PARTICLE ACCELERATION AND KINEMATICS IN SOLAR FLARES

A Synthesis of Recent Observations and Theoretical Concepts

PARTICLE ACCELERATION AND KINEMATICS IN SOLAR FLARES

MARKUS J. ASCHWANDEN
Lockheed Martin, Advanced Technology Center, Palo Alto, California, USA

Reprinted from *Space Science Reviews*, Volume 101, Nos. 1–2, 2002

KLUWER ACADEMIC PUBLISHERS
DORDRECHT / BOSTON / LONDON

A.C.I.P. Catalogue record for this book is available from the Library of Congress

ISBN: 1-4020-0725-6

Published by Kluwer Academic Publishers,
P.O. Box 990, 3300 AZ Dordrecht, The Netherlands

Sold and distributed in North, Central and South America
by Kluwer Academic Publishers,
101 Philip Drive, Norwell, MA 02061, U.S.A.

In all other countries, sold and distributed
by Kluwer Academic Publishers,
P.O. Box 322, 3300 AH Dordrecht, The Netherlands

Printed on acid-free paper

Printed in the Netherlands

Deployment of the *Compton Gamma Ray Observatory (CGRO)* in April 1991, photographed from the space shuttle *Atlantis*. The 8 *Large-Area Detectors (LAD)* of the *Burst and Transient Source Experiment (BATSE)* are located at the 8 corners of the spacecraft. The cylindrical detectors on the top are *COMPTEL* and *EGRET*, while *OSSE* is on the far side. The *BATSE* detectors provided for the first time sufficient photon statistics to measure electron time-of-flight delays in solar flares. [Courtesy of CGRO Team].

Preface

Over the last decade we entered a new exploration phase of solar flare physics, equipped with powerful spacecraft such as Yohkoh, SoHO, and TRACE that provide us detail-rich and high-resolution images of solar flares in soft X-rays, hard X-rays, and extreme-ultraviolet wavelengths. Moreover, the large-area and high-sensitivity detectors on the *Compton GRO* spacecraft recorded an unprecedented number of high-energy photons from solar flares that surpasses all detected high-energy sources taken together from the rest of the universe, for which CGRO was mainly designed to explore. However, morphological descriptions of these beautiful pictures and statistical catalogs of these huge archives of solar data would not convey us much understanding of the underlying physics, if we would not set out to quantify physical parameters from these data and would not subject these measurements to theoretical models. Historically, there has always been an unsatisfactory gap between traditional *astronomy* that dutifully describes the morphology of observations, and the newer approach of *astrophysics*, which starts with physical concepts from first principles and analyzes astronomical data with the goal to confirm or disprove theoretical models. In this review we attempt to bridge this yawning gap and aim to present the recent developments in solar flare high-energy physics from a physical point of view, structuring the observations and analysis results according to physical processes, such as particle acceleration, propagation, energy loss, kinematics, and radiation signatures. The discovery of electron time-of-flight delays and magnetic trapping delays made with CGRO provides us for the first time accurate parameters to apply relativistic particle kinematics to solar flares. Although this review fills an entire volume, we scratched only the surface of particle physics in such complex plasma environments as magnetic reconnection processes, and I hope to stimulate readers to embark on more comprehensive physical modeling of these unique goldmines of data the Sun provides us every day.

Markus J. Aschwanden
Palo Alto, California
2001 October 13

PARTICLE ACCELERATION AND KINEMATICS IN SOLAR FLARES

A Synthesis of Recent Observations and Theoretical Concepts (Invited Review)

MARKUS J. ASCHWANDEN

Lockheed Martin, Advanced Technology Center, Solar and Astrophysics Laboratory, Dept. L9-41, Bldg.252, 3251 Hanover St., Palo Alto, CA 94304, USA; e-mail: aschwanden@lmsal.com

(Received 1 March 2002)

Abstract. We review the physical processes of particle acceleration, injection, propagation, trapping, and energy loss in solar flare conditions. An understanding of these basic physical processes is inexorable to interpret the detailed timing and spectral evolution of the radiative signatures caused by nonthermal particles in hard X-rays, gamma-rays, and radio wavelengths. In contrast to other more theoretically oriented reviews on particle acceleration processes, we aim here to capitalize on the numerous observations from recent spacecraft missions, such as from the *Compton Gamma Ray Observatory (CGRO)*, the *Yohkoh Hard X-Ray Telescope (HXT)* and *Soft X-Ray Telescope (SXT)*, and the *Transition Region and Coronal Explorer (TRACE)*. High-precision energy-dependent time delay measurements from *CGRO* and spatial imaging with *Yohkoh* and *TRACE* provide invaluable observational constraints on the topology of the acceleration region, the reconstruction of magnetic reconnection processes, the resulting electromagnetic fields, and the kinematics of energized (nonthermal) particles.

Keywords: solar flares, particle acceleration, particle kinematics, nonthermal particles, hard X-ray emission, soft X-ray emission, radio emission

Abbreviations: BATSE – Burst And Transient Source Experiment (instrument aboard *CGRO*);
CGRO – Compton Gamma Ray Observatory;
ETH – Eidgenössische Technische Hochschule (Zurich, Switzerland), radio observatory;
EUV – Extreme ultra-violet;
HXT – Hard X-Ray Telescope (instrument aboard *Yohkoh*);
HXRBS – Hard X-Ray Burst Spectrometer (instrument aboard *SMM*);
SMM – Solar Maximum Mission;
SXT – Soft X-Ray Telescope (instrument aboard *Yohkoh*);
TRACE – Transition Region and Coronal Explorer;
VLA – Very Large Array (Socorro, New Mexico).

Space Science Reviews **101:** 1–227, 2002.

Table of Contents

1. Introduction

Particle kinematics applied to solar flare processes is a relatively new analysis method that became feasible over the last few years, thanks to the availability of high-precision timing measurements combined with high-resolution spatial imaging. High-precision timing measurements (down to a few microseconds) can be performed with high-sensitivity hard X-ray detectors (e.g. with the *Compton Gamma Ray Observatory, CGRO*) or with fast-sweeping, frequency-agile radiotelescopes. In addition, high-resolution imaging of solar flares (down to a few arcseconds) became readily available in hard X-rays (from *Yohkoh/HXT*), soft X-rays (from *Yohkoh/SXT*), extreme-ultraviolet [from the *Solar and Heliospheric Observatory (SoHO) Extreme-ultraviolet Imaging Telescope (EIT)*, or the *Transition Region and Coronal Explorer (TRACE)*], and in radio wavelengths [e.g. from the *Very Large Array (VLA), Owens Valley Radio Observatory (OVRO)*, or *Nobeyama Radio Observatory (NRO)*]. The combination of timing and imaging measurements provide the necessary physical parameters (distance $\mathbf{x}$, velocity $\mathbf{v}$, time t) to conduct quantitative data analysis on the kinematics of relativistic particles produced in solar flare processes. While particle physicists in laboratories made fundamental discoveries of new particles by applying the relativistic kinematics to recorded trajectories and energy spectra in acceleration-and-collision experiments, we solar physicists work the other way round, by remote-sensing of particle trajectories and thick-target collisions, and then we invert the timing with relativistic kinematics to infer the characteristics and location of the unknown "accelerator machines" in solar flare sites.

In this review we focus on a rather special aspect of solar flares, namely on the spatial and energy-dependent timing of nonthermal particles, which is relevant to probe the basic physics of particle acceleration and kinematics. We review first (in Section 2) a comprehensive set of solar flare models in order to characterize the relevant magnetic configurations $\mathbf{B}(\mathbf{x})$ that are crucial to determine the spatial trajectories $\mathbf{x}(t)$ of particle orbits. Topological changes of magnetic field lines during reconnection processes are of over-riding importance to deduce the electromagnetic fields that accelerate charged particles and control their propagation. So far, particle acceleration in solar flares happened in a black box that was not accessible to observations, but the new high-resolution images from *TRACE* taken in narrow-band temperature filters start to give us a sensitive tracer of individual pre-reconnection and post-reconnection magnetic field lines, from which we can, in principle, reconstruct the topological changes during localized reconnection processes in flares. In Section 3 we summarize all geometric constraints we have obtained about the acceleration region in flares from hard X-ray and radio observations, including methods such as direct imaging, time-of-flight localization, propagation delays, directionality, and coincidence measurements of electron beams in hard X-rays and radio. Because particle acceleration in solar flares is apparently not a static process, but rather occurs in a highly dynamic environment of rapidly-changing field lines during magnetic instabilities, we review next (in Section 4) dynamic processes that play a role; flare trigger mechanisms and bursty magnetic reconnection processes such as tearing and coalescence instabilities. We review relevant observations that provide us information on the spatial and temporal intermittency of magnetic reconnection processes: statistics of fast time scales, wavelet analysis of time series, spatio-temporal correlations, the variability of radio bursts, and the spatial fine structure of flare arcades. Armed with all observational constraints on the acceleration region we have obtained so far, regarding magnetic topology, geometry, and dynamics, we take now a fresh look at the theoretically known acceleration mechanisms (in Section 5), such as electric DC-field acceleration, wave turbulence (or stochastic) acceleration, and shock acceleration. We discuss the theoretical concepts of these acceleration mechanisms, recent flare models where they have been applied, and review some relevant numerical simulations that yield particle orbits and energy spectra of accelerated particles. Next we proceed to the kinematics of the energized relativistic particles (in Section 6), which includes particle orbits in the stages of acceleration, injection into field lines escaping the acceleration region, propagation in form of free-streaming along adiabatic orbits, as well as propagation in magnetic mirror traps, collisional deflection, and precipitation to energy loss sites. After we have obtained a basic description of the particle motion in terms of physical parameters, i.e. the time-dependent location $\mathbf{x}(t)$ and velocity $\mathbf{v}(t)$, we can model the timing and spectra of radiative signatures in various wavelengths, including nuclear deexcitation lines in gamma rays (Section 7), bremsstrahlung in hard X-rays (Section 8), and various radio emissions (Section 9), such as gyrosynchrotron emission, beam-driven plasma emission, or losscone-driven electromagnetic emission. The energy-dependent timing of these radiative signatures, which contains very specific kinematic information on the processes of particle acceleration and propagation, can even be used in non-imaging observations from other astrophysical plasmas at large (e.g. extragalactic gamma-ray bursts), if sufficient photon statistics (by high-throughput detectors) is provided. To test theoretical models of high-energy particle physics in astrophysical plasmas, however, solar flares provide us the only data with imaging information.

2. Magnetic Topology of Acceleration Regions

Virtually all solar flare models involve magnetic reconnection in one or the other form. Magnetic reconnection changes the topology of the magnetic field by reconfiguring the connectivities between opposite magnetic polarities. A fundamental characteristic of magnetic field lines in the solar corona is their connectivity back to the solar surface or out into interplanetary space, which we call *closed* and *open field lines*. Closed field lines orinignate in a natural way from emerging magnetic dipoles, while open field lines result after the local dipole field has been broken up by a magnetic reconnection process with the interplanetary magnetic field. Consid-

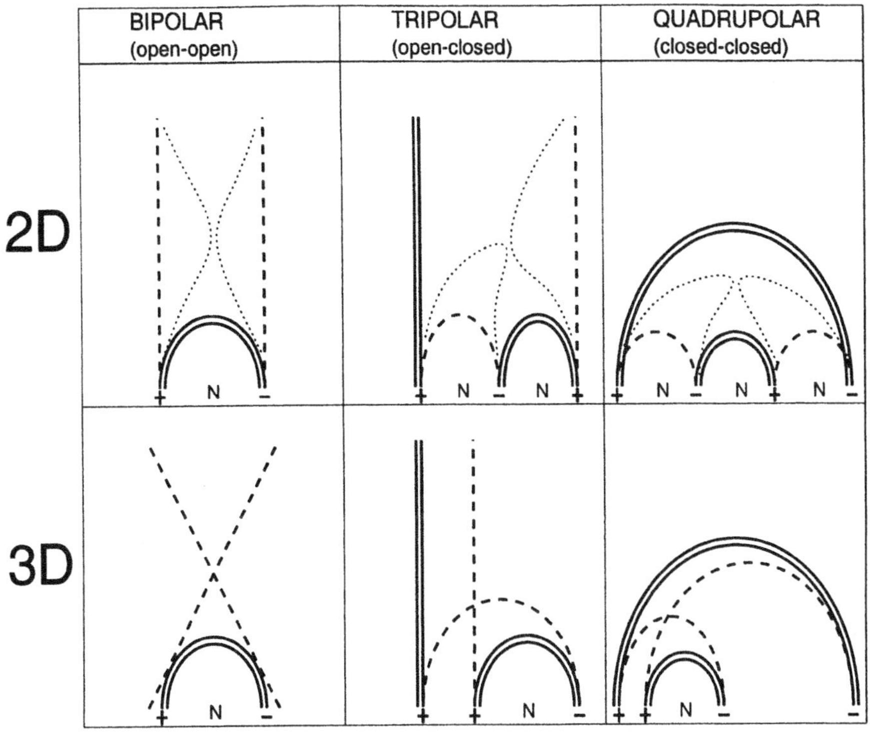

Figure 1. Classification of X-type magnetic reconnection topologies: (1) *bipolar* models have reconnection between two open field lines, (2) *tripolar* models have reconnection between an open and a closed field line, and (3) *quadrupolar* models have reconnection between two closed field lines. The pre-reconnection field lines are marked with dashed linestyle, at the time of reconnection with dotted linestyle, while the post-reconnection field lines, as they occur after relaxation into a near-potential field state, are outlined with solid double linestyle. 2D versions, invariant in the third dimension (forming arcades), are shown in the upper row, while 3D versions are captured in the lower row. The pre-reconnection field lines (dashed) are located behind each other in the 3D versions, but approach each other in the image plane during reconnection. Note that the number of neutral lines (marked with symbol N), extending into the third dimension, is different in the corresponding 2D and 3D cases.

ering *X-type reconnection* cases (which can be described by the interaction of two intersecting field lines in a 2D plane), we have three basic combinations between closed and open field lines, i.e. interactions between (1) open and open, (2) open and closed, and (3) closed and closed field lines. These three cases can also be distinguished by the number of poles rooted on the solar surface, which vary from two to four, and thus we name these three classes of models: (1) bipolar, (2) tripolar, and (3) quadrupolar models. The pre-reconnection configuration of these three classes is shown in Fig. 1. During magnetic reconnection, the connectivity between opposite polarities is switched and the new-configured field lines snap back into a lower energy state. Thus, in the post-reconnection phase we expect a relaxation into a near-potential field configuration. The outcome of this reconfiguration changes the topology in the dipolar case, where a transition from two open field lines to a closed field line occurs, while the topology remains equivalent in the other two cases (open-and-closed, closed-and-closed field lines). Each of the three multi-pole ($N=2,3,4$) topologies can be realized in a 2D geometry, requiring N-1 neutral lines for N magnetic poles (Fig.1, top row), as well as in a 3D geometry, with a single neutral line only (Fig.1, bottom row). The 2D geometries are appropriate for the interpretation of flares that show postflare arcades, while 3D geometries are more suitable for flares that involve interacting loops.

In the following we make use of this classification scheme to group different flare models according to the topology of the underlying magnetic reconnection processes. As a working hypothesis, we assume in this review that particle acceleration sites in solar flares are generally associated with magnetic reconnection regions, or with reconnection-driven shocks. This working hypothesis reflects the growing evidence and relevance of magnetic reconnection processes we believe to have observed in solar flares over the last decade. Of course there exist also alternative particle acceleration mechanisms that happen without reconnection, e.g. acceleration in quasi-static electric fields (V-events in the Earth's aurora) or in stationary shocks (gradual *Solar Energetic Particle (SEP)* acceleration in interplanetary space), but they are not the focus of this review.

To enable a quantitative physical picture in the framework of magnetic reconnection, we start with an even more restricted working assumption (I), i.e. that the acceleration regions are located on surfaces where reconnected magnetic field lines snap back during and after reconnection into a lower energy state of a relaxed, near-potential field. These regions can be characterized by "cusp" geometries as indicated in Fig. 2 (hatched areas). The justification for this working hypothesis is simply that the electric fields capable of rapid particle acceleration are mainly created by the dynamic electromagnetic forces during magnetic field reconfigurations, while static electric fields, possibly existing before and after flares, seem to have negligible efficiency in producing nonthermal particles (based on the absence of hard X-ray and nonthermal radio emission outside of impulsive flare episodes). Another important constraint is that particles are accelerated and propagate in collisionless plasmas only, while they suffer energy loss and become thermalized in collisional plasmas. Therefore, the possible sites of particle acceleration depend crucially whether the reconnection sites are located in the corona (Fig. 2 top), near the chromosphere or transition region (Fig. 2 middle), or in the deeper chromosphere or photosphere (Fig. 2, bottom). The transition region is an approximate boundary between collisionless plasma (in the corona) and collisional plasma (in the chromosphere), and thus represents a lower boundary for acceleration sites (assumption II). Based on these two working hypotheses we expect in the case of coronal X-type reconnection (Fig. 2 top row) acceleration sites in the upward- and downward-oriented cusp regions, as well as in the (frontal) fast-shock and (sideward) slow-shock regions of the upward and downward reconnection outflows (Fig. 2 top row); in the case of X-point reconnection in the transition

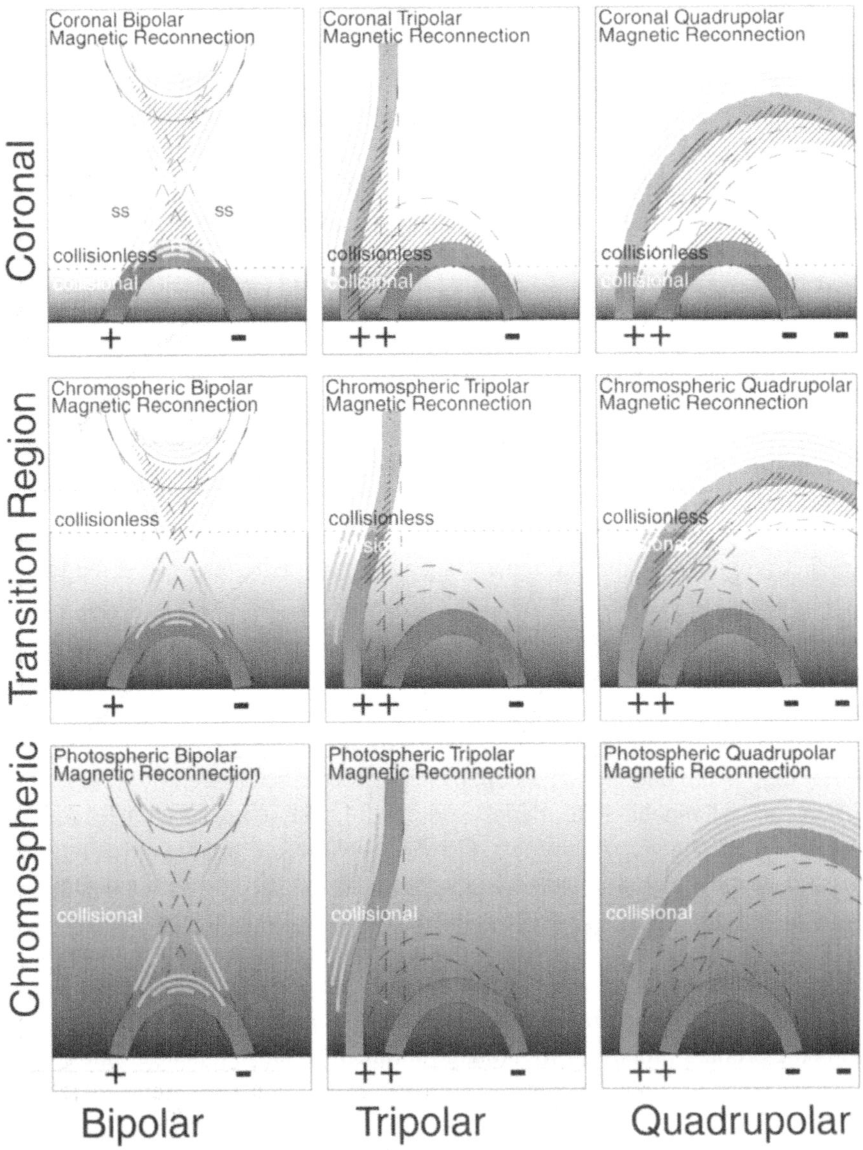

Figure 2. The topology of acceleration regions is outlined for three types of 3D magnetic reconnection processes (bipolar, tripolar, and quadrupolar), and for three different locations of the magnetic reconnection point with respect to the collisionless/collisional boundary (above, near, and below transition region). The greyscale indicates the relative density, with dark for collisional plasma and white for collisionless plasma. Pre-reconnection field lines are marked with dashed lines, post-reconnection field lines with solid lines, the intermediate area is defined as acceleration region (hatched) if it is located in a collisionless zone. Additional acceleration regions are located in upward/sideward directed shocks (marked with grey thick curves) in front of the upward and downward relaxing field lines. Also the locations of fast shocks (FS) and slow shocks (SS) in downward reconnection outflows are indicated (top left panel).

region (Fig. 2 middle row) we expect primarily acceleration regions associated with the upwardly-directed cusp and shock; but we expect none for chromospheric reconnection processes (Fig. 2 bottom row) because it is completely embedded in a collisional regime. The topology of these acceleration regions is important to understand the occurrence or absence of hard X-ray emission (mainly produced by downward propagating electrons) and of radio emission (mainly produced by upward propagating or trapped electrons), and possible correlations between the two. In the following we discuss the three classes of bipolar, tripolar, and quadrupolar reconnection scenarios in more detail in the context of specific flare models.

2.1. Bipolar Reconnection Models

Magnetic reconnection between two open field lines has in principle two possible outcomes. If the two field lines have oppositely-directed magnetic fields, a single bipolar loop is produced between the two footpoints (Figs. 1 and 2 left), while the symmetric counterpart in upward direction contains a "concave" disconnected loop that probably escapes into interplanetary space. Alternatively, if the two field lines have parallel-directed magnetic fields, current coalescence might occur, or reconnection that reconfigures two open field lines into two less-twisted open field lines. Because bipolar soft X-ray loops are always observed in flares, the latter processes (which start and end with open field lines) are probably irrelevant for flare processes. We therefore concentrate only on the former case, where a single bipolar loop is the outcome of an oppositely-directed reconnection process.

The class of bipolar flare models has continuously evolved over the last three decades. In the model of Sturrock (1966), a helmet-streamer configuration was assumed to exist at the beginning of a flare, where the tearing-mode instability (induced by footpoint shearing) near the Y-type reconnection point triggers a flare, accelerating particles in downward direction and producing shock waves and plasmoid ejection in upward direction. However, the (preflare) driver that leads up to tearing and reconnection is not specified in this model. Hirayama (1974) explains the preflare process with a rising prominence above a neutral line (between oppositely-directed open magnetic field lines), which carries an electric current parallel to the neutral line and induces a magnetic collapse on both sides of the current sheet after eruption of the prominence. The magnetic collapse is accompanied by lateral inflow of plasma into the opposite sides of the current sheets, which drives then Petschek-type reconnection and forms closed field lines beneath the X-point that eventually become the soft X-ray-bright flare loops filled with heated plasma. Kopp and Pneuman (1976) refined this scenario further and predicted a continuous rise of the Y-type reconnection point. The observation of cusp-shaped soft X-ray loops in the famous "candle-flame" (long-duration) flare on 1992-Feb-21 (Tsuneta et al., 1992) provided strong observational evidence of the now called Carmichael-Sturrock-Hirayama-Kopp-Pneuman (CSHKP) reconnection model, in the sense that the hot plasma illuminates the cusp geometry expected for X-type reconnection topologies. On the other side, the observed long duration of hot cusps might not be consistent with fast reconnection and could be interpreted alternatively (e.g. by upward diffusion of hot plasma in a high-β environment, see Gary 2001). Nevertheless, the standard scenario received additional support by Masuda's discovery of above-the-loop-top hard X-ray sources (Masuda et al., 1994a,b), which are thought to confine the primary acceleration site of hard X-ray-producing electrons. The details of the downward reconnection outflows were described in the model of Tsuneta (1996) and simulated by Yokoyama and Shibata (1996). Observational evidence for supra-arcade downflows were first shown by McKenzie (2000), while observational evidence of a lateral inflow was reported by Yokoyama et al. (2001). Specific acceleration models in the reconnection outflow and associated fast shocks between the cusp region and the soft X-ray flare loop top have been developed by Tsuneta and Naito (1998), Somov and Kosugi (1997), and in terms of slow shocks by Blackman (1997).

In Fig. 3 we show a collection of *Yohkoh/SXT* and *TRACE* observations that illustrate flares with bipolar structures, which can appear as single loops or as arcades of loops along a neutral line. The latter geometry is traditionally called "double-ribbon flare". The neutral line does not need to be a straight line, it could be a curved linear feature, which leads to curved arcades which also have been dubbed "slinky", e.g. in the famous Bastille-day (2000-Jul-14) flare. Despite of the variety of morphologies, all these flares shown in Fig. 3 essentially can be described by bipolar loops and are believed to be the endresult of a bipolar reconnection process of the CSHKP-type model.

2.2. Tripolar Reconnection Models

The reconnection between an open field line (which has one unipolar footpoint on the Sun) and a closed field line (with bipolar footpoints) obviously involves three magnetic poles (Figs. 1 and 2 middle column), so we call this class of flares consequently *tripolar reconnection models*, although this term is not widely used in solar flare literature. However, variants of this type of magnetic reconnection in tripolar geometries appeared in solar flare literature in the context of emerging-flux models (Heyvaerts, Priest, and Rust 1977) and after the discovery of soft X-ray plasma jets with *Yohkoh* (Shibata et al., 1992). The observation of long straight soft X-ray jets were taken as evidence of plasma flows along open field lines, a fact that constitutes a flare-like process between a closed and an open field line. Such a process is also envisioned for so-called "explosive events" observed in EUV (Chae, 1999; Chae et al., 1999). Although it is presently unclear whether there exists a fundamental difference between "explosive events", "microflares", and "jets" in soft X-rays and EUV, the observation of radio type III bursts associated with soft X-ray jets (Aurass et al. 1995; Kundu et al., 1995) indicates flare-like properties such as acceleration of nonthermal particles, which may well happen in tripolar reconnection processes. High-resolution observations of EUV jets with *TRACE*

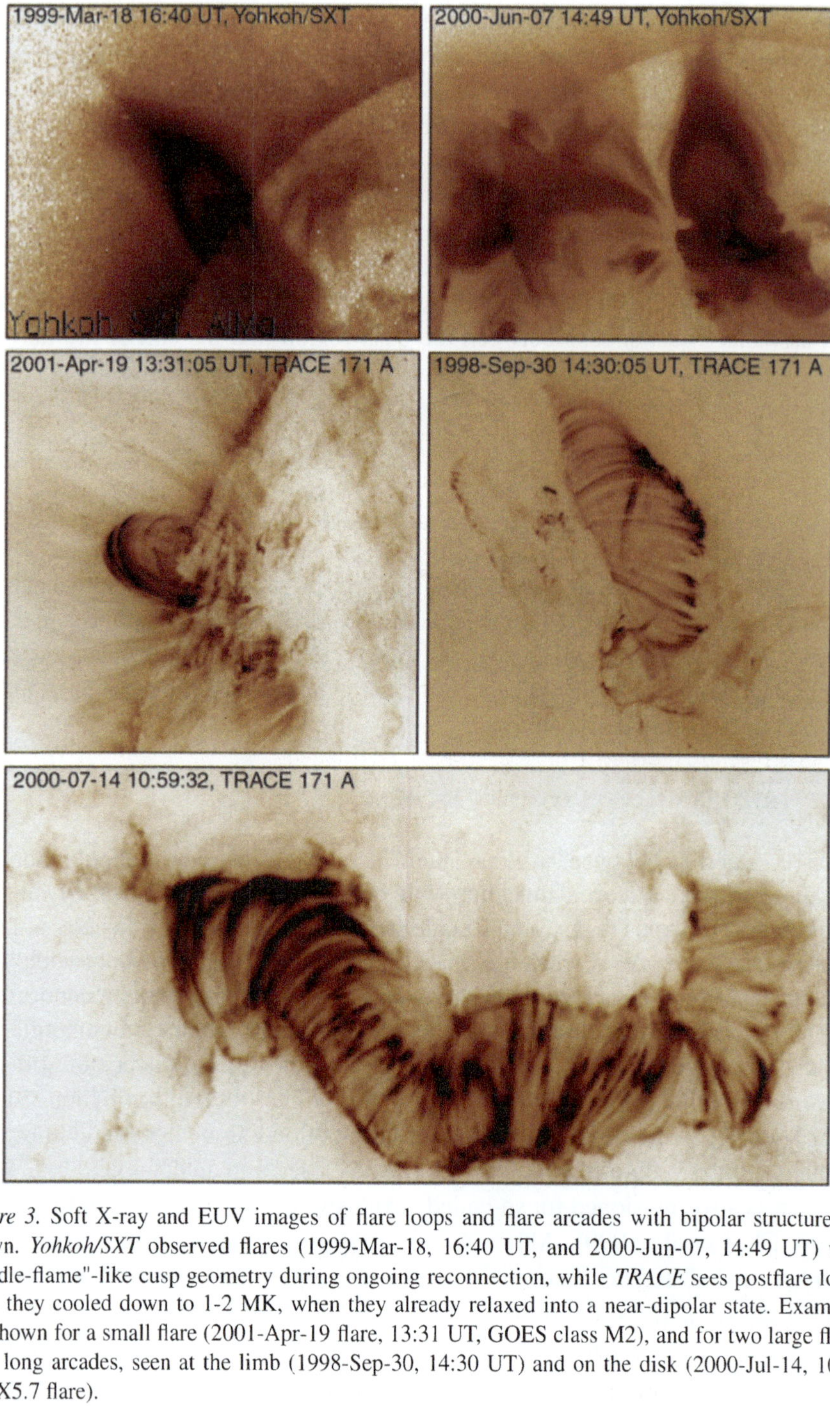

Figure 3. Soft X-ray and EUV images of flare loops and flare arcades with bipolar structure are shown. *Yohkoh/SXT* observed flares (1999-Mar-18, 16:40 UT, and 2000-Jun-07, 14:49 UT) with "candle-flame"-like cusp geometry during ongoing reconnection, while *TRACE* sees postflare loops once they cooled down to 1-2 MK, when they already relaxed into a near-dipolar state. Examples are shown for a small flare (2001-Apr-19 flare, 13:31 UT, GOES class M2), and for two large flares with long arcades, seen at the limb (1998-Sep-30, 14:30 UT) and on the disk (2000-Jul-14, 10:59 UT, X5.7 flare).

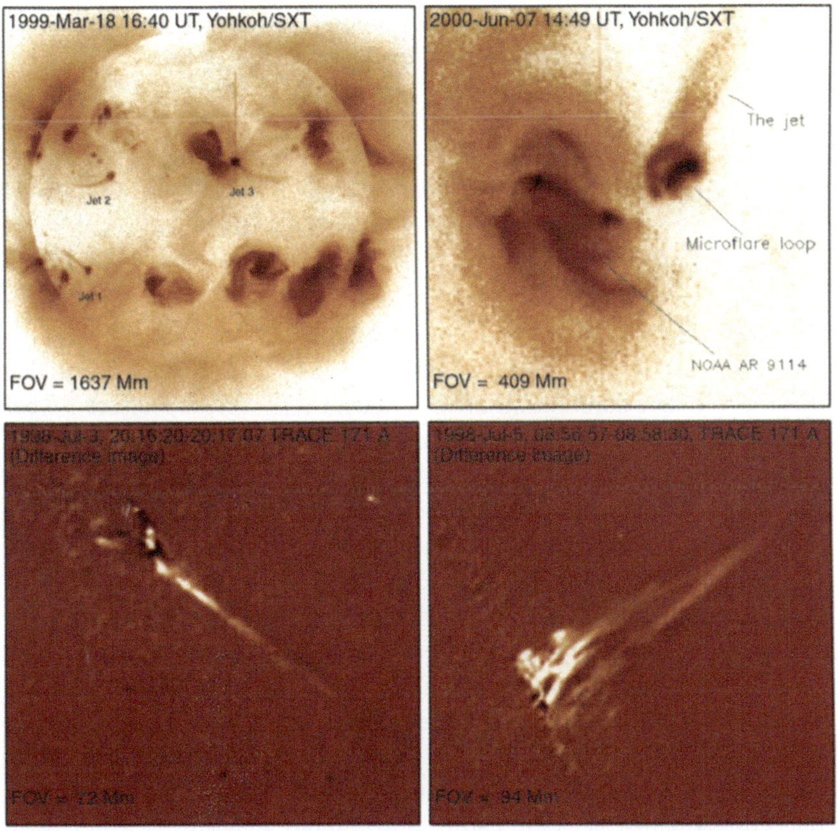

Figure 4. Soft X-ray and EUV images of flare loops with tripolar structure or jets are shown. *Yohkoh/SXT* observed triple jets on 1998-Jul-11 12:57 UT at three different locations on the Sun, a rare coincidence (top left). An enlarged view of the jet of 2000 Aug 7, 09:06 UT, observed with *Yohkoh* shows clearly a microflare loop at its feet (top right). Two high-resolution *TRACE* images of tripolar flares with jets are shown as difference images (bottom row), one corresponds to a GOES C2.1 flare (1998-Jul-3, 20:14 UT, bottom left), the other to a GOES C1.1 flare (1998-Jul-5, 08:56 UT, bottom right). Note that tripolar reconnection preferentially produces small flares.

were interpreted as evidence of reconnection-driven plasma flows (Alexander and Fletcher, 1999), produced by a magnetic reconnection process between an emerging or cancelling small-scale loop and an open field line of the ambient corona, according to the model of Shibata et al. (1992). The most recent high-resolution observations of flares by *TRACE* suggest that magnetic reconnection processes between small-scale closed loops and large-scale open field lines, what we call tripolar configuration here, may be part of a more complex geometry that involves 3D reconnection at a coronal nullpoint (Fletcher et al. 2001), see Section 2.4.

Examples of such tripolar reconnection events are shown in Fig. 4. Every soft X-ray or EUV jet shows that a small bipolar loop is involved at its footpoint, usually associated with a small flare, also called "microflare" or "active region transient brightening" (ARTB) in soft X-rays, or "explosive event" in EUV.

2.3. Quadrupolar Reconnection Models

We define a third category of flare models that involve the interaction of two closed field lines. Such flare models are also called *"interacting-loop models"* or *"quadrupolar reconnection models"*. Some first ideas about interacting flare loops were proposed by Gold and Hoyle (1960), where the currents of two oppositely-twisted loops (with parallel magnetic fields) are thought to annihilate and to dissipate energy. This scenario was, however, criticized by Melrose (1997) by the argument that the resulting induction times (hours) would be much longer than typically observed flare durations (minutes). The ideas of current coalescence between current-carrying loops were further developed by Sakai (see references in Sakai and De Jager, 1996), who subdivided the loop-loop interactions into three categories, depending on the relative angle between the two current-carrying loops (leading to I, Y, and X-type reconnection geometries for parallel, oblique, or perpendicular loop orientations (Sakai and De Jager, 1991).

A classical flare scenario was developed by Heyvaerts, Priest, and Rust (1977), who assumed the formation of a current sheet between the antiparallel magnetic fields of an emerging flux tube and a pre-existing overlying large-scale loop (or open field line). Observational evidence for interacting dipolar flare loops was first reported from SMM observations by Machado et al. (1988). Because the overlying large-scale loop necessarily has one footpoint close to one footpoint of the reconnecting small-scale loop, earlier observations dubbed this configuration as *"three-legged structure"* (Hanaoka, 1996; Nishio et al., 1997), although strictly-speaking it corresponds to a quadrupolar structure. However, it was shown that the quadrupolar geometry consistently fits those "three-legged" or "interacting-loop" flares (Aschwanden et al. 1999b). The 3-dimensional reconstruction of two such quadrupolar flares is shown in Fig. 5. Other interactions between quadrupolar flare loop geometries can be found in Rudawy et al. (2001). Quadrupolar topologies have also been employed in a number of other studies, e.g. in Baum and Bratenahl (1980), Antiochos (1998), or Titov and Demoulin (1999).

A generalization of quadrupolar reconnection between two interacting flare loops is the extension into the third dimension, i.e. quadrupolar reconnection between to interacting flare arcades (Fig.1 top right panel). Such a model has been developed by Uchida et al. (1994, 1999a,b) and Hirose et al. (2001). The 2D model of quadrupolar reconnection requires three (parallel) neutral lines, which seems to be less common than the simplest 3D model of quadrupolar reconnection (Fig.1 bottom right panel), which requires only one single neutral line.

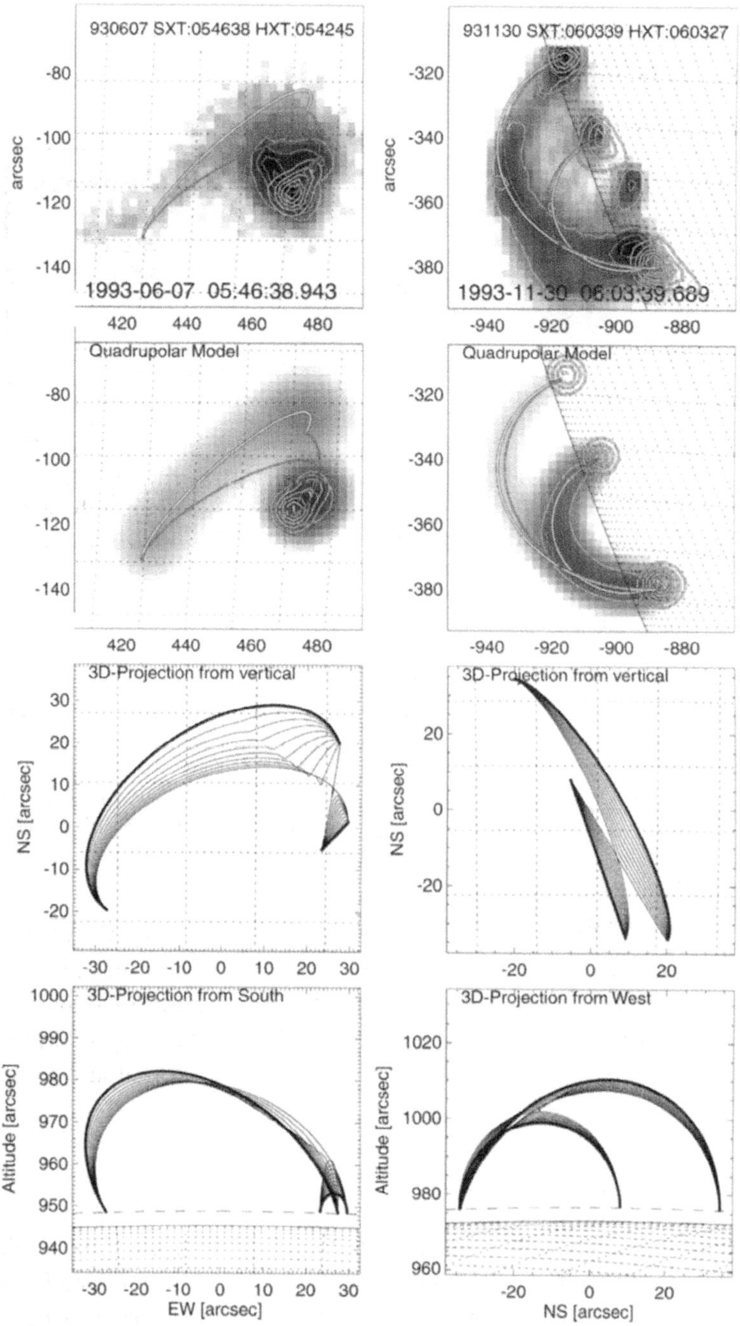

Figure 5. 3D reconstruction of interacting loops in the two flares of 1993-Jun-07, 05:46 UT, and 1993-Nov-30, 06:03 UT. The top panels show the soft X-ray maps (in greytone) and hard X-ray maps (contours) recorded with *Yohkoh*. The second row shows the simulated 3D model with the same projection, based on the inferred 3D coordinates of the interacting loops, for which the 3D solution is projected from vertical (third row) and from the side (forth row). [from Aschwanden et al., 1999b].

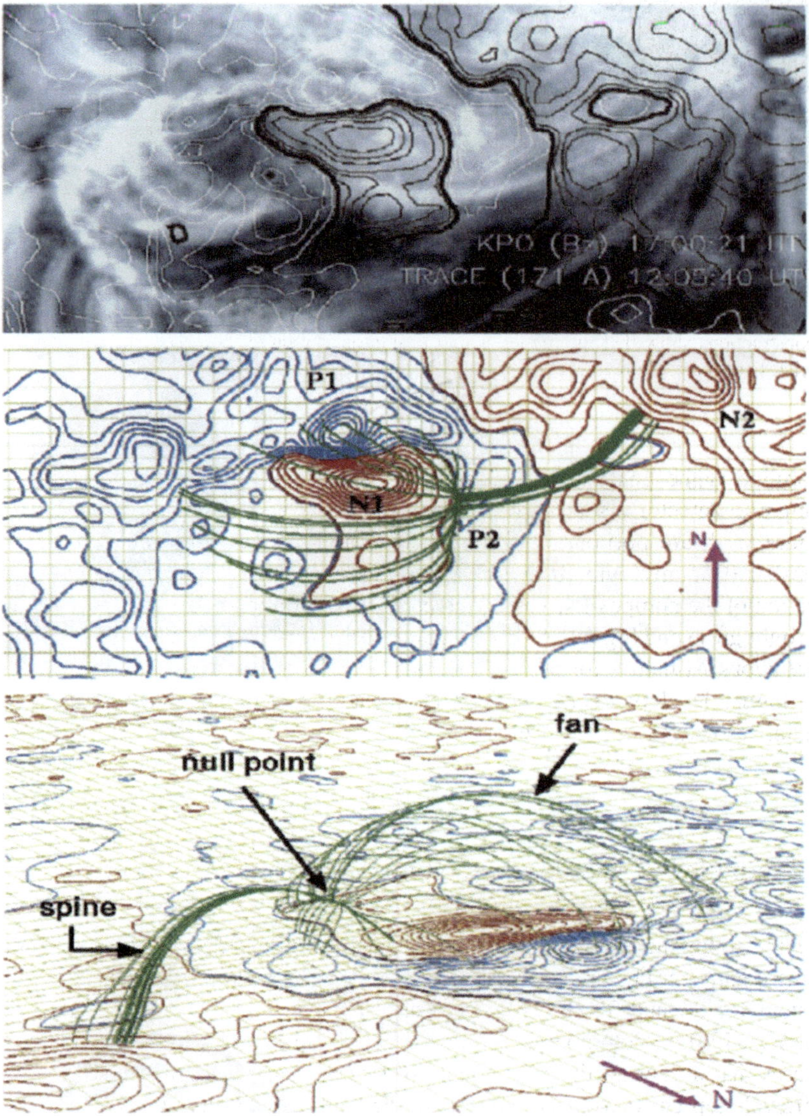

Figure 6. A 3D nullpoint with its associated spine field lines and fan surface is inferred for the first Bastille-day (1998-Jul-14, 12:55 UT) flare by Aulanier et al. (2000). *Top:* The magnetic field (with contours at $B = \pm 20, 50, 100, 250, 400, 600$ G) from a KPNO magnetogram is overlaid on a *TRACE* 171 Å image with a FOV of 203×104 Mm. The neutral lines are indicated with thick black lines. *Middle:* Extrapolated magnetic field lines are shown that closely trace out a fan-like separatrix surface above the δ-spot (P1-N1) and end in a 3D nullpoint (P2), which is connected through spine field lines to the leading polarity in the west (N2). *Bottom:* A 3D view is shown from a different viewing angle (from north-west).

2.4. Spine and Fan Reconnection Models

While most previous flare models with magnetic reconnection were restricted to X-type configurations and current sheet geometries (with 2D nullpoints), recent flare studies involve also three-dimensional (3D) nullpoints [see, e.g., Priest and Forbes (2000), Chapter 8, for an introduction]. In such reconnection models, magnetic reconnection can occur at 3D nullpoints, saddle points (also called *bald patches*), along singular lines, or quasi-separatrix layers, dubbed *spine reconnection, fan reconnection, or separator reconnection.* Separator reconnection occurs in 3D if two nullpoints exist (where the separator connects the two nullpoints) and dominates the other 3D-nullpoint reconnection cases (fan and spine reconnection) if at least two nullpoints exist (Galsgaard et al. 2000). Observational evidence for such types of 3D-reconnection to occur in solar flares has been brought forward only recently. Such new magnetic topologies have also implications for the geometry and electromagnetic fields of particle acceleration regions, and thus should be seriously studied in this context.

One of the first suggestions for the presence of a 3D nullpoint has been inferred from *SoHO/EIT* observations in an active region (Filippov, 1999). However, despite of the fact that a saddle-like or hyperbolic magnetic configuration is considered as a necessary condition for magnetic field-line reconnection, no heating or flaring was observed in this case. Magnetic field extrapolations of the first Bastille-day flare (1998-Jul-14) led Aulanier et al. (2000) to the conclusion that magnetic reconnection occurred along a *spine field line* and the *fan surface* associated with a coronal nullpoint (Fig. 6). The existence of a coronal 3D nullpoint was also inferred in the 1999-May-3 flare by Fletcher et al. (2001), based on the emergence of a positive magnetic flux concentration inside a separatrix dome above the surrounding ring of negative magnetic polarity (Fig. 75). Relaxation of the stressed magnetic field was interpreted in terms of spine or fan reconnection during this flare. Upflows of heated plasma was also observed along the spine field line, which can be considered as consequence of the reconnection-driven heating along the spine field line. These examples offer convincing evidence that 3D reconnection are not only a theoretical construct, but indeed occur in real solar flares.

3. Geometry of Acceleration Region

The magnetic topology in solar flare models, which we discussed in the previous Section, constrains in principle the location and geometry of acceleration regions, because the propagation of charged particles is guided by the magnetic field lines. In principle, when the sites of their collisional interactions are observed (e.g. the chromospheric thick-target sites at the footpoints of flare loops as seen in hard X-rays), the trajectories of free-streaming particles can be traced back to the acceleration region, using the knowledge of the magnetic topology. In the following we review results on the geometry of flare acceleration sites obtained with different methods: magnetic topology constraints (Section 3.1), direct detection of coronal hard X-ray sources (Section 3.2), electron time-of-flight localization (Section 3.3), conjugate-footpoint constraints (Section 3.4), remote-footpoint time delays (Section 3.5), bi-directional electron beams (Section 3.6), hard X-ray and radio-coincident electron beams (Section 3.7), and start frequencies of radio bursts (Section 3.8). Although most of these methods require models of the electron kinematics, which is discussed later on (in Section 6), we summarize in this Section the results concerning the geometry of the acceleration region. Of course, each of the 8 different methods makes somewhat different assumptions, so the mutual comparison of the results provides a powerful consistency test.

3.1. Magnetic Topology Constraints on the Acceleration Region

A magnetic reconnection model describes the topological change of magnetic field line connectivities. In X-type reconnection at least, for every pair of reconnecting field lines, there is only one possibility of connectivity exchange. Say, the first field line is anchored at the (positive) $1+$ and (negative) $1-$ polarity at the boundaries, and a second field line at $2+$ and $2-$, the only possible reconfiguration with opposite polarities for each reconfigured field line is a new field line between $1+$ and $2-$, and a second new field line between $2+$ and $1-$. The forces that accelerate particles,

$$m \frac{d\mathbf{v}}{dt} = q(\mathbf{E} + \frac{1}{c}\mathbf{v} \times \mathbf{B}) \ , \qquad (1)$$

are given by the direct and induced electric fields in the reconnection region itself (i.e. at X-points, current sheets, nullpoints), as well as caused by the time-varying magnetic field $\mathbf{B}(\mathbf{x}, t)$ between pre-reconnection and post-reconnection field line locations. Newly-reconfigured field lines relax into a near-dipolar state, which implies a shortening of the length of the field line (from a cusp-like s' to a circular-like shape s), and thus a reduction of the magnetic energy density,

$$\Delta W = \int_{cusp} \frac{B^2(s')}{8\pi} ds' - \int_{relaxed} \frac{B^2(s)}{8\pi} ds \approx \frac{< B >^2}{8\pi}(s' - s) > 0 \qquad (2)$$

which is available as free energy to heat and accelerate particles. The effect of field line shrinkage is also discussed in the context of postflare loop expansions (Forbes and Acton, 1996) or flare implosions (Husdon 2000). We can therefore quantify, in first order, the geometry of the acceleration region by the 3D surfaces that are subtended between the initial and final (relaxed) position of the field line during the post-reconnection phase. The efficiency of particle acceleration is probably highest near the reconnection points due to the fast changes of the electromagnetic fields, and is likely to fall off gradually with increasing distance to the reconnection point. A detailed map of the acceleration region requires the reconstruction of the electromagnetic fields (Section 5) and particle orbits (Section 6). Thus the area covered by topological changes of the magnetic field lines during reconnection outlines just the approximate maximum spatial extent of the particle acceleration region. What is usually observable is the position of the post-reconnection field lines (i.e. postflare loops), because they become illuminated only after being filled with upflowing plasma by the chromospheric evaporation process. However, if all footpoints of the reconnecting field lines are known, the position of the pre-reconnection field lines can be reconstructed by switching the connectivities between the magnetic polarities.

Following this principle, we show in Fig. 7 the expected locations of acceleration regions: for the cases of X-type reconnection (bipolar loops, arcades, tripolar, quadrupolar) and for 3D nullpoints with fan and spine reconnection. In the cases of X-type reconnection, the geometry of the acceleration region can be characterized by a curved surface with the shape of a cusp, where the field line has initially a cusp at the X-point and relaxes into a near-circular segment between the new-connected polarities. The positions of the relaxed postflare loops is marked with thick lines in Fig. 7. In the cases of 3D nullpoints, the new-configured field line moves through the fan separatrix surface (fan reconnection) or through the spine separatrix line (spine reconnection) during the reconnection process (which both intersect at a coronal nullpoint here), and relax into a more dipole-like state afterwards. The acceleration region is located beneath the fan surface and coronal nullpoint in the case of *fan reconnection* (Fig. 7 middle right), while it is located near the spine separatrix line above the coronal nullpoint in the case of *spine reconnection* (Fig. 7 bottom right). Because particle acceleration sites have to be located in collisionless or low-collisional plasmas, we envision that only *coronal 3D nullpoints* are relevant for flares. 3D-reconnection models that involve *chromospheric nullpoints* cannot be efficient particle accelerators, because the accelerating fields are located in the highly-collisional chromosphere.

3.2. Direct Hard X-ray Detection of the Acceleration Region

The most reliable localization of a particle acceleration source would be, of course, a direct detection of some radiative signature from particles in the stage of acceleration. However, free-streaming particles accelerated in an electric field do not emit much radiation in a collisionless plasma, as the corona was believed to be. However, Masuda's discovery of hard X-ray emission at coronal positions *above* the soft-X-ray bright flare loops changed our minds. Masuda et al. (1994a) discovered an *above-the-loop-top* hard X-ray source at energies of $\gtrsim 20 - 50$ keV with *Yohkoh/HXT* in about 10 flares (four examples are shown in Fig. 8), besides the well-known (usually double-footpoint) chromospheric footpoint sources. Initially it was not clear how electrons can emit collisional bremsstrahlung in such low plasma densities as measured above flare loops, typically $n_e \lesssim 10^8 - 10^9$ cm^{-3}. An interpretation in terms of thermal hard X-ray emission was ruled out based on the required temperatures $T \approx 200$ MK, for which there was no evidence from

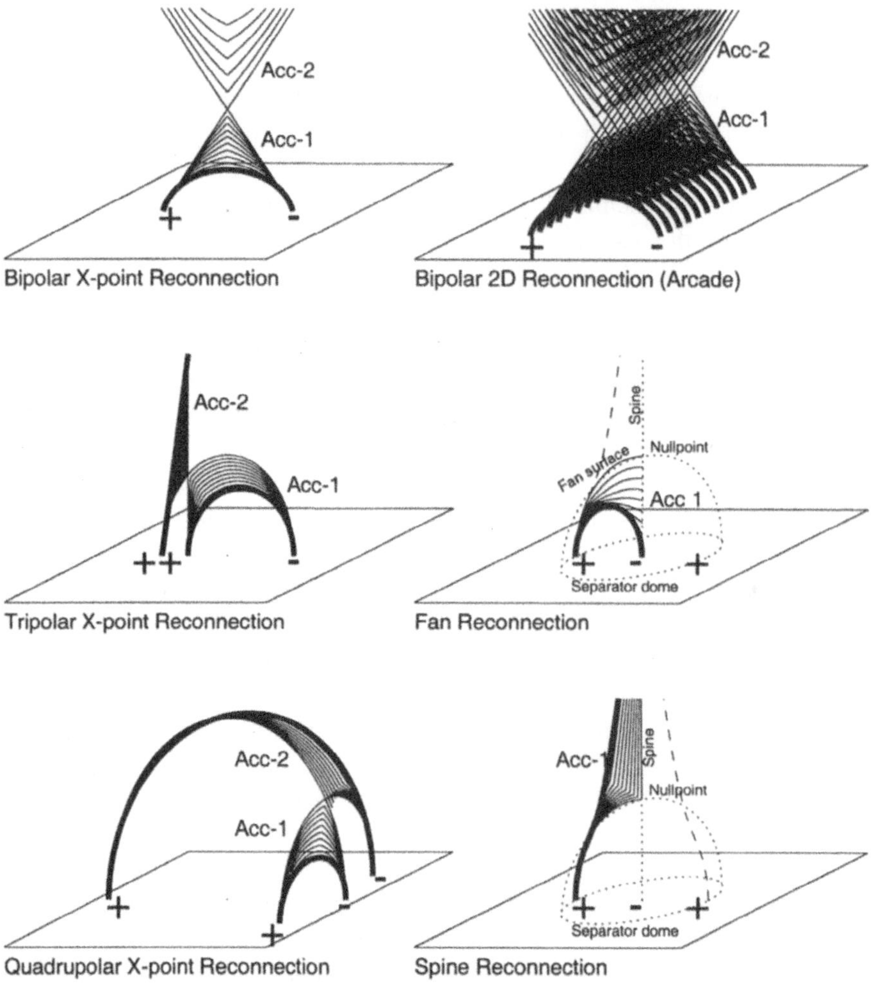

Figure 7. The geometry of the acceleration region inferred from magnetic topology constraints for X-type reconnection models (bipolar loops, arcades, tripolar, quadrupolar) and reconnection models with 3D nullpoints (fan reconnection, spine reconnection). The acceleration region is outlined by the area over which a new-configured field line relaxes (marked as a time sequence of thin lines) after reconnection into a near-dipole state (with the final state marked with a thick line). For X-type reconnection models there are always two acceleration regions (denoted with Acc-1 and Acc-2), below and above the reconnection point. The reconnection models with 3D nullpoints are characterized by a fan surface that forms a separator dome and intersects the solar surface with a ring of the same magnetic polarity, surrounding an area with opposite magnetic polarity, located below a coronal nullpoint. The pre-reconnection field lines are indicated with dashed lines.

other soft X-ray instruments (*Yohkoh/SXT* and *BCS*). Also the time variability of the looptop hard X-ray emission was too rapid to be consistent with thermal cooling times, and thus required a nonthermal interpretation (Hudson and Ryan, 1995). Therefore, a plausible explanation is collisional bremsstrahlung from trapped electrons, which are directly fed in from the accelerator in the cusp region beneath the reconnection point. The location of the coronal hard X-ray source was measured to be about 10" (7250 km) above the soft X-ray loop, and in slightly higher altitudes in images taken in the higher ($\gtrsim$ 50 keV) energy bands (Masuda et al., 1994b). This location is fully consistent with the cusp geometry in bipolar reconnection models (Fig. 7), and thus the coronal hard X-ray emission has to be emitted relatively close to the acceleration region associated with the reconnection point, rather than from a secondary trap somewhere else in the corona.

The spectrum of the coronal hard X-ray source in the 1992-Jan-13 flare was analyzed in detail by Alexander and Metcalf (1997). To overcome the problem of analyzing a weak source that is suppressed by the much (ten times) brighter footpoint sources nearby, they employed the photometrically more accurate pixon method. Their conclusions were that the coronal hard X-ray source: (a) has an impulsive temporal profile (similar to the footpoint sources), (b) has a nonthermal spectrum, (c) has a very hard spectrum or a low-energy cutoff in the electron injection spectrum, and (d) the loop-top and footpoint hard X-ray sources are produced by two distinct particle populations. All this information indicates that the coronal hard X-ray emission is produced directly by nonthermal electrons in or near the acceleration region, and thus makes these observations to one of the most direct witnesses of the acceleration process itself. Apparently, electrons are not simply accelerated in large-scale electric fields, because the required background density for free-streaming electrons would be $n_e \approx 3 \times 10^{11}$ cm^{-3} (Fletcher, 1995) to $n_e \approx 10^{12}$ cm^{-3} (Wheatland and Melrose, 1995) to produce $\gtrsim$ 30 keV hard X-ray emission, which is about 2-3 orders of magnitude higher than what is observed at the locations of Masuda's hard X-ray sources. A more likely acceleration mechanism is stochastic acceleration, which allows the electrons to stay sufficiently long in a local trap to produce detectable amounts of nonthermal bremsstrahlung in the low-density coronal plasma. An alternative idea is particle trapping by wave scattering in MHD turbulent regions above flare loops (Jakmiec et al. 1998; Jakimiec 1999).

Obviously, coronal hard X-ray sources are most easily observed for limb flares, because the separation between footpoint, looptop, and above-the-looptop sources is least confused for such a view. This is the reason why Masuda's sample of 10 flares includes only limb flares. However, there are also examples found on the solar disk, e.g. the 1991-Nov-19, 09:29 UT, flare shown in Fig. 8 (top right). An even better opportunity are flares with occulted footpoints (Masuda, 2000), but they are rare. The hard X-ray spectra of occulted flares were found to be softer and the nonthermal broadening was found to be smaller (Mariska et al. 1996), clearly indicating different physical conditions above the looptop than at

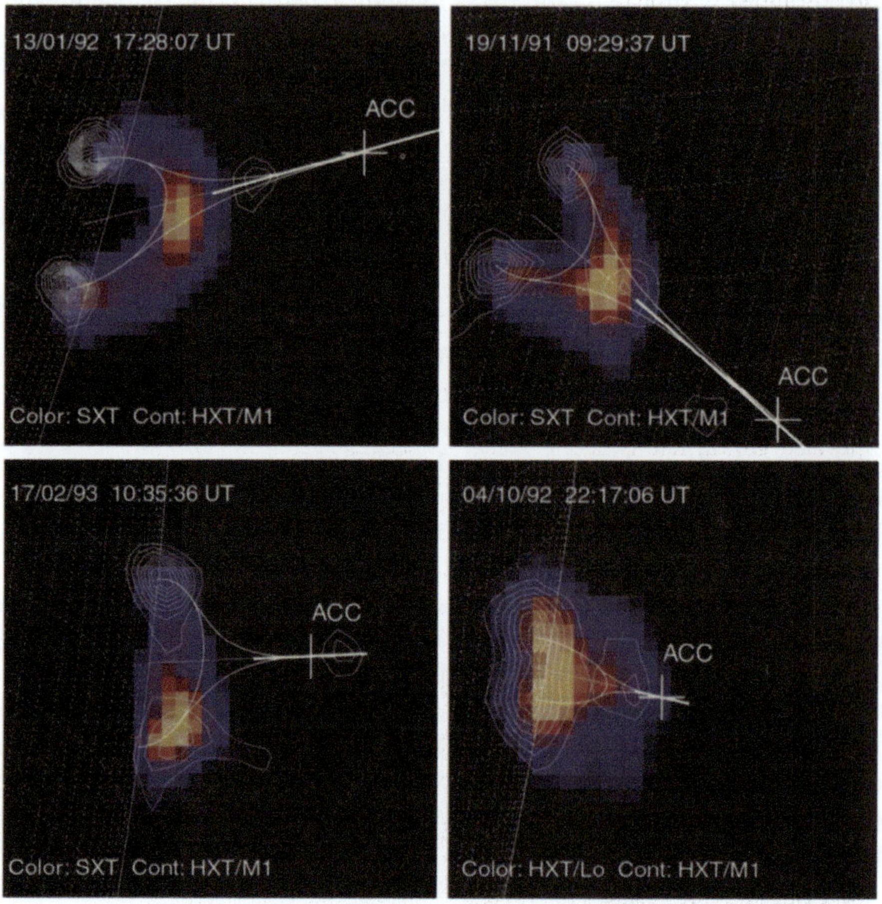

Figure 8. The geometry of the acceleration region inferred from direct detections of above-the-loop-top hard X-ray sources with *Yohkoh/HXT* (contours) and simultaneous modeling of electron time-of-flight distances based on energy-dependent time delays of 20-200 keV hard X-ray emission measured with *BATSE/CGRO* (crosses marked with ACC). Soft X-rays detected with *Yohkoh/SXT* or thermal hard X-ray emission from the low-energy channel of *Yohkoh/HXT/Lo* is shown in colors, outlining the flare loops.

the footpoints. The hard X-ray time profiles of occulted flares were also found to exhibit spiky emission occasionally, perhaps a signature of turbulent flare kernels (Tomczak 2001). Earlier stereoscopic multi-spacecraft measurements of occulted flares yielded maximum heights of $h \lesssim 30,000$ km above which impulsive hard X-ray emission was identified (Kane et al., 1979; Kane et al., 1982a; Hudson, 1978; Kane, 1983). A record height of $\gtrsim 200,000$ km was determined for an occulted flare (1984-Feb-16, 09:12 UT) at hard X-ray energies of $\gtrsim 5$ keV, by combined stereoscopic measurements with the *ICE* and ISEE-3 spacecraft (Kane et al., 1992). We will see in the following Section that typical altitudes of acceleration regions are found in a range of $h_{acc} \approx 5000 - 35,000$ km based on electron time-of-flight measurements, which is fully consistent with the height limit of $h \lesssim 30,000$ km in earlier stereoscopic measurements. The most extreme case with a stereoscopic height of $h \approx 200,000$ km could be associated with thin-target emission from trapped electrons in a large-scale flare loop (e.g. in a quadrupolar geometry, see Fig. 7).

3.3. Time-of-Flight Localization of the Acceleration Region

A new method to determine the geometric location of the acceleration region are electron time-of-flight measurements between the acceleration region and the collisional stopping region, which is the chromosphere in the thick-target model. In the simplest scenario, electrons are accelerated at some coronal altitude h_{acc} and propagate along the magnetic field lines down to the chromosphere, where they are stopped at height h_{stop} in the transition region or upper chromosphere. Because hard X-ray photons at different energies are caused by electrons with different kinetic energies or velocities, say v_1 and v_2, there will be an electron time-of-flight (TOF) difference $\Delta t = t_2 - t_1$ in the arrival time of electrons, if they are simultaneously injected and propagate over the same distance $L = h_{acc} - h_{stop}$,

$$t_2 - t_1 = \left(\frac{L}{v_2} - \frac{L}{v_1} \right) . \tag{3}$$

Such time-of-flight differences could be recently measured from fast time structures in different hard X-ray spectral channels with the *Burst and Transient Source Experiment (BATSE) Large Area Detectors (LAD)* onboard the *Compton Gamma Ray Observatory (CGRO)* (Aschwanden et al., 1995a, b; 1996b, c, d). In reality, there are a number of effects that need to be considered in high-precision measurements, which require various assumptions that can only be fully justified in hindsight. The full derivation of the underlying electron kinematics and conversion into hard X-ray bremsstrahlung will be reviewed later on (Section 8), but we list here the major assumptions that have been made in the localization of the acceleration region with the electron time-of-flight method:

1. Electrons at different energies are simultaneously injected (or accelerated) at some coronal height h_{acc} into a magnetic field line, during each of the fast ($\lesssim 1$ s) pulses typically seen in the impulsive flare phase. This assumption can be weakened to a criterion where the difference in acceleration or injection time is much smaller than the propagation time in the free-streaming part or the coronal field line, $t_{acc} \ll t_{prop}$.
2. Electrons are free-streaming in the coronal part of the magnetic field line, i.e. they propagate in a collisionless plasma, where the theory of adiabatic particle orbits can be applied.
3. The pitch-angle correction of the parallel velocity component $v_{\parallel} = v \cos \alpha$ to the magnetic field line is estimated in the limit of large magnetic mirror ratios

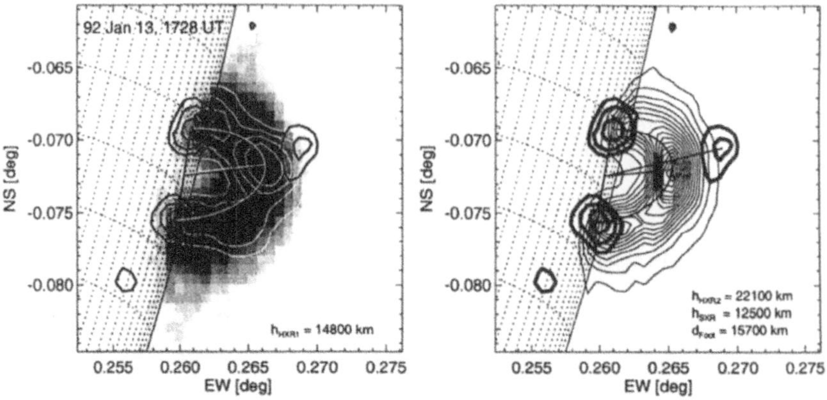

Figure 9. Yohkoh/HXT (1728:04-1728:40 UT) images of the 1992-Jan-13 *"Masuda"* flare. *Left:* SXT image (grey scale), *HXT/Lo* 14-23 keV image (white contours), and *HXT/M1* 23-33 keV image (black contours). *Right: SXT* image (thin contours) and *HXT/M1* image (thick black contours). The heliographic positions of the centroids of the *HXT/M1* footpoints are 85.7 W/15.2 S and 87.7 W/16.4 S. Elliptical loop geometries are fitted through the two footpoints and the brightness maxima near the loop top. The measured heights refer to the *HXT/Lo* loop (h_{HXR1}, left), the *HXT/M1* loop-top source (h_{HXR2}, right), and the *SXT* loop (h_{SXR}, right), d_{foot} is the projected footpoint separation [from Aschwanden et al. 1996b].

($R \gg 1$), yielding a correction factor of $q_\alpha = t_0/t_B \le 0.64$ between the parallel transit time t_0 and the bounce time t_B for mirroring particles. This correction factor defines the projected time-of-flight distance $L' = L \times q_\alpha$ in relation to the total length L of the helical particle trajectory. (Appendix A in Aschwanden et al., 1996b).

4. The helical twist of the magnetic field line is estimated from a safety margin below the kink instability (less than $\approx 2.5\pi$) and an average loop thickness-to-length aspect ratio of ≈ 0.2, which yields a helical twist correction factor of $q_H = 0.85$. This factor is a second correction to the projected time-of-flight distance, $L' = L \times q_\alpha \times q_H \approx L \times 0.64 \times 0.85 = L \times 0.54$. (Appendix B in Aschwanden et al., 1996b).
5. The hard X-ray flux time profiles $F_X(t)$ consist of two components with different energy-dependent timing: (a) fast time structures ($\lesssim 1$ s) are caused by a modulated electron injection function, and are preserved for directly-precipitating electrons, i.e. for electrons with small pitch angles $\alpha \le \alpha_{losscone}$, and (b) a smoothly and slow-varying component of the time profile is produced by the convolution of the electron injection function with the trapping time of electrons injected at large pitch angles $\alpha > \alpha_{losscone}$. To measure electron time-of-flight differences, the fast component (a) needs to be subtracted from the slowly-varying component (b), which has an opposite energy-dependent timing. The two components can be separated with a lower-envelope subtraction or with a deconvolution of an exponential trapping time distribution.

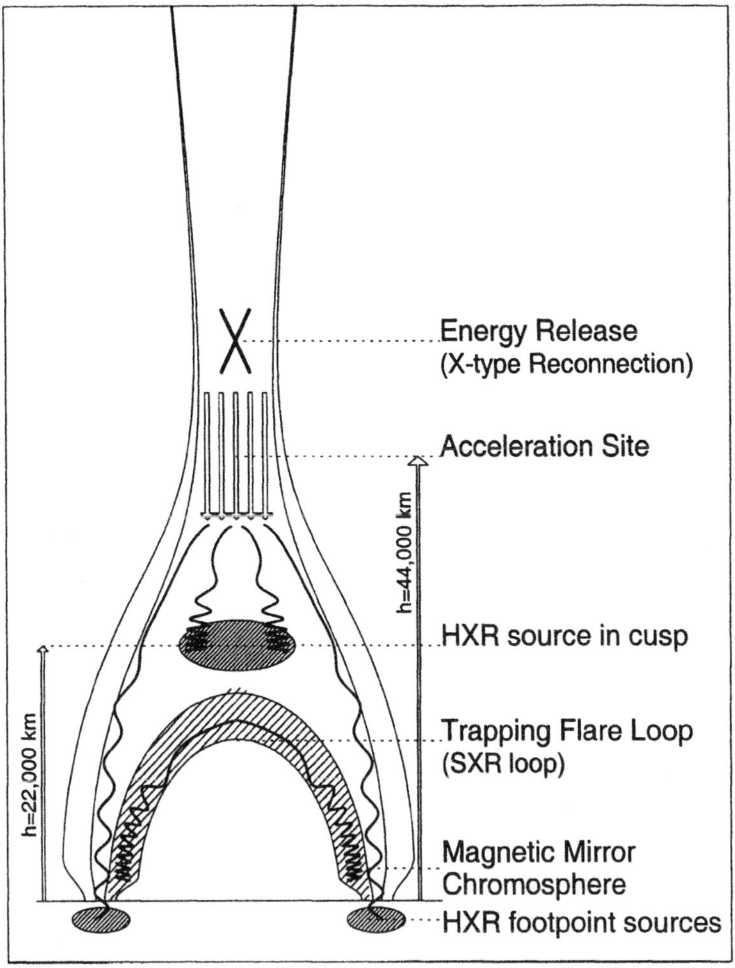

Figure 10. Synthesized geometry of the relative heights of particle acceleration source, coronal and chromospheric hard X-ray sources based on direct imaging from *Yohkoh/HXT*, *SXT* and from electron time-of-flight measurements with Compton GRO [from Aschwanden et al. 1996b].

6. The electron energy loss time t_{loss} in the chromospheric thick-target region is much smaller than the propagation time t_{prop} in the free-streaming part of the coronal field line, i.e. $t_{loss} \ll t_{prop}$.

A first application of this electron time-of-flight method was conducted for the famous Masuda flare (1992-Jan-13, 17:29 UT) by Aschwanden et al. (1996b), for which Masuda already discovered a coronal hard X-ray source associated with the acceleration site. The geometry of footpoint, looptop, and above-the-looptop hard X-rays sources for this flare are shown in Fig. 9. The height of the centroid of the soft X-ray-bright (Be 119) flare loop is measured to $h_{SXR} = 12,500$ km (Fig. 9 right), the height of the low-energy 14-23 keV (mostly thermal) hard X-ray flare loop is slightly higher $h_{HXR1} = 14,800$ km (Fig. 9 left), and Masuda's above-

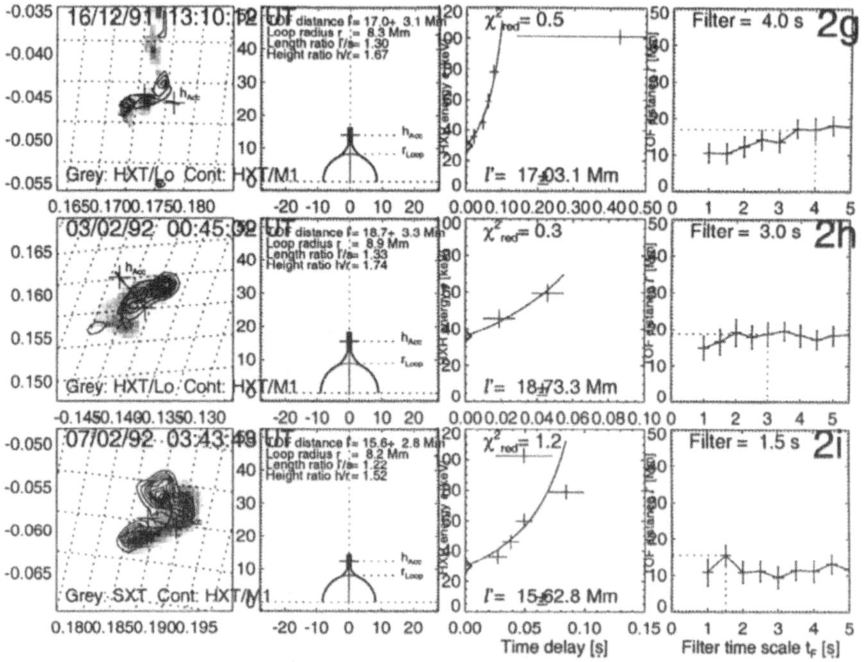

Figure 11. Geometric reconstruction of electron time-of-flight distance between acceleration site and chromospheric footpoints with a cusp flare model. *Left:* Overlays of *Yohkoh SXT* (grey), *HXT* (contours) maps and projected TOF distance. *Second column:* Projection of cusp geometry and TOF distance into vertical flare loop plane. *Third column:* Energy-dependent hard X-ray time delay measurements with *CGRO* (crosses with error bars) and fit of TOF difference (curve). *Right column:* Variation of TOF distance L' measurements with different highpass filters, the maximum is taken as least contaminated TOF measurement [from Aschwanden et al. 1996d].

the-looptop hard X-ray source at 23-33 keV is even higher, $h_{HXR2} = 22,100$ km. Compared with these directly measured altitudes, the location of the acceleration site inferred from the electron time-of-flight differences is even higher, i.e. $h_{TOF} = 44,000$ km (Aschwanden et al., 1996b). This leads to the following picture shown in Fig. 10. An important result is that the altitude of the TOF-inferred acceleration site is substantially higher than Masuda's coronal hard X-ray source, which suggests that the Masuda sources are not exactly cospatial with the acceleration site, but rather located in a trapping region in the cusp below the acceleration site. From the four examples shown in Fig. 8, the same relation is observed in two other flares (1991-Nov-19 and 1992-Oct-4), but an exception is found for one flare (1993-Feb-17). One could argue that the pitch angle ($q_\alpha = 0.64$) and helical twist correction factors ($q_H = 0.85$) are overestimated, but smaller pitch angles and less helical twist would increase the TOF distance and the inferred altitude of the acceleration site up to a factor of $\lesssim 2$. If this were the case, all Masuda hard X-ray sources would definitly be located substantially lower than the TOF-inferred acceleration site.

The same technique of electron time-of-flight measurements has been applied to all flares observed jointly with *Yohkoh* and *BATSE/CGRO* (Aschwanden et al., 1996c,d), including some 100 flare events with a wide range of (primary) flare loop sizes, with radii of $r = 3000 - 25,000$ km. To determine the geometric location of acceleration sites, the measured projected time-of-flight distance L' has to be mapped into the flare loop plane. A parameter of interest for geometric flare models is the ratio of the altitude h of the acceleration site to the flare loop radius r. The flare loop radius r can simply be measured from the footpoint separation, after de-projecting the heliographic footpoint coordinates with the line-of-sight projection angle. A cusp-like field line that connects the footpoints with the acceleration site can simply be composed of two circular segments with opposite curvature, joined together with a steady tangent. The length of such a field line, which corresponds to the projected time-of-flight distance L', is then

$$L' = \frac{r}{2}\left(1 + \frac{h^2}{r^2}\right) \arctan\left(\frac{2hr}{h^2 - r^2}\right) . \quad (4)$$

Some examples of such geometric reconstructions of acceleration heights with the TOF method are shown in Fig. 11, based on double footpoint sources observed with *Yohkoh/HXT*, soft X-ray images of the flare loops observed with *Yohkoh/SXT*, and hard X-ray time delay measurements carried out with *BATSE/CGRO* data in the energy range of $\approx 20 - 120$ keV. In Fig. 11 we show the coaligned maps (left panels), the vertical geometry of the TOF distance (second column), the energy-dependent time delays (in the order of $\lesssim 100$ ms) and the fits of the TOF model (third column), and the robustness of the result (TOF distance L') as function of the highpass filter for fast hard X-ray pulses (right column). The optimum filter time scale (typically $\approx 1 - 4$ s) was chosen where the TOF distance L' reaches a maximum, because a less optimum separation of the smoothly-varying flux component (caused by trapped electrons) which has generally an opposite sign in the energy-dependent delays, would add a contamination of trap delays to the TOF delays and this way reduce the inferred TOF distance L'.

The statistics of 42 suitable events, for which both the TOF distance L' and the flare loop radius r could be determined, reveals a remarkable result, shown in Fig. 12: There is a strong correlation between the TOF distance L' and the flare loop half length $s = r \times (\pi/2)$, revealing a linear relation with a ratio of

$$L'/s = 1.43 \pm 0.30 . \quad (5)$$

This correlation clearly demonstrates a physical relation between the independently measured spatial and temporal parameters, and moreover, implies a scaling law between the location of the acceleration site (which is supposedly close the the reconnection point), and the relaxed flare loop size after reconnection. This corroborates the scale-free property of reconnection geometries (Fig. 7), where the length ratio of a newly reconnected (cusp-like) field line to the relaxed (dipole-like) field line is essentially constant for different sizes of the system. One could

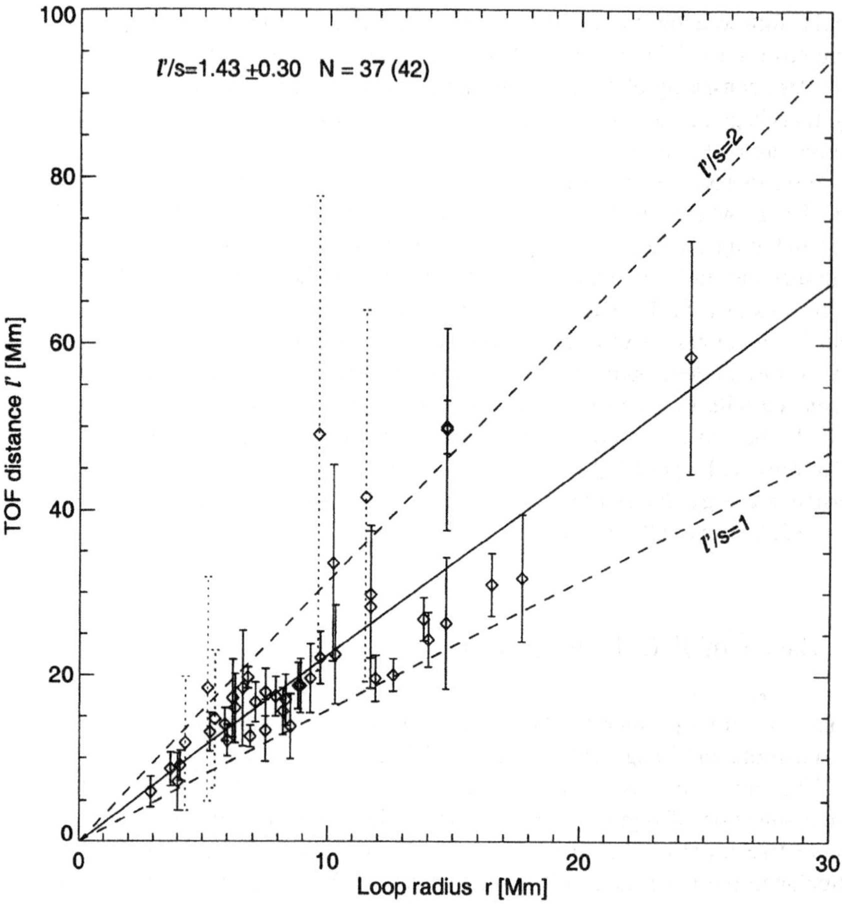

Figure 12. Scaling law of TOF distance L' versus loop radius r in 42 flare events. The average ratio of the TOF distance L' and loop half length s is indicated with a solid line: the dashed lines indicate a loop half-length ($L'/s = 1$) or a full loop length ($L'/s = 2$).

argue that the accuracy of this result may be hampered by the separation method of oppositely-signed energy-dependent time delays dt/dE in the analyzed time profiles. In order to test the robustness of the scaling law, a full kinematic model of directly-precipitating and trap-plus-precipitating electrons was developed (Aschwanden, 1998a) and the time-of-flight analysis was repeated with a deconvolution method of the two electron components. A quite similar result was found with this different method (Aschwanden et al., 1998b, 1999a),

$$L'/s = 1.6 \pm 0.6 \,. \tag{6}$$

This scaling law can be applied to the various reconnection geometries shown in Fig. 7. The fact that the scaling law is closer to $L/s' \approx 1.5$, which roughly corresponds to the length of the cusp-shaped field lines, rather than $L/s' \approx 1.0$, which is the value for the relaxed, dipole-like field line, indicates that the particle

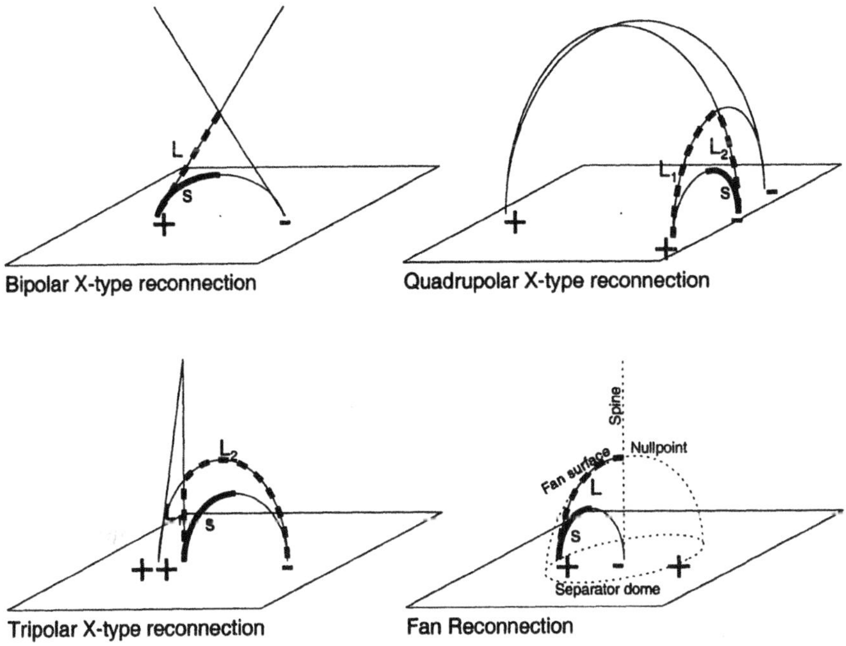

Figure 13. TOF distance L from acceleration region near the reconnection point (dashed line) and half length s of a postflare field line in relaxed dipole-like state, for four different magnetic reconnection models.

acceleration site is closely located near the reconnection point, rather than near the soft X-ray-bright flare loop. The location of the acceleration region significantly above the SXR-bright flare loops makes a lot of sense, because the densities in the soft X-ray bright flare loops are much higher ($n_e \approx 10^{10} - 10^{12}$ cm^{-3}, Aschwanden et al., 1998b; $n_e \approx (0.2 - 2.5) \times 10^{11}$ cm^{-3}, Aschwanden et al., 1997b) than in the cusp above, $n_e \approx (0.6 - 10) \times 10^9$ cm^{-3}, Aschwanden et al., 1997b). If the acceleration site would be located inside the dense soft X-ray bright flare loops, as assumed in earlier simple flare models (e.g. see Fig. 1 in Dennis and Schwartz (1989), the collisional energy loss would be so high that the electrons could not be accelerated to high energies or could not propagate to the footpoints, where the brightest hard X-ray emission is generally seen from the thick-target bremsstrahlung of precipitating electrons.

The ratios of the time-of-flight distance to the relaxed postflare loop half length is also consistent with an analysis of the field line shrinkage in two bipolar flares with "candle-light" morphology, where a shrinkage of 20% and 32% was found (Forbes and Acton, 1996), which corresponds in our definition to a ratio of $L'/s = 1.25$ and $L'/s = 1.5$, a range that is expected in dipole-field models with a vertical (Syrovatskii-type) current sheet. Other models constrained by the quite narrow opening angle of reconnection outflows yield ratios of $L'/s \approx 2 - 4$ (Tsuneta 1996).

Let us now consider the expected ratios of electron time-of-flight distances L' to the half length s of the relaxed postflare loops in the various magnetic reconnection models shown in Fig. 7. The corresponding distances of L' and s are outlined in Fig. 13. For bipolar reconnection, the ratio L'/s can have any value above unity, depending on the angle between the reconnecting field lines. The same is true for tripolar reconnection, if we define the average time-of-flight distance by the average of the two asymmetric segments, $L' = (L'_1 + L'_2)/2$. For quadrupolar reconnection, it can be shown with elementary geometry, that the ratio varies between $L'/s = \sqrt{2}$ for equal-sized loops (for vertical loop planes and reconnection at the top of the smaller loop), and $L'/s = (1 + 2/\pi)$ for extremely different loop sizes (see Eq. 18 in Aschwanden et al. 1999b). For 3D nullpoints, only the case of fan reconnection produces a postflare loop with closed field line topology, which has the ratio $L'/s = 2$. Spine reconnection produces an open postflare fieldline (Fig. 7), perhaps manifested as a soft X-ray jet (see also Fig. 75). So, we find the following ranges of ratios:

$$L'/s = 1...\infty \quad \text{for Dipolar X}-\text{type Reconnection} \tag{7}$$

$$L'/s = 1...\infty \quad \text{for Tripolar X}-\text{type Reconnection} \tag{8}$$

$$L'/s = 1.41 - 1.64 \quad \text{for Quadrupolar X}-\text{type Reconnection} \tag{9}$$

$$L'/s = 2 \quad \text{for 3D Fan Reconnection} \tag{10}$$

In summary, it can be said that the electron time-of-flight measurements are fully consistent with a location of particle acceleration at or near the reconnection point in all X-type reconnection models, located about a factor of 1.5 higher than the soft X-ray-bright flare loop in the average. The time-of-flight measurements are also compatible with fan reconnection models, if the acceleration region is spread out inside the separator dome. Of course, this location constrains only the centroid of the acceleration region, while the spatial extent is not much constrained, but can be bracketed by the cusp area of the relaxing field lines in the post-reconnection phase.

3.4. Conjugate Footpoint Constraints

While electron time-of-flight measurements mainly provide 1-dimensional constraints on the altitude of the acceleration site, relative time delays between conjugate footpoint hard X-ray emission yield additional information on the horizontal offset or asymmetry of the acceleration site with respect to the vertical symmetry axis of the flare loop.

Sakao (1994) measured the simultaneity of hard X-ray emission from two conjugate footpoints by cross-correlating the hard X-ray fluxes from both footpoints (F_1, F_2). The cross-correlation time delays were found to be near-simultaneous in 6 cases out of 7 investigated flares, with uncertainties of $\pm(0.1 - 0.5)$ s. Using the geometry indicated in Fig. 14 we obtain the following relation between the

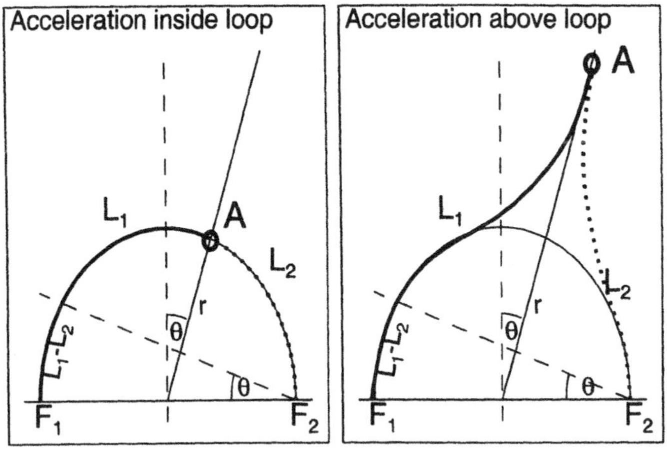

Figure 14. Geometry of electron path trajectories (with lengths L_1, L_2) from a common acceleration site (A) to two conjugate footpoints (F_1, F_2). The inclination angle θ of an asymmetric location A is related to the path difference by $(L_2 - L_1) = 2r\theta$. The two cases illustrate the geometry for an acceleration source inside a semi-circular flare loop (left panel) and for an acceleration site above the flare loop (right panel) [from Aschwanden 1998d].

asymmetry angle θ and the travel times $t_1 = L_1/\mathrm{v}$ and $t_2 = L_2/\mathrm{v}$ for particles propagating with speed v from a common acceleration site A to two conjugate footpoints F_1 and F_2:

$$\theta = \frac{\mathrm{v}}{2r} \frac{(t_1 - t_2)}{q_\alpha q_H} \tag{11}$$

with r the semi-circular loop radius, $q_\alpha \approx 0.64$ is a correction factor for the average pitch angle α, and $q_H \approx 0.85$ a correction factor for helical twist of the magnetic field line along the trajectory (Aschwanden et al., 1996b). The path difference $(L_1 - L_2)$ does not depend on the height of the acceleration source, or on the assumption whether acceleration takes place inside or outside of the loop. In Table I we list the upper limits of the travel time differences $(t_2 - t_1)$ measured by Sakao (1994), the loop radii r of the corresponding flares, and calculate the upper limit for the asymmetry angles θ for three different speeds: (a) for electrons that produce 33-53 keV hard X-ray emission in the *HXT/M2* channel ($E_{kin} \approx 2\epsilon_{HXR} = 66$ keV, v/c=0.46), (b) for 0.1-1 MeV protons (v/c=0.015-0.046), and (c) for conduction fronts (v$\approx$1000 km/s). The results show that the simultaneity limits do not require a highly symmetric acceleration site for electrons (in contrast to widely-held statements), but would require an unplausibly high symmetry for protons or conduction fronts, which can thus be ruled out as energy carriers to produce hard X-ray emission at the footpoints.

In Fig. 15 we show the time profiles of the conjugate footpoint sources for 14 different flares, including Sakao's original selection. The two conjugate time profiles have often a striking high degree of correlation, down to the time resolution of *HXT* (0.5 s). This fact tells as another important clue about the acceleration

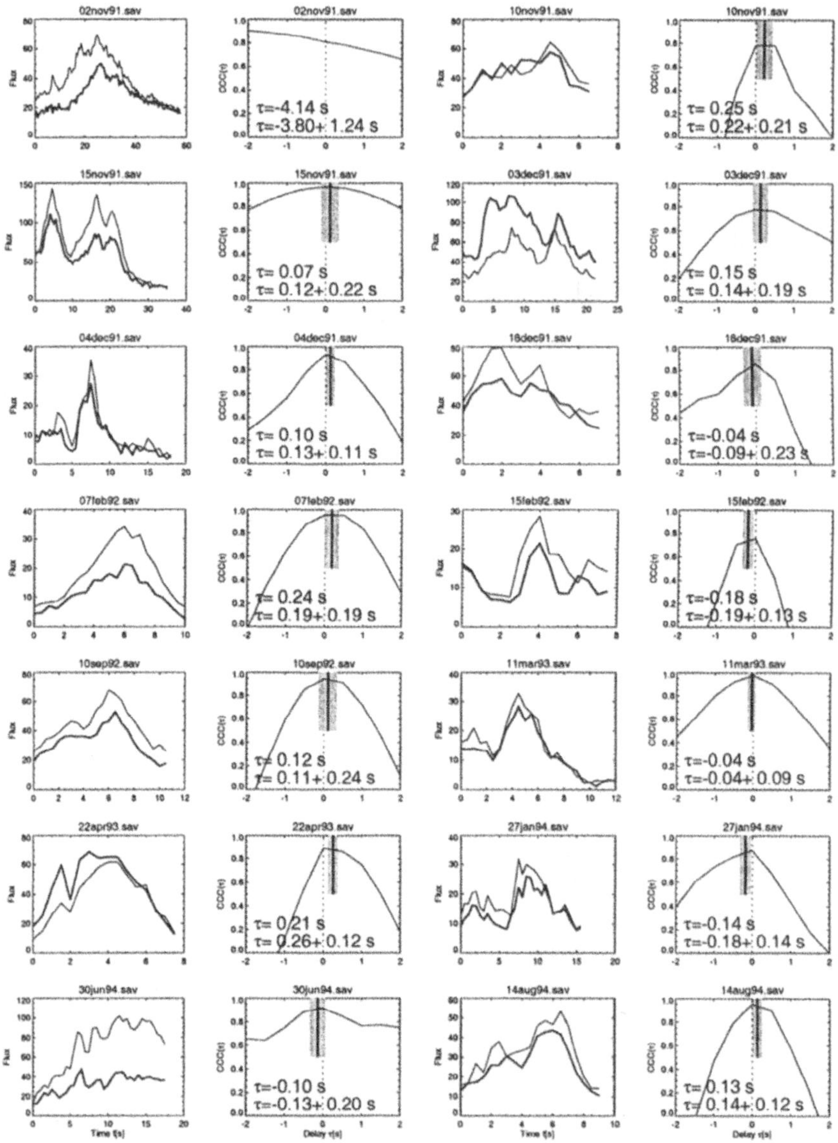

Figure 15. Hard X-ray time profiles $F_1(t)$ and $F_2(t)$ from magnetically conjugate footpoint sources, extracted from *Yohkoh/HXT* images for 14 flares (first and third column; courtesy of Taro Sakao). The two conjugate time profiles are cross-correlated and the cross-correlation coefficient $CCC(\tau)$ and relative time delay τ and uncertainty σ_τ (assuming a dynamic range of 1:10 in *HXT* images) is shown (second and forth column) [from Aschwanden 2000a].

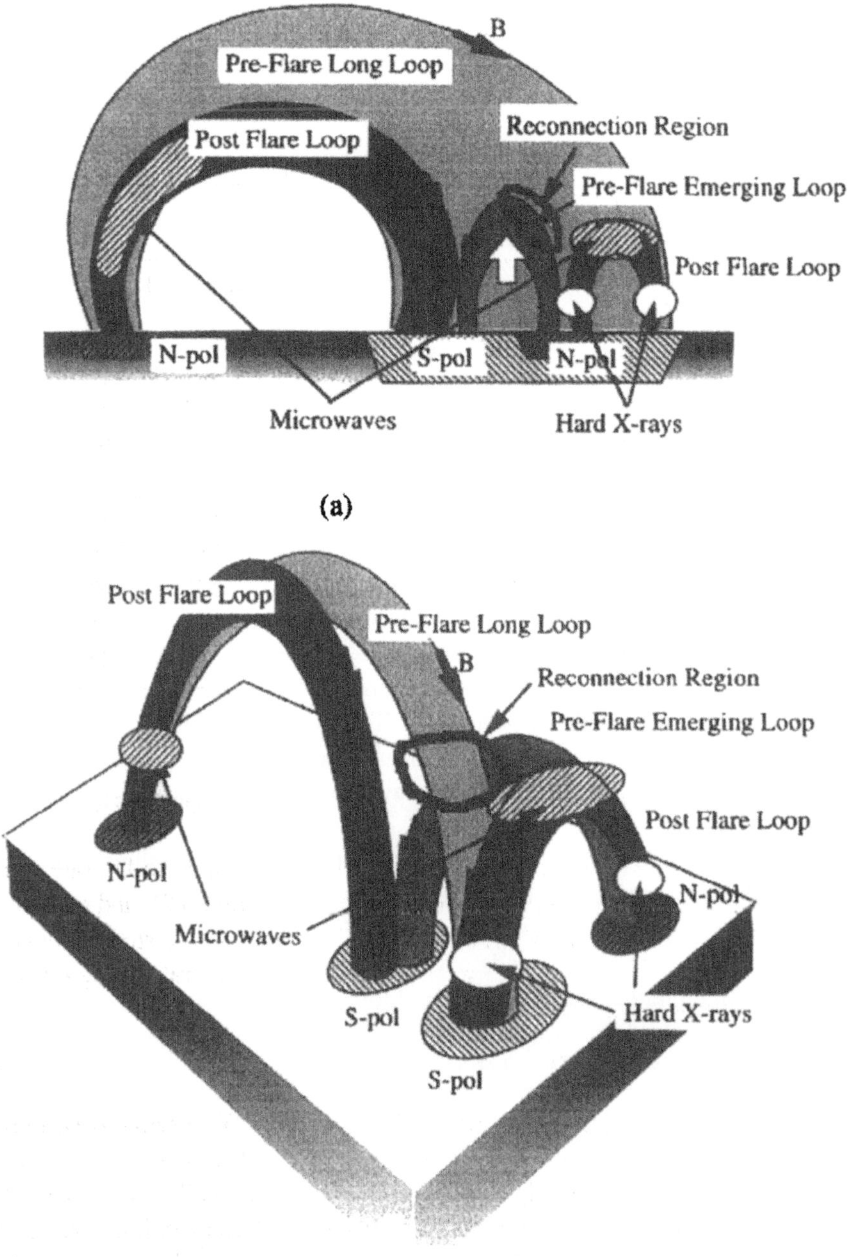

Figure 16. Flare loop configuration in quadrupolar reconnection scenario (Nishio et al. 1997), (see also Fig. 7) showing the locations of primary hard X-ray emission and radio microwave emission. The correlated radio emission in the primary flare loop and near the remote footpoint requires an interaction with oppositely-directed injection of energetic electrons.

TABLE I

Upper limits on asymmetry angle of acceleration site for the flare events of Sakao (1994)

Flare Date	Simultaneity $(t_1 - t_2)$ [s]	Loop radius r[Mm]	Electron asymmetry θ	Proton asymmetry θ	Conduction front asymmetry θ
91/11/15	± 0.1	5.6	$\pm 38^0$	$\pm 1 - 4^0$	$\pm 0.3^0$
91/12/03	± 0.2	4.7	$> 90^0$	$\pm 3 - 9^0$	$\pm 0.7^0$
92/02/07	± 1.0	6.4	$> 90^0$	$\pm 10 - 33^0$	$\pm 2.4^0$
91/12/16	± 0.2	9.7	$\pm 44^0$	$\pm 1 - 4^0$	$\pm 0.3^0$
92/09/10	± 0.3	8.0	$\pm 80^0$	$\pm 2 - 8^0$	$\pm 0.6^0$
91/11/10	± 0.5	15.7	$\pm 67^0$	$\pm 2 - 7^0$	$\pm 0.5^0$
91/11/02	-1.5	11.8	$> 90^0$	$\pm 8 - 27^0$	$\pm 2.0^0$

region. It means that pulses of accelerated particles are not independently produced on field lines at both sides, but rather every acceleration episode produces a stream of particles that “bifurcates” to both footpoints. We can not tell whether the bifurcation happens inside or outside the acceleration region. A natural scenario is that acceleration or injection of particles is controlled on a freshly reconnected field line near the reconnection point in the cusp, with simultaneous injection into opposite directions to conjugate footpoints. If particle acceleration would happen outside the cusp, they would be topologically connected only to one footpoint, and thus the correlation of conjugate hard X-ray footpoint modulation could not be explained.

Conclusions on the degree of simultaneity obtained from correlation of conjugate footpoint time profiles should be taken with caution. Accurate time-of-flight measurements require a decomposition of directly-precipitating from trapped-precipitating electrons, which demands asymmetric trapping models. The timing of the observed precipitating fluxes at each footpoint is therefore non only determined by pure time-of-flight differences, but also the trapping times in asymmetric loops, an effect that is not included in Sakao's analysis.

3.5. Remote Footpoint Delays

Quadrupolar reconnection was found to occur mainly in flares with strongly unequal loop sizes (Hanaoka, 1996; Nishio et al., 1997), with typical loop size ratios of $r_2/r_1 = 1.5...4.1$ (Aschwanden et al., 1999b). The smaller of the two interacting loops always dominates the total hard X-ray and soft X-ray emission, while the secondary, larger loop exhibits only weak soft X-ray emission, radio emission, and occasionally hard X-ray emission at the remote footpoint (Fig. 16). Hanaoka (1999) managed to determine the timing separately for the microwave emission at the remote footpoint with respect to hard X-ray emission in the primary small-scale flare loop and found time delays with a mean of $\Delta t \approx 465 \pm 60$ ms for 4 different flare events. Taking the distance of the remote footpoint to the primary flare loop into account, $L = 67,000 \pm 8,000$ km, he evaluated a mean propagation speed of v$\approx$ 228, 000 km/s (β=v/c=0.76), which is typical for relativistic electrons ($E = 275$ keV). Similar delays for correlated radio emission at remote flare footpoints were previously measured by Nakajima et al. (1985), Lang and Willson (1989), and Willson, Lang, and Gary (1993). Although the previous measurements lacked the spatial resolution to resolve the quadrupolar magnetic topology, it is likely that they involve flares with small-scale and large-scale loops that interact by a quadrupolar reconnection process. Nakajima et al. (1985) suggests that in each of his five analyzed events the distant microwave burst was produced by electrons with energies of 10-100 keV which were channeled along a huge loop from the main flare site to the remote location. In two events, he finds evidence that the electron stream from the primary flare site triggered a sympathetic flare within the remote region. Lang and Willson (1989) observed that energetic electrons propagated along a 260,000 km long transequatorial loop to a remote location. Willson et al. (1993) witnesses electron propagation with a speed of β=v/c$\geq$0.4 over a similar large distance during a flare simultaneously observed with *BATSE/CGRO* and the *VLA*.

These observations do not only confirm the magnetic topology of quadrupolar reconnection models, but constrain also the extent of the acceleration region to a fairly compact region at the interaction point of the reconnection region. The high correlation of primary hard X-rays and secondary radio emission measured by Hanaoka (1996) implies also synchronized acceleration episodes that bifurcate energetic particles in opposite direction, downward to the footpoints of the primary flare loop, and upward along huge secondary loops to a remote footpoint. This scenario further implies that the accelerating fields are not located one-sided in one outflow region of the reconnection volume, but rather symmetrically configured to allow ejection of accelerated particles into opposite directions. This constraint calls for some symmetry of the accelerating fields, which is easily warranted in many reconnection geometries, but cannot be accounted for by parallel acceleration in uni-directional electric fields.

3.6. Bi-Directional Electron Beams

The most stringent geometry constraints of acceleration regions in solar flares probably comes from the startfrequencies of bi-directional electron beams, because they bracket the vertical extent of an accelerating field region. While it was earlier not clear how far away from the acceleration region an electron beam starts to produce radio (type III) emission, the discovery of bi-directional, oppositely-directed electron beams provide a rigorous limit on this minimum distance. Early estimates of the minimum distance x_{min} which the electron beam must move from its point of injection before it becomes unstable toward growing Langmuir waves were estimated

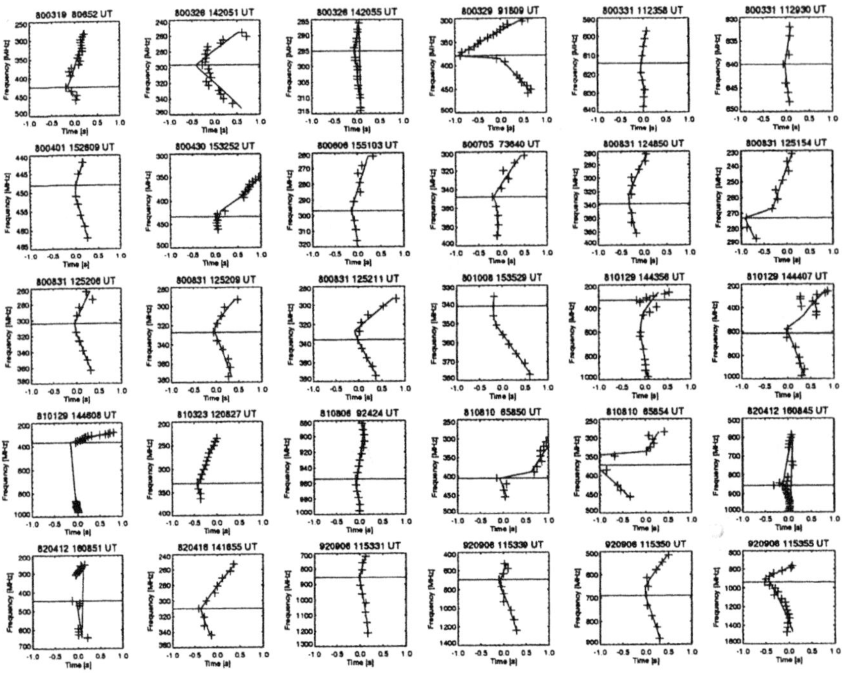

Figure 17. The frequency-time drift rate of 30 oppositely drifting pair bursts is shown from the analysis in Aschwanden et al. (1995c). The time window is always 2 s wide, and the start frequency (of earliest detection) is marked with a horizontal line. The data were obtained with the radio observatory of *ETH* Zurich, mostly in the 300-1000 MHz frequency range, with 3-10 MHz frequency resolution and 0.1 s time resolution. Most burst pairs coincide at the start frequency within $\lesssim 0.1$ s, a few are offset in time.

to be of order $x_{min} \approx 75,000$ km (Kane et al., 1982b). This minimum distance of detection for upward, downward, and bi-directional electron beams have been studied in more detail by Robinson and Benz (2000).

Bi-directional electron beams, i.e. a simultaneously starting pair of an upward-propagating type III burst (with a negative frequency-time drift rate $df/dt < 0$) and a downward-propagating *reverse-drift* (RS) burst (with a positive frequency-time drift rate $df/dt > 0$) were detected mostly in decimetric frequency ranges (Aschwanden et al., 1993a; Aschwanden et al., 1995c; Melendez, 1998; Xie et al., 2000). A collection of 30 observations of such bi-directional radio bursts is shown in Fig. 17. Surprisingly, no frequency gap has been detected between the two starting frequencies. Using the frequency resolution of the radio instruments (3-10 MHz at 300-1000 MHz), an upper limit of the spatial separation h_{sep} between the start frequencies of the two oppositely-directed radio burst sources (assuming plasma emission, i.e. the plasma frequency scales with $f \propto \sqrt{n_e}$) has been derived from

$$h_{sep} \approx \frac{\Delta n_e}{n_e} \lambda = \frac{2 \Delta f}{f} \lambda \ , \qquad (12)$$

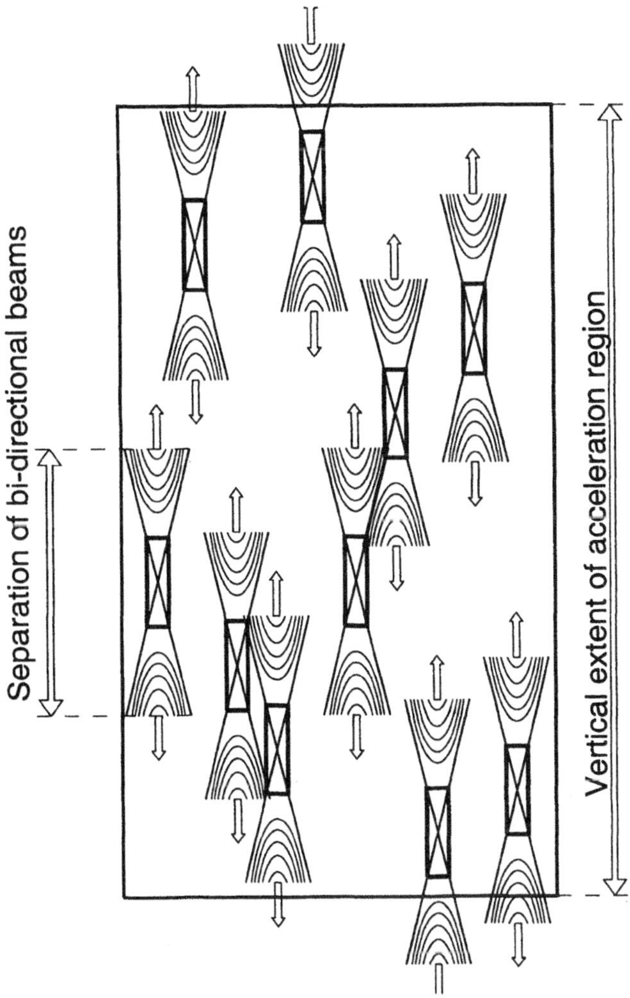

Figure 18. The start frequencies of bi-directional electron beams are shown with respect to the vertical geometry of the acceleration region. The frequency separation between simultaneously ejected electron beams in upward and downward direction yields an upper limit on the vertical extent of a current sheet in a magnetic reconnection region, while the distribution of all start frequencies yields a measure of the vertical extent of the entire flare acceleration region.

yielding values of $h_{sep} \approx 1000$ km for a density scale height of $\lambda \approx 50,000$ km (Aschwanden et al., 1995c). The density in these active regions corresponds to $n_e \approx 10^9 - 10^{10}$ cm^{-3}, assuming fundamental plasma emission in the frequency range of $f \approx 300 - 1000$ MHz. Xie et al. (2000) find bi-directional bursts over an even wider range of 250-2900 MHz, implying electron densities up to $n_e \lesssim 10^{11}$ cm^{-3}. They find also smaller frequency separations, down to 2 MHz, implying compact acceleration regions as small as $\lesssim 100$ km.

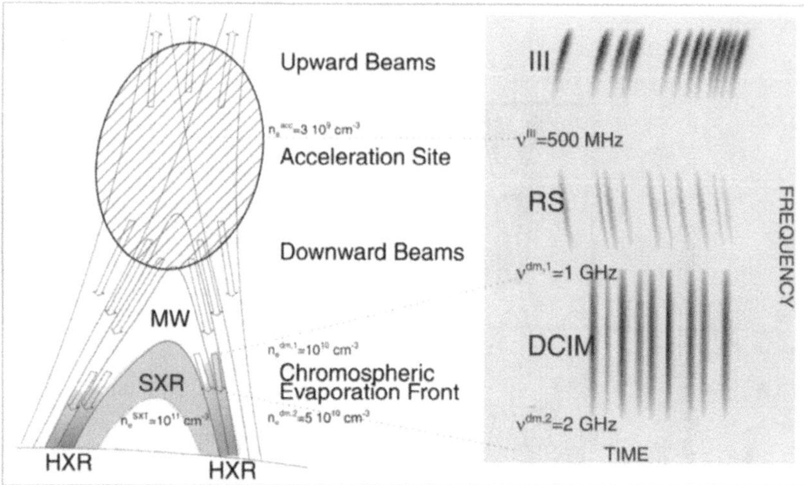

Figure 19. Diagram of a flare model inferred from the magnetic topology constraint of simultaneously detected upward and downward electron beams, in radio and hard X-rays. The diagram on the right illustrates a dynamic radio spectrum with radio bursts indicated in the frequency-time plane. The acceleration site is located in a low-density region (in the cusp) with a density of $n_e^{acc} \approx 10^9$ cm^{-3} from where electron beams are accelerated in upward (type III) and downward (RS bursts) directions. Downward-precipitating electron beams produce pulses of chromospheric thick-target bremsstrahlung emission, possibly intercepting chromospheric upflows. Those loops that have already been filled with heated chromospheric plasma brighten up in soft X-rays and have higher densities of $n_e^{SXR} \approx 10^{11}$ cm^{-3} than the acceleration region. There is a filling delay of soft X-ray loops, during which the magnetic reconnection point rises higher, widening the hard X-ray emitting footpoints [from Aschwanden and Benz 1997].

This separation distance $h_{sep} \approx 100-1000$ km can be interpreted as upper limit of the vertical extent of an elementary acceleration region, e.g. located in a symmetric current sheet geometry (Fig. 18). However, startfrequencies of subsequent radio bursts during the course of a flare do not occur all at an identical frequency, they are clustered in a frequency range of $\Delta f/f \approx 0.2 - 0.5$, suggesting a much larger vertical extent of the entire acceleration region (Fig. 18), in the order of 5,000-50,000 km (Aschwanden et al., 1993a; 1995c; Melendez, 1998; Melendez et al. 1999; Xie et al., 2000).

3.7. Hard X-ray and Radio-Coincident Electron Beams

Another topolocigal constraint of the acceleration region comes from triple coincidences, i.e. two radio bursts (in upward and downward direction) and a simultaneous hard X-ray pulse (which is produced by thick-target bremsstrahlung of downward-precipitating electrons). Detailed correlations between radio bursts and hard X-ray pulses have been studied since the *Solar Maximum Mission (SMM)* era, when hard X-ray data from the *Hard X-Ray Burst Spectrometer (HXRBS)* with a time resolution of 0.1 s became available. Single coincidences of radio and hard

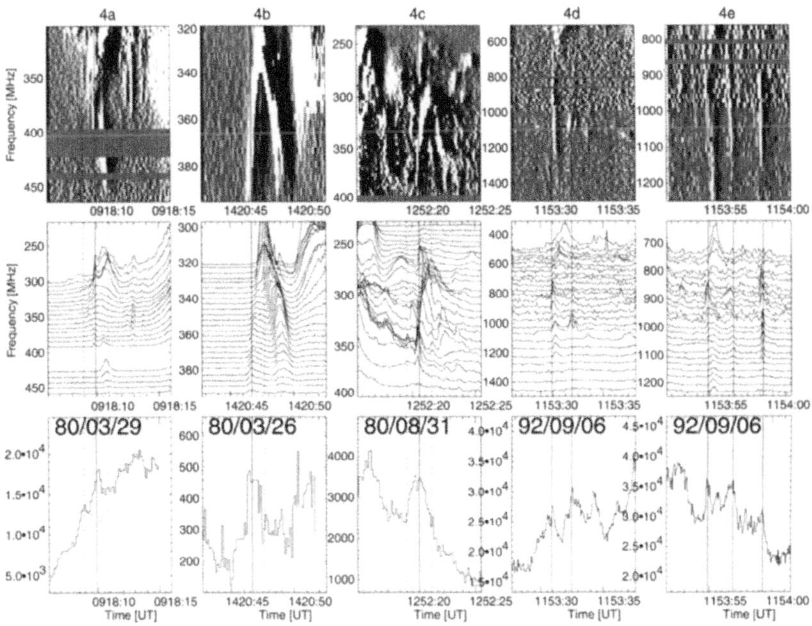

Figure 20. Examples of radio and hard X-ray coincident bursts: (4a,4b) oppositely-drifting burst pair; (4b,4c) type U bursts; (4d,4e) reverse-slope-drifting bursts. The hard X-ray data are from *SMM/HXRBS* (4a-4c) and from *BATSE/CGRO* (4d-4e). The radio data were recorded from *ETH* Zurich, shown as dynamic spectra (top row) and as time profiles at multiple frequencies (middle row) [from Aschwanden et al. 1995c].

X-ray bursts as good as $\lesssim 0.1$ s were interpreted as common acceleration events, occurring in altitudes of $h \approx 40,000$ km (Dennis et al., 1984), which is a typical height for cusps above soft X-ray flare loops. The existence of such short delays was used to rule out slow acceleration processes, e.g. production of type III bursts by cross-field drifts (Dennis et al., 1984). The detailed timing of radio bursts with hard X-ray pulses observed with *BATSE/CGRO* with a time resolution of 64 ms revealed many detailed one-to-one correlations within $\lesssim 500$ ms (Aschwanden et al., 1993a; 1995c,d; see examples in Fig. 20), but the start of the radio bursts occurred always slightly delayed (by a few 100 ms) to the start of the hard X-ray pulses.

In summary, from these correlation and timing studies the following was concluded about the nature of the particle acceleration source:

1. Electron beams are simultaneously accelerated in upward and downward direction.
2. Acceleration occurs in intermittent pulses with typical durations of $\lesssim 1$ s.
3. An upper limit of the acceleration time of nonthermal electrons is given by the pulse duration, i.e. $t^{acc} \lesssim 1$ s.

4. The temporal pattern of intermittent acceleration pulses is neither periodic nor strictly random, suggesting some memory function (or nonlinear evolution) between subsequent acceleration pulses.
5. The acceleration source is spatially highly fragmented, based on the scatter of start frequencies.
6. The mean height of the acceleration source, as well as its vertical extent, vary over a wide range, say $h^{acc} \approx 5000 - 50,000$ km.
7. The vertical extent of an elementary acceleration structure (determined from the separation distance between upward and downward starting electron beams) is much smaller than the vertical extent of the entire acceleration region, in the order of $\Delta h \approx 100 - 1000$ km.

These observations do not allow to discriminate between specific theoretical acceleration mechanisms, but they provide information on the spatial and temporal organization of accelerating fields.

3.8. Electron Density in Acceleration Sites

The most prominent features observed in imaging data from flares are the hard X-ray footpoints (e.g. seen with *Yohkoh/HXT*), where nonthermal particles bombard the chromosphere, and the soft X-ray-bright dipolar-like flare loops (e.g. seen with *Yohkoh/SXT*) that bridge the footpoints or chromospheric flare ribbons (seen in Hα). Radio images show also gyrosynchrotron emission of trapped electrons in or above the soft X-ray flare loops. Given all this imagery, it is still not clear where the particle acceleration site is located with respect to these features. From the topology of magnetic reconnection events, the discovery of Masuda's above-the-loop-top sources, and from electron time-of-flight measurements we expect that the primary particle acceleration site is located *above* the soft X-ray flare loops, but there is still some uncertainty about the detailed spatial structure in the cusp region. An additional diagnostic of the acceleration site are comparisons of electron density measurements in radio and soft X-rays.

The observation of soft X-ray flares often show single or multiple loops, for which the transverse loop diameter w can be measured. It is then reasonable to assume that these loops have a cylindrical symmetry, which yields an estimate of the line-of-sight depth $z \approx w$, so that the emission measure EM can be used to infer the electron density n_e in such loops,

$$EM = \int n_e^2(z)dz \approx < n_e >^2 w . \tag{13}$$

If the soft X-ray source has a filling factor of less than unity, the so determined electron density represents only a lower limit. On the other side, if radio type III or RS bursts are observed in the same flare region, the plasma frequency f_p measured

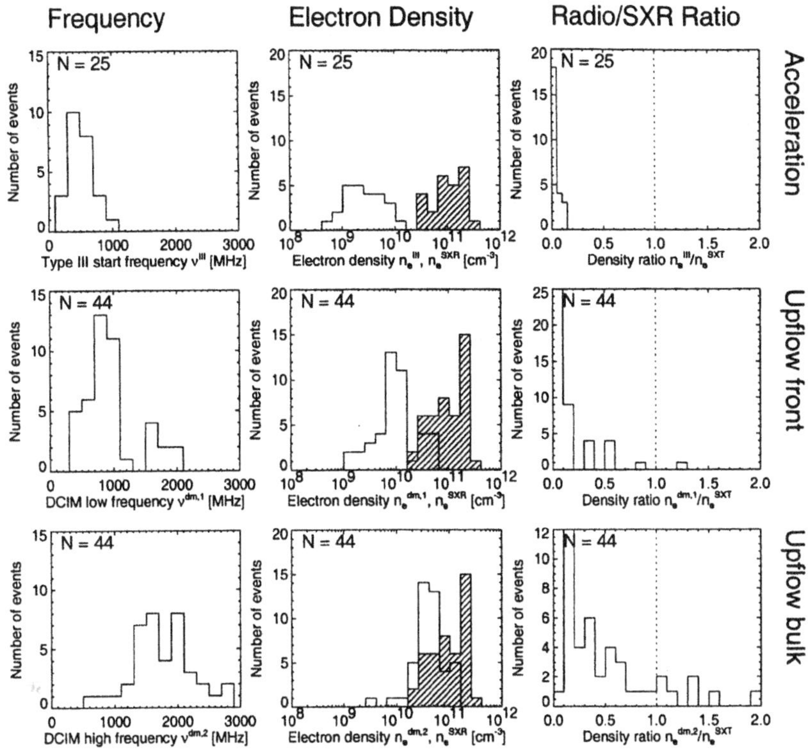

Figure 21. Statistics of radio frequencies (left column), electron densities (middle column), and the ratio of radio-inferred to soft X-ray-inferred densities (right column). The open histograms refer to the radio-inferred values, the hatched histograms to the soft X-ray-inferred values. The values are measured at the start frequencies of metric type III bursts (top row), at the low frequency of decimetric bursts (middle row), and at the high frequency of decimetric bursts (bottom row). The frequency regimes of the three rows are identified with the locations of the acceleration site (top), the upflow front (middle), and the bulk upflow (bottom) of the chromospherically evaporated plasma (see scenario in Fig. 19).

by their radio spectrum is a direct function of the electron density n_e along their path,

$$f_p = \sqrt{\frac{n_e e^2}{\pi m_e}} = 8980\sqrt{n_e} \quad [Hz] , \tag{14}$$

with e the elementary charge and m_e the mass of an electron. Measuring the soft X-ray emission measure and simultaneous start frequencies of near-cospatial radio bursts thus provides us with a density diagnostic of the acceleration site with respect to the flare loop. The only uncertainty is whether plasma emission is generated at the fundamental or harmonic (changing the density by a factor of $\sqrt{2}$) and whether there is a significant minimum distance before an electron beams starts to produce observable radio emission. The latter uncertainty is, however, completely removed for bi-directional electron beams. Using this method of density diagnostic during 14 flares it was found that the soft X-ray-inferred electron densities ($n_e^{SXR} = (0.2 - 2.5) \times 10^{11}$ cm^{-3}) were always significantly higher than the radio-inferred electron densities in acceleration sites, $n_e^{acc} = (0.6 - 10) \times 10^9$ cm^{-3}, inferred from the plasma frequency at the separatrix between upward and downward-accelerated electron beams (Aschwanden and Benz, 1997). Histograms of the distributions of both densities are shown in Fig. 21. Thus the ratio of densities in the acceleration region amounts only to a fraction of $n_e^{acc}/n_e^{SXR} = 0.005 - 0.13$. This is again a strong corroboration that the acceleration site is fully detached from the soft X-ray bright (post-)flare loops, located in a low-density region fully separated from the chromospheric evaporation-filled flare loops, leading to the scenario shown in Fig. 19.

A summary of observational constraints and diagnostics on the geometry of acceleration regions in flares is given in Table II.

4. Dynamics of Acceleration Region

While we focused so far on the magnetic topology (Section 2) and the spatial geometry (Section 3) of the acceleration region, we focus in this Section on the temporal and dynamic properties. This includes the dynamic processes of flare triggers and drivers (Section 4.1) and current sheet dynamics (Section 4.2), such as tearing mode instability, magnetic island formation, coalescence instability, which all together constitute a non-steady, bursty, and intermittent reconnection regime. Related observations of fast time structures (Section 4.3), spectral fragmentation (Section 4.4), and spatial fine structure in flare loop arcades (Section 4.5) are then discussed in the context of the foregoing theoretical concepts of current sheet dynamics. Because particle acceleration can happen in all these phases, it is instructive to review first the dynamics of these processes, before we deal with the actual acceleration mechanisms (in Section 5).

4.1. Dynamics of Flare Triggers and Drivers

A solar flare is the occurrence of a magnetic instability, which requires a trigger or driver before the magnetic field configuration loses its equilibrium and evolves catastrophically into a new equilibrium of a lower energy state. A physical model of the trigger or driver mechanism is important for us to understand what type of magnetic reconnection topology is initiated, which determines the currents and electromagnetic fields set up in the preflare phase for particle acceleration during the impulsive phase. Table III shows a list of possible flare driver mechanisms that have been proposed in literature. The dynamic processes are grouped into (1) *preflare drivers*, which store and build up energy in the preflare phase, (2) *flare triggers*, which represent disturbances that launch catastrophic changes in a marginally stable system, and (3) *flare drivers*, which feed energy into a dissipating system. The processes listed in Table III are just examples and by no means complete.

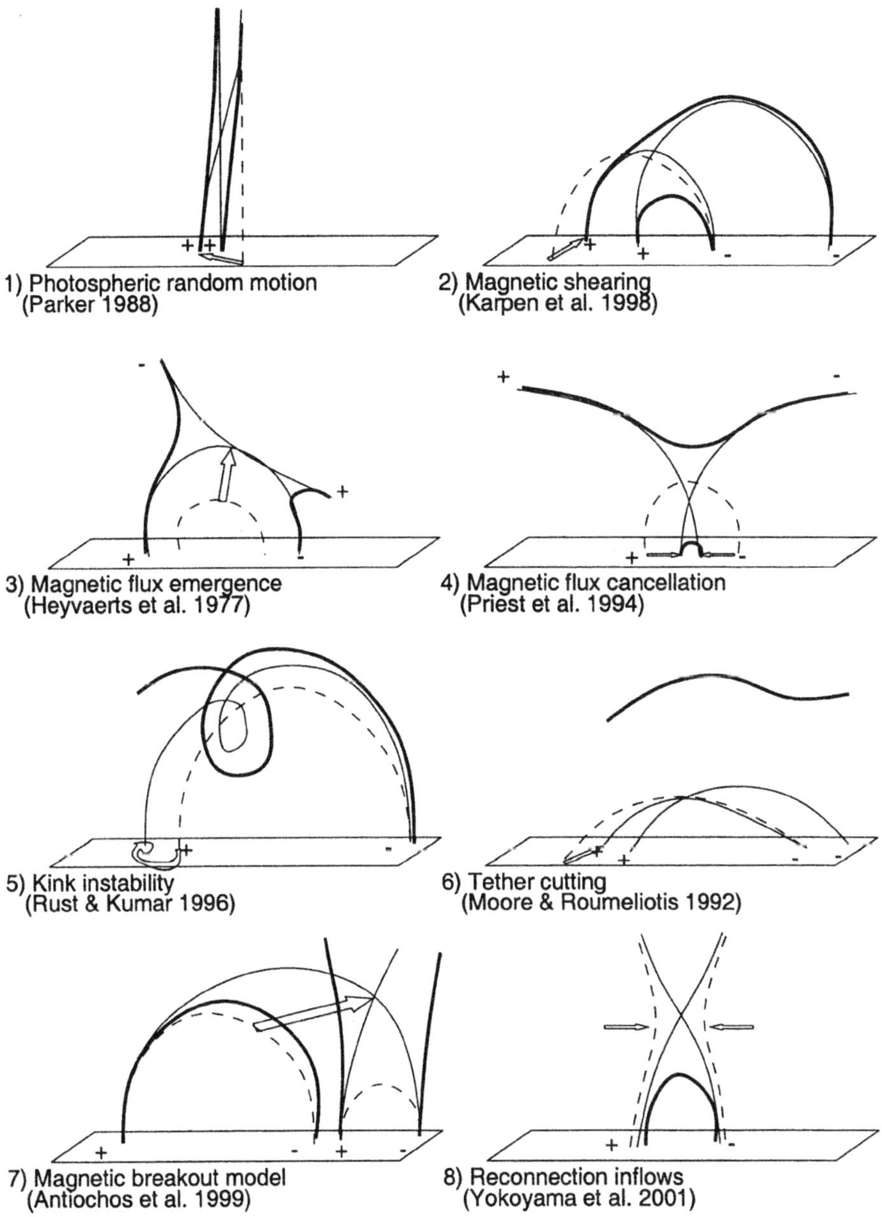

Figure 22. Flare trigger and driver mechanisms (see Table III). The driving force is indicated with an arrow, the pre-reconnection field lines with dashed line style, at the time of reconnection with solid line style, and the relaxed post-reconnection field lines with thick solid line style.

TABLE II
Summary of Observational Constraints on Acceleration Region Geometry

Acceleration Region Diagnostic	Observational constraint	Typical Values	Refs.
Altitude	Stereoscopic spacecraft	$h \approx 2500 - 30,000$ km	1)
	Electron time-of-flight	$h \approx 5000 - 35,000$ km	2)
Locus relative to SXR	Coronal HXR sources	≈ 7000 km above SXR	3)
	Electron time-of-flight	$L/s \approx 1.4 \pm 0.3$	2)
	Conjugate HXR timing	symmetric midpoint	4)
	SXR density + type III freq.	low density region outside SXR	5)
Vertical Extent	Start frequency range III,RS	$\Delta h = 5000 - 50,000$ km	6)
Directionality	HXR+Remote HXR correlations	upward/downward	7)
	Type III+RS coincidence	upward/downward	6)
	Type III+HXR coincidence	upward/downward	8)
Spatial organization	Type III start freq.distrib.	spatially fragmented, $\Sigma(\delta h)$	6)
	Type III+RS freq. separation	$\delta h \approx 100 - 1000$ km	6)
	Type III maximum rate	$\delta h \approx 500 - 5000$ km	9)
	metric spikes bandwidth	$\delta h \approx 3000$ km	10)
	arcade loop separation	$\delta h \approx 1800$ km	11)
Temporal organization	Type III rate	intermittent, 1-10 Hz	6)
	HXR pulses	intermittent, 1-10 Hz	8)
Acceleration time	HXR pulse duration	$t_{acc} \lesssim 0.1 - 1$ s	8)
Electron density	Type III+RS start frequency	$n_e = (0.6 - 10) \times 10^9$ cm^{-3}	5)

References: 1) Kane et al. 1979, 1982a; Kane 1983, Hudson 1978, 2) Aschwanden et al. 1996c, 1996d, 3) Masuda et al. 1994a, 1994b, 4) Sakao 1994, Aschwanden 2000a, 5) Aschwanden and Benz 1997, 6) Aschwanden et al. 1993a, 1995c, Melendez 1998; Melendez et al., 1999, Xie et al. 2000, 7) Hanaoka 1999, 8) Dennis et al. 1984, Aschwanden et al. 1993a, 1995c, 1995d, 9) Aschwanden, Benz, and Kane 1990, 10) Benz, Csillaghy, and Aschwanden 1996, 11) Aschwanden and Alexander 2001.

4.1.1. *Photospheric random motion*

The random motion of photospheric flows is ultimately driven by the convectional motion of granules, mesogranules, and supergranules, with upflows in the center of granules and downflows at the outer boundaries of the granules, associated with the photospheric network. This random motion has the effect of shuffling the footpoints of coronal field lines around, which become tangled and braided. This scenario was proposed by Parker (1988) to trigger nanoflares in the corona, whenever the tangential discontinuities exceed some critical angle. The resulting magnetic reconnection occurs then between two almost parallel magnetic field lines, and the free energy released by this process was estimated to be of order $\approx 10^{24}$ erg (Parker, 1988). Due to the smallness of this energy amount, this energy

TABLE III

Flare Trigger and Driver Mechanisms

Physical Mechanism	Observational manifestations	Refs.
PREFLARE DRIVERS:		
– Photospheric random motion	photospheric network, nanoflares	1)
– Magnetic shearing	magnetic field along neutral line	2)
– Magnetic flux emergence	new dipoles in magnetograms,	3)
– Magnetic flux cancellation	disappearing dipoles in magnetograms	4)
FLARE TRIGGERS:		
– Kink instability	twisted soft X-ray loops, sigmoids, filament eruption	5)
– Tether cutting	arch filaments, filament fibrils, Hα crossing fibrils	6)
– Magnetic breakout model	multipolar magnetic topologies, δ-spots	7)
FLARE DRIVERS:		
– Reconnection Inflows	converging coronal flows in EUV	8)

References: 1) Parker 1988, 2) Hagyard et al. 1984, 1990, Karpen et al. 1991, 1995, 1998, 3) Heyvaerts et al. 1977, Kurokawa 1987, 1991, Nitta et al. 1994, Leka et al. 1996, Shibata et al. 1996, Schmieder et al. 1997, Canfield and Reardon 1998, Fan 1999, 4) Wang and Shi 1993, Priest, Parnell, and Martin 1994, Jiang and Wang 2001, Zhang et al. 2001, 5) Cheng 1977, Hood and Priest 1979, Rust and Kumar 1996, Canfield, Hudson, and McKenzie 1999, 6) Van Ballegooijen and Martens 1989, Moore and Roumeliotis 1992, Canfield et al. 1994, Canfield and Reardon 1998, Moore et al. 2001, 7) Antiochos 1998, Antiochos, DeVore, and Klimchuk 1999, Aulanier et al. 2000, 8) Yokoyama et al. 2001, Blackman 1997.

release process was dubbed *nanoflaring*, which could account by its cumulative effect for continuous coronal heating, rather than for individual flare events. Nevertheless, in the context here we can ask whether this process is capable of particle acceleration. With a simple energy argument we can estimate the maximum free energy per particle,

$$E = \frac{B_\perp^2}{8\pi} \frac{1}{n_e} \approx 1.5 - 25 \text{ keV} \tag{15}$$

where $B_\perp \approx 25$ G is the transverse magnetic field component estimated by Parker (1988), and $n_e = (0.6 - 10) \times 10^9$ cm^{-3} the electron density in acceleration regions inferred from bi-directional radio bursts (Aschwanden and Benz, 1997). This process is therefore capable of accelerating nonthermal electrons. However, because photospheric random motion is ubiquitous on the solar surface, such nonthermal electrons should be produced copiously all the time (not only during flares) in coronal fields above active regions.

4.1.2. *Magnetic shearing*

Oppositely-directed photospheric flows on both sides of a neutral line shear the overlying coronal magnetic field. Measuring this shear, defined as angular difference in the photosphere between the extrapolated potential magnetic field and the observed vector magnetic field, Hagyard et al. (1984) found that locations exceeding a critical shear angle ($\gtrsim 80^0 - 85^0$) triggered co-spatial flare events. Neutral lines with longer sheared segments were found to be more likely to produce large flares with gamma rays (Hagyard, Venkatakrishnan, and Smith, 1990). Thus, highly-sheared magnetic fields seem to be more efficient in accelerating particles to high energies. Symmetric and asymmetric shearing of a quadrupolar configuration was studied with hydrodynamic simulations by Karpen et al. (1991), but it was found that this topology does not form coronal current sheets. However, when enhanced resistivity is assumed in the magnetic X-point of a quadrupolar configuration, shearing was found to lengthen the initial X-point into a vertical current sheet, which gradually reconnects and undergoes tearing (Karpen et al., 1995). In particular, for asymmetric shearing of unequal dipoles the effects of reconnection-driven current filamentation and magnetic island formation are most prominent (Karpen et al., 1998). Magnetic shearing in coronal zones with enhanced resistivity thus seems to be a very favorable condition to trigger magnetic reconnection and particle acceleration.

4.1.3. *Emerging Flux*

Subphotospheric magnetic fluxtubes that emerge through the photosphere and chromosphere push against the pre-existing overlying coronal field, leading to heating of the intervening current sheet, increase of resistivity, and eventual rapid reconnection and particle acceleration (Heyvaerts et al., 1977). The ultimate source of emerging magnetic flux is thought to be at the bottom of the convection zone, in the tachocline and in convective overshoot layers (e.g. Fisher et al. 1991). The subphotospheric rise of kink-unstable magnetic flux tubes, which emerge as twisted Ω-tubes, produce compact magnetic bipoles similar to those seen in δ-spots (Fan et al., 1999), which are most flare-prone. Evidence for current-carrying emerging flux was demonstrated by Leka et al. (1996) from potential-field extrapolations of vector magnetograms. Observational evidence for emerging flux in the preflare phase was found in form of arch filaments in Hα and increased vertical flux in vector magnetograms (Canfield and Reardon, 1998). Successive emergence of twisted magnetic flux ropes was found to be an essential criterion to produce major flares (Kurokawa, 1987, 1991). However, the relation between emerging flux and flare events is somewhat indirect, vigorous flux emergence is often observed hours before the flare (Nitta et al., 1994). While emerging flux definitely contributes to enhanced flaring, the detailed trigger that leads to the rapid instability at the beginning of the impulsive flare phase is not known. The emerging flux model works as a preflare driver in tripolar (Shibata et al., 1996) and quadrupolar (Nishio et al., 1997) reconnection geometries, as well as in 3D-nullpoint topologies, e.g.

where the emergence of a new magnetic polarity surrounded by opposite polarity creates a coronal nullpoint encompassed by a fan-surface separator dome (Fig. 7). Evidence for the latter scenario was found from the intersection of computed quasi-separatrix layers with observed Hα flare ribbons (e.g. Gorbachev and Somov 1989; Schmieder et al., 1997).

4.1.4. *Cancelling Flux*

The opposite process to magnetic flux emergence, i.e. the submergence of a magnetic fluxtube, which appears as a disappearance of a dipolar feature in a photospheric magnetogram has also been associated with the role of a flare trigger (Wang and Shi, 1993; Jiang and Wang, 2001; Zhang et al., 2001). A converging flux model that triggers magnetic reconnection and explains X-ray bright points associated with cancelling magnetic features was proposed by Priest et al. (1994). In a recent work it was found that the only obvious magnetic change during the 2000-Jul-14 flare was magnetic flux cancellation at many sites in the vicinity of the erupting filament, where all the initial disturbance in the filament and the initial brightening around the filament took place at the cancellation sites (Zhang et al., 2001). It suggests that slow magnetic reconnection in the chromosphere, manifested as flux cancellation, leads up to the global instability responsible for filament destabilization in major flares.

4.1.5. *Kink Instability and Filament Eruption*

Flares are often accompanied by a filament eruption. It was suggested that the eruption is caused by the kink instability, which sets in when the amount of magnetic twist in the flux tube exceeds a critical value, typically $2\pi - 6\pi$ (Hood and Priest, 1979; Cheng, 1977). Observational evidence for helically kinked magnetic flux ropes was shown in loop bundles from *Yohkoh/SXT* data (Rust and Kumar, 1996). Such twisted loop bundles have a macroscopic S-shape, also called *sigmoids*, which were found to be good predictors of eruptive flares (Canfield et al., 1999).

4.1.6. *Tether Cutting*

The destabilizing step when quadrupolar reconnection between highly-sheared field lines above the neutral line trigger the loss of equilibrium and eruption of a filament has been coined *tether cutting* by Moore and Roumeliotis (1992). This step is part of a more comprehensive model on the formation and eruption of a filament by Van Ballegooijen and Martens (1989), where the initial trigger (preceding the tether-cutting phase) of the filament eruption is flux cancellation at the neutral line beneath the filament, transferring gradually more magnetic flux to the helical field of the filament, until it loses equilibrium. Applications of this *tether-cutting* model to flare observations can be found in Canfield et al. (1994), Canfield and Reardon (1998), and Moore et al. (2001).

4.1.7. *Magnetic Breakout Model*

An extension of sheared-arcade flare models (producing the well-known double-ribbon flares) is the participation of an overlying flux system (e.g. a parallel-oriented secondary arcade) that plays a crucial role in triggering the eruption (Antiochos et al., 1999). In this multipolar magnetic topology, reconnection between the sheared arcade and the adjacent arcade removes the unsheared field above the low-lying primary arcade and allows thereby this core flux to burst open, leading to filament eruption, flaring, and CME. Aulanier et al. (2000) generalized the 2D-quadrupolar reconnection version of Antiochos et al. (1999) to the case of 3D reconnection, involving a coronal nullpoint above a δ-spot. He defines: "a magnetic breakout is the opening of initially low-lying sheared fields, triggered by reconnection at a nullpoint that is located high in the corona and that defines a separatrix enclosing the sheared fields" (see Fig. 6).

4.1.8. *Reconnection Inflows*

First evidence of reconnection inflows in a solar flare was discovered by Yokoyama et al. (2001), who identified converging flows with a speed of v=5 km s^{-1} in EUV toward the magnetic X-point above a cusp-shaped soft X-ray flare loop. They inferred a reconnection rate (Alfvén Mach number) of $M_A = 0.001 - 0.03$. The momentum of the inflowing mass or the difference in magnetic pressure high up in the corona represents a driver of the magnetic reconnection process. An interesting consequence for particle acceleration scenarios is that the rate of accelerated particles in the fast shocks of the downstreaming outflows is directly related to the angle of the inflow field and the velocity of the slow shocks (Blackman, 1997).

In conclusion, we can associate with each flare driver and trigger mechanism certain type of magnetic reconnection topologies (as defined in Fig. 7). These associations can be used either as diagnostic of the reconnection topology for an observed flare trigger mechanism, or to infer an unknown flare trigger mechanism from an observed magnetic topology. In the context of this review, the knowledge of the magnetic reconnection topology in a particular flare is the most essential prerequisite to understand the electromagnetic fields that determine particle acceleration and kinematics.

4.2. Dynamics of Magnetic Reconnection

In the last Section on flare triggers and drivers we described the forces that build up non-potential magnetic energy until the system becomes unstable and loses equilibrium. The energy build up phase involves relatively slow processes (on time scales of hours to days before a flare), while the loss of equilibrium and primary energy release happens on relatively short time scales (seconds to minutes) that make up the impulsive flare phase, which is the most relevant phase for particle acceleration. We are therefore interested in fast magnetic reconnection processes (on

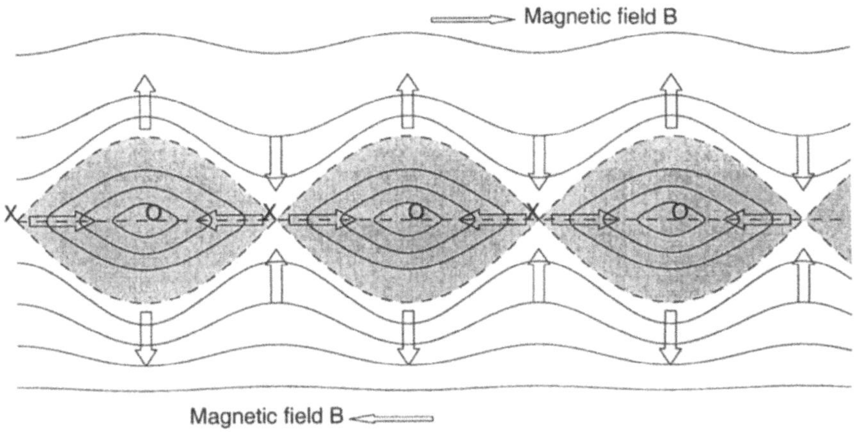

Figure 23. Magnetic island formation by tearing mode instability in magnetic reconnection region. Magnetically neutral X and O points are formed at the boundary between regions of oppositely directed magnetic field, with plasma flow in the directions indicated by arrows. [after Furth et al., 1963].

time scales of seconds to minutes), and thus do not consider steady-state or quasi-stationary magnetic reconnection processes here (for an overview see Sections 4 and 5 in Priest and Forbes, 2000). The dynamics of steady reconnection processes can basically be described by orthogonally-directed inflows and outflows through the reconnection (or magnetic diffusion) region, where the flows are constrained by the law of mass conservation. Early concepts where the diffusion region during the magnetic reconnection process is described with long current sheets (Sweet, 1958a,b; Parker, 1957; 1963) were found to be too slow to explain the rapid energy release during solar flares. The process can be speeded up by reducing the diffusion region to short X-points (Petschek, 1964), which is considered as one of the possible mechanisms for solar flares. What we really need, to explain the bursty and intermittent pulses (on time scales of seconds or subseconds) of particle acceleration seen in hard X-rays and radio wavelengths during the impulsive flare phase, are *unsteady magnetic reconnection modes* (see Sections 6 and 7 in Priest and Forbes, 2000; Kliem, 1995). We describe in the following a few such unsteady reconnection modes that are thought to be most relevant for particle acceleration in solar flares, such as the tearing instability, the coalescence instability, and their combined dynamics (i.e. the regime of *bursty reconnection*). There are also other reconnection types, such as X-type collapse (Dungey, 1953; Priest and Forbes, 2000, p.205), resistive reconnection in 3D (e.g. Schumacher, Kliem, and Seehafer, 2000; Priest and Forbes, 2000, p.230), or collisionless reconnection (e.g. Drake, Biskamp, and Zeiler 1997; Haruki and Sakai, 2001a, b), which have not yet been applied to solar flares, but have been discovered in the Earth's magnetotail (Øieroset et al. 2001).

4.2.1. *Tearing Mode Instability and Magnetic Island Formation*

In current sheet formations, resistive instabilities can occur, where the magnetic field lines can move independently of the plasma due to the non-zero resistivity (opposed to the *frozen-flux theorem* for zero resistivity). In magnetic reconnection regions with high magnetic Reynolds numbers ($R_m = \tau_d/\tau_A$), where the outward diffusion (on a time scale of $\tau_d = l^2/\eta$, with $2l$ the width of the current sheet and $\eta = (\nu\sigma)^{-1}$ the magnetic diffusivity) is much larger than the Alfvénic transit time $\tau_A = l/v_A$ (with $v_A = B_0/\sqrt{\mu\rho_0}$), i.e. $\tau_d \gg \tau_A$, three different types of resistive instabilities can occur: gravitational, rippling, and tearing mode (Furth et al., 1963). The tearing mode, which has a wavelength greater than the width of the sheet ($kl < 1$), has a growth time τ_G^{tear} of

$$\tau_G^{tear} = [(kl)^2 \tau_d^3 \tau_A^2]^{1/5} \tag{16}$$

for wavenumbers in the approximate range $(\tau_A/\tau_d)^{1/4} < kl < 1$. (e.g. see derivations in Furth et al. (1963), Priest (1982, p.272), White (1983), or Sturrock (1994, p.272). Thus, the mode with the longest wavelength has the fastest growth rate,

$$\tau_{G,min}^{tear} = [\tau_d \tau_A]^{1/2} . \tag{17}$$

The tearing mode produces magnetic islands in 2D (see Fig. 23), or magnetic flux ropes in 2.5D, respectively. These structures saturate in the nonlinear phase of the tearing mode (if coalescence is not permitted) and their subsequent diffusion at the diffusive timescale τ_d is extremely slow (since $R_m \gg 1$ in the corona). The energy release of the tearing-mode instability occurs during the process of island formation. Tearing modes have been applied to solar flares in a number of theoretical studies (e.g. Sturrock 1966; Heyvaerts et al., 1977; Spicer 1977a,b; 1981; Somov and Verneta, 1989; Kliem, 1990), and numerical MHD simulations have been performed by Biskamp and Welter (1989). Kliem (1995) estimated the growth time of the tearing mode for coronal conditions ($n_e = 10^{10}$ cm^{-3}, $T = 2.5 \times 10^6$ K, B=200 G, with smallest current sheets half widths of $l \approx 7 \times 10^3$ cm), which yields $\tau_G^{tear} \approx 0.4$ s. This time scale is comparable with the duration of elementary time structures observed in form of hard X-ray pulses and radio type III bursts, which provide an upper time limit on the responsible acceleration processes. So, the tearing mode has the right time scale to modulate particle acceleration in the reconnection region. Because the tearing mode has a threshold current density orders of magnitude below the threshold of kinetic current-driven instabilities, it will occur first. Continued shearing and tearing may reduce the width of the current sheet until the threshold of a kinetic instability is reached (Kliem, 1995).

4.2.2. *Coalescence Instability*

While the tearing mode leads to filamentation of the current sheet, the resulting filaments are not stable in a dynamic environment. If two neighbored filaments approach each other and there is still non-zero resistivity, they enter another instability, the *coalescence instability*, which merges the two magnetic islands into a

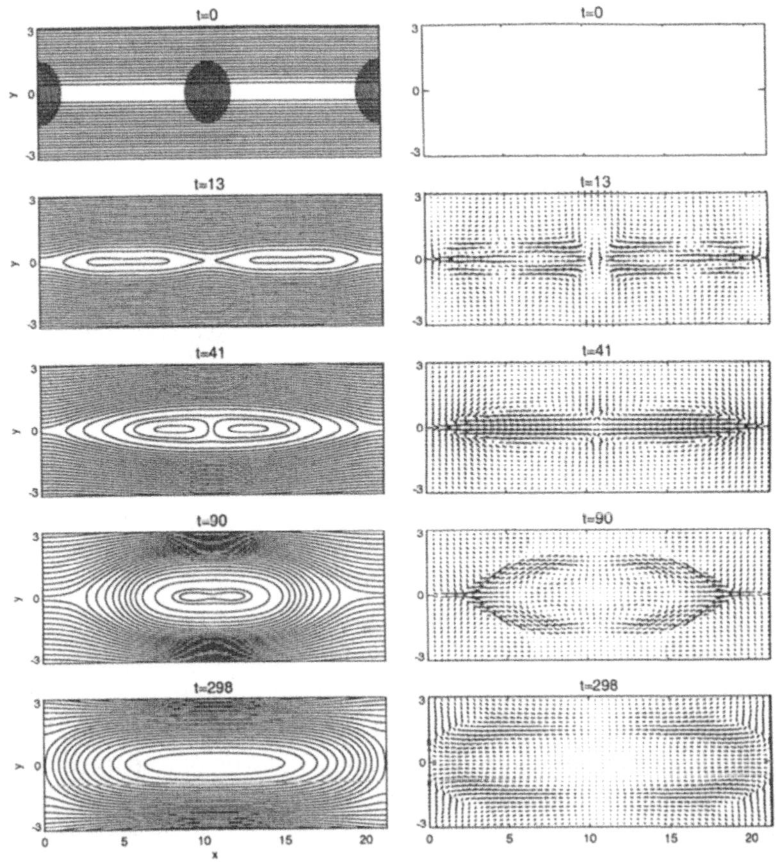

Figure 24. MHD simulation of the coalescence instability for a Lundquist number of $S = 1000$ and a plasma-$\beta = 0.1$. The magnetic field is shown in the left panels, the velocity field in the right panels. The initial resistivity perturbation is shown shaded [from Schumacher and Kliem 1997].

single one (Pritchett and Wu, 1979). For a recent MHD simulation see Schumacher and Kliem (1997), shown in Fig. 24. The coalescence instability completes the collapse in sections of the current sheet, initiated by the tearing mode instability, and thus releases the main part of the free energy in the current sheet (Leboef et al., 1982). There is no complete analytical description of the coalescence instability, but numerical MHD simulations (Pritchett and Wu, 1979; Biskamp and Welter, 1979; 1989; Leboef et al., 1982; Tajima et al. 1987; Schumacher and Kliem, 1997) show that the evolution consists of two phases: first the pairing of current filaments as an ideal MHD process, and then a resistive phase of pair-wise reconnection between the approaching filaments. The characteristic time scale of the ideal phase is essentially the Alfvénic transit time through the distance λ_{coal} between the approaching current filaments,

$$\tau_{coal} = \zeta^{-1} \frac{\lambda_{coal}}{v_A} , \qquad \zeta = \frac{u_{coal}}{v_A} \approx 0.1 - 1 \tag{18}$$

where u_{coal} is the velocity of the approaching filaments. For coronal conditions (say $n_e = 10^{10}$ cm^{-3}, B=200 G, $\lambda_{coal} = 1000$ km) we estimate coalescence times of $\tau_{coal} \approx 0.2 - 2.0$ s, which is again a typical time for the observed modulation of hard-X pulses and type III electron beams.

4.2.3. *Dynamic Current Sheet and Bursty Reconnection*

In praxis, the two previously described processes of tearing instability and coalescence instability occur iteratively, leading to a scenario of *dynamic current sheet evolution*, also known as *impulsive bursty reconnection* (Leboef et al., 1982; Priest, 1985a; Tajima et al. 1987; Kliem, 1988; 1995). A long current sheet is first subject to tearing that creates many filaments, while rapid coalescence clusters and combines then groups of closely-spaced filaments, these are then again unstable to secondary tearing, to secondary coalescence, and so forth. MHD simulations reproduce this iterative chain of successive tearing and coalescence events (Malara, Veltri, and Carbone, 1992; Kliem, Karlický, and Benz, 2000). An example of such a numerical simulation from the study of Kliem et al. (2000) is shown in Fig. 25 (magnetic field evolution) and Fig. 26 (temporal variation of reconnection rate).

This type of dynamic current sheet evolution has the attractive feature that it can reproduce fast modulation of particle acceleration on the right (subsecond) time scales observed in hard X-rays and radio. Let us review three key studies (Tajima et al., 1987; Karpen et al., 1995; Kliem et al., 2000), where numerical MHD simulations of this process have been directly applied to solar flares.

Tajima et al. (1987) performed numerical MHD simulations of the nonlinear coalescence instability between current-carrying loops and derived also an analytical model of the temporal evolution of the electromagnetic fields (see also two comprehensive reviews on this subject by Sakai and Ohsawa (1987) and Sakai and De Jager (1996), and references therein). A remarkable result of these studies is that they obtained for the electric fields and particle acceleration rate a pulsating time profile (with double-peak substructures, see Fig.50), which closely resembled the observed hard X-ray time and microwave profiles, which contained a quasi-periodic sequence of 7 peaks (with a period of $\approx$ 10 s) in this particular flare. The quasi-periodic dynamics of the coalescing loops was reproduced by relaxational oscillations of a nonlinear system obtained from a self-similar ansatz for the variation of the MHD variables during coalescence and driven by the interplay of the $\mathbf{j}\times\mathbf{B}$ force and the hydrodynamic pressure response. A very interesting consequence of this model is the ability to reproduce quasi-periodic particle acceleration, as it was commonly observed at radio wavelengths. Recent *TRACE* observations reveal also oscillating loops after flares and filament eruptions (Aschwanden et

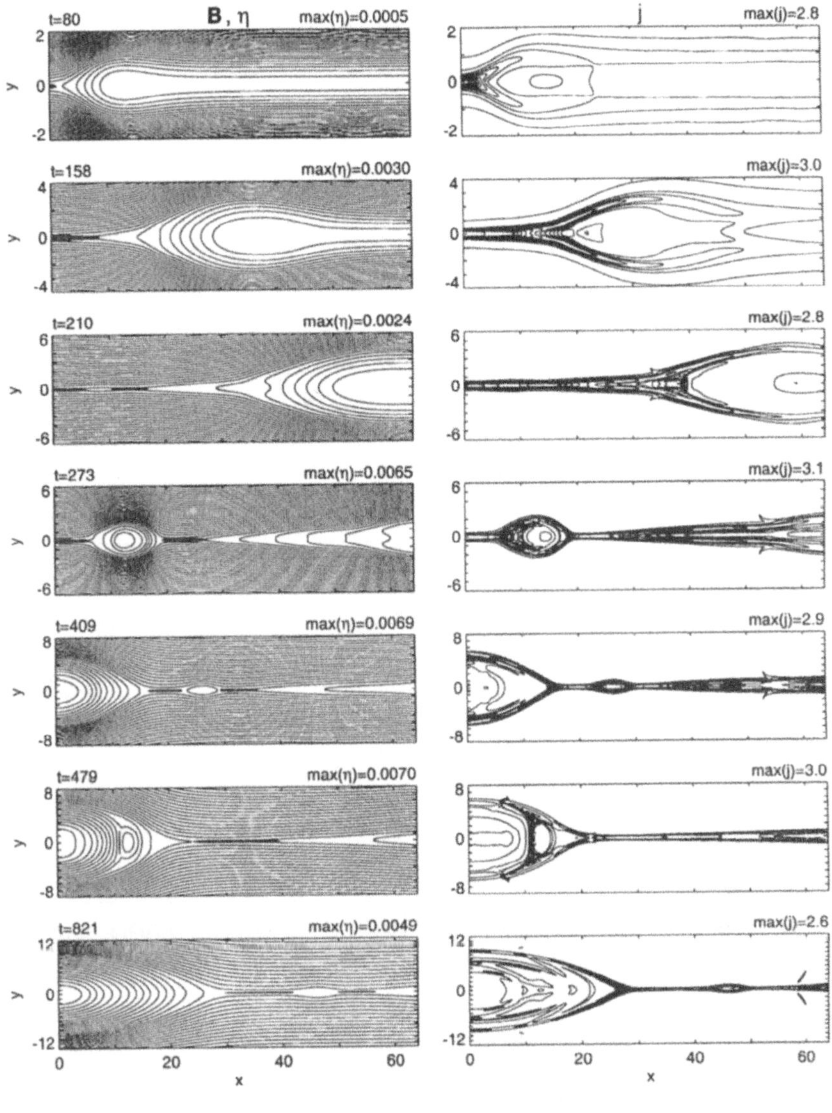

Figure 25. Two-dimensional MHD simulation of dynamic magnetic reconnection, showing the magnetic field (left panels) and current density (right panels). Regions with anomalous resistivity are shown shaded in the magnetic field plot (at y=0) [from Kliem, Karlický, and Benz, 2000].

al., 1999c), which may be governed by similar MHD dynamics and could explain modulated particle acceleration. On the other side, recent flare observations with *TRACE* show also a high degree of spatial structuring, amounting up to some 100 postflare loops in larger flare events. This suggests a spatio-temporal relation between subsequent peaks of flare time profiles and individual flare loops, rather than an oscillatory time pattern of a single dynamic system. However, although oscilla-

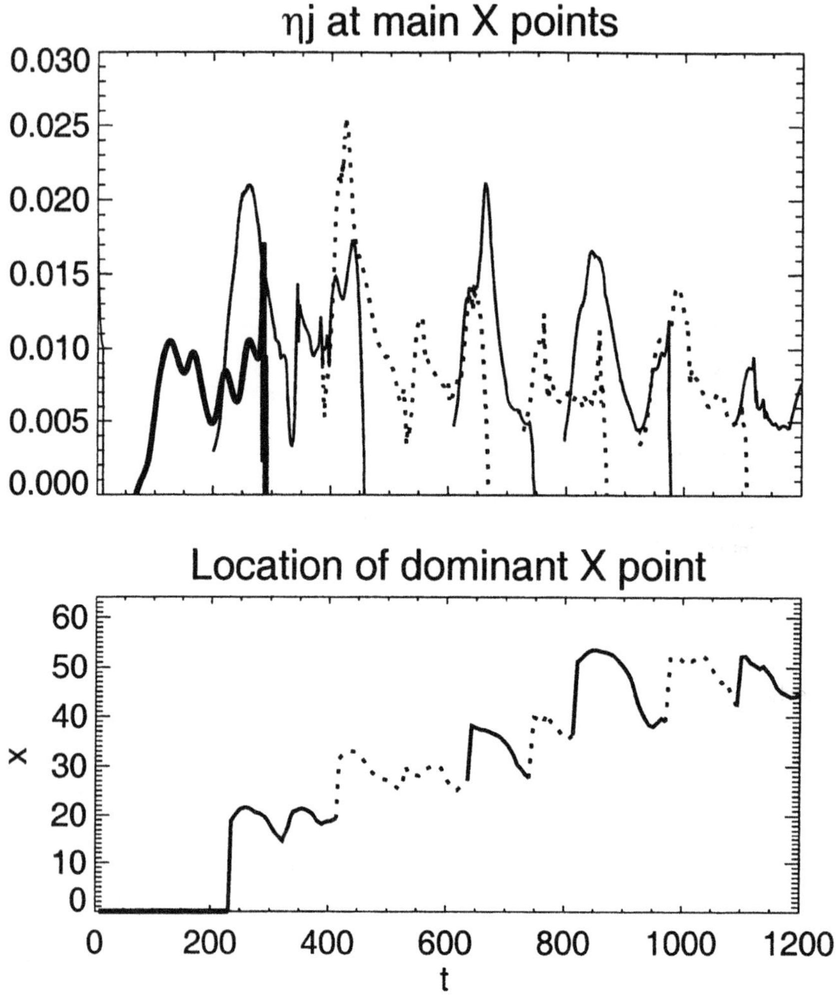

Figure 26. Electric field at the main X points (reconnection rate) vs. time, from simulation shown in Fig. 25. [from Kliem, Karlický, and Benz, 2000].

tions of coalescing loops may not explain most of the numerous and (non-periodic) rapidly fluctuating time structures seen in hard X-rays and radio, they provide a physically sound model for some of the flare loops that have been observed to oscillate after the impulsive flare phase.

Karpen et al. (1995) performed 2.5-dimensional numerical MHD simulations of shear-driven magnetic reconnection in a double arcade with quadrupolar magnetic topology. For strong shear, the initial X-point was found to lengthen upward into a current sheet, which reconnects gradually for a while but then began to undergo multiple tearing. Several magnetic islands develop in sequence, move toward the ends of the sheet, and disappear through reconnection with the overlying and underlying field (Fig. 27). A second study with similar quadrupolar configuration

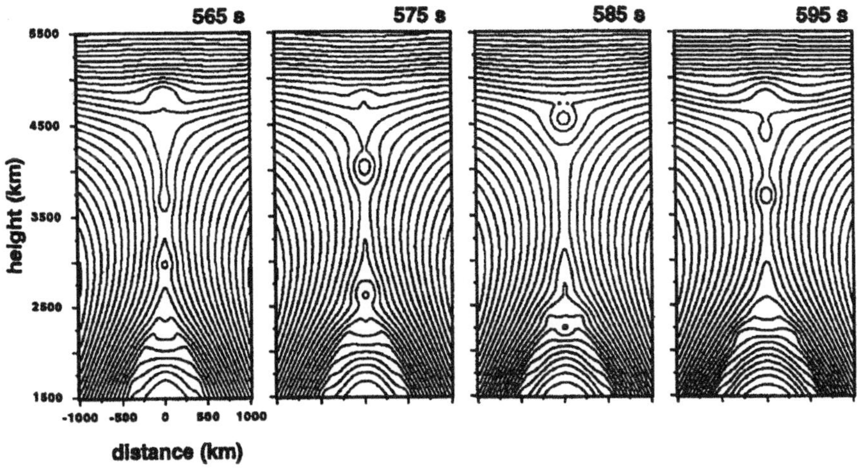

Figure 27. Magnetic field lines near the reconnection region at four different times (565, 575, 585, 595 s) during a strong-shear MHD simulation by Karpen et al. (1995). Note the tearing along the vertical current sheet (first frame), which forms two magnetic islands (second frame), which are ejected from the sheet and merge with the flux systems above or below the sheet (third frame), followed by another tearing plus magnetic island formation (forth frame). [from Karpen et al. 1995].

was performed, but with asymmetric shear in dipoles with markedly unequal field strengths (Karpen et al., 1998). Similar intermittency was found in the shear-driven magnetic reconnection process, and the simulations moreover show that each dissipated magnetic island leaves a footprint in form of fine filaments in the overlying separatrix layer (Fig. 28). This dynamic behavior is essentially identical with the pattern of repeated tearing and coalescence, first investigated by Leboef et al. (1982) and dubbed *impulsive bursty reconnection* by Priest (1985b). In Fig. 28 there are also some other dynamic processes present: (a) a thin region along the slowly rising inner separatrix is compressed, (b) a downflow with $v \approx 30$ km s^{-1}, (c) followed by an upflow along the same field lines. Although these simulations by Karpen et al. (1995; 1998) are carried out with parameters corresponding to chromospheric conditions, it demonstrates that magnetic reconnection in sheared flare arcades occurs in a bursty and intermittent mode, and not in a quasi-stationary Sweet-Parker or Petschek mode. The physical origin of this intermittent reconnection dynamics is most essential to understand the rapidly-varying time structures of accelerated particles.

The most recent work on *impulsive bursty reconnection* applied to solar flares was carried out by Kliem et al. (2000). Fig. 25 shows the evolution of tearing, magnetic island formations, magnetic islands coalescence, secondary tearing, and so forth. Particle acceleration occurs through the electric fields associated with the X-points, which occurs in bursts, as shown in Fig. 26. Particle acceleration is expected to be more efficient in such a multiple X-point reconnection process, because the particles stay longer trapped in the meander-like orbits around the O-points (Ambrosiano et al. 1988; Kliem 1994; Kliem, Schumacher, and Shklyar

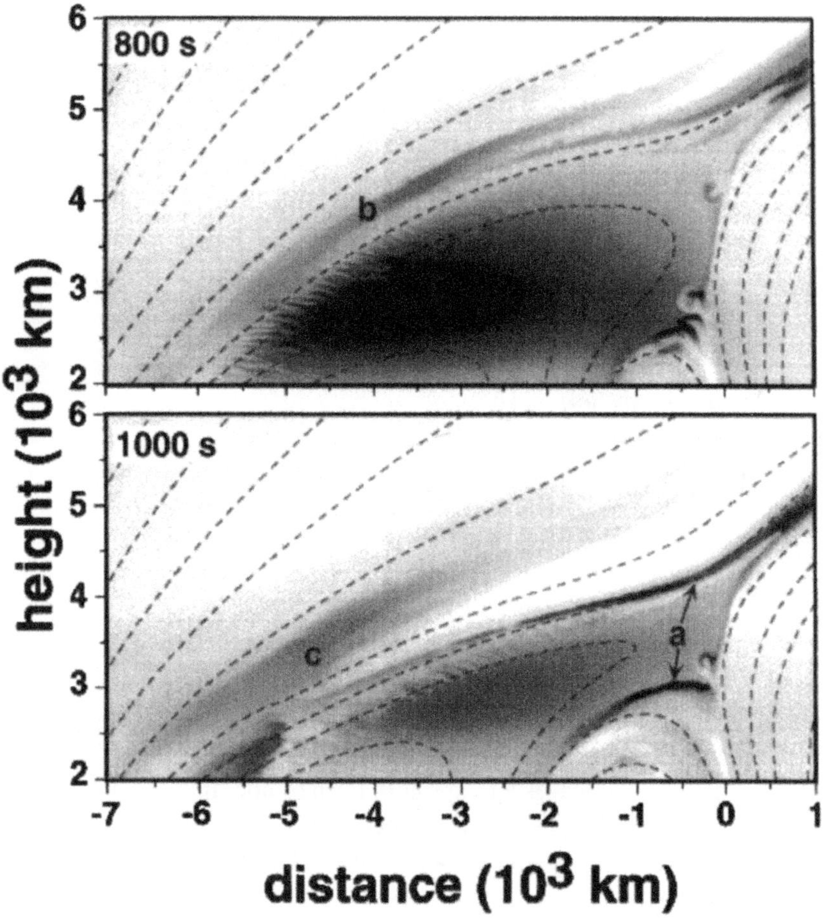

Figure 28. Mass density difference ratio (grey scale) and projection of magnetic field lines into the image plane (dashed lines) at 800 s and 1000 s in the vicinity of the reconnection region, during a MHD simulation of a sheared arcade. The location **a** corresponds to a thin compressed region along the slowly rising inner separatrix, **b** to a narrow downflow falling outside of the left outer separatrix, and **c** indicates a broader upflow that follows along the same field lines. Note that the dimensional scales of the numerical simulation do not scale one-to-one to solar flares. [from Karpen et al. 1998].

1998), and thus are longer exposed to the accelerating fields, opposed to single X-point (Petschek mode) configurations, where they can escape quickly. Kliem et al. (2000) scaled the physical parameters of the numerical MHD simulation to solar conditions to estimate the acceleration time scales of radio-emitting electrons. The time intervals between subsequent peaks in the reconnection rate were found to be $t_R \approx 200\tau_A \approx 13\delta_x/v_A$ (Fig. 26), using an average distance of $\delta_x \approx 15\, l_{CS}$ between neighbored X-points, where l_{CS} is the current sheet half width. Translating this timescale $t_R \approx 13\delta_x/v_A$ to solar flare conditions ($n_0 = 10^{10}$ cm^{-3}, $T_0 = (2.5-9) \times 10^6$ K, $B_0 = 70$ G), and estimating δ_x from the mean free path length λ_{mfp}

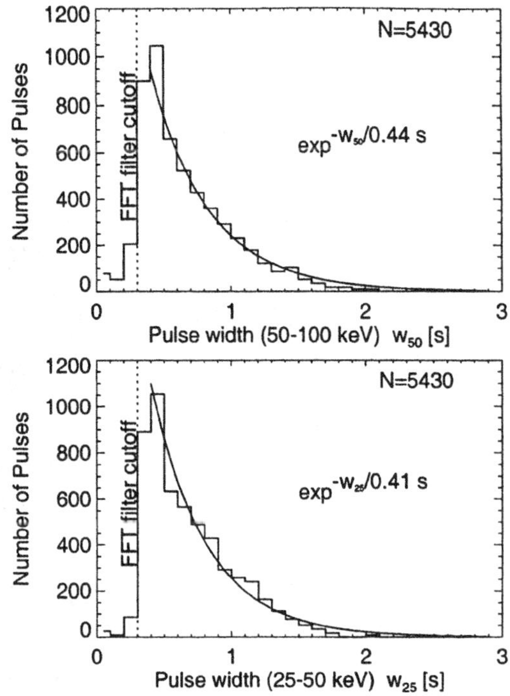

Figure 29. Distribution of hard X-ray pulse durations detected in 25-50 keV and 50-100 keV channels from *BATSE/CGRO*. The cutoff of detected pulse widths at $\lesssim 0.3$ s is caused by the Fourier (FFT) filter used for structure detection [from Aschwanden et al. 1995a].

(which implies $t_R \approx n_0^{-1/2} T_0^2 B_0^{-1}$), they find typical time intervals of $t_R = 0.4 - 4$ s between subsequent radio bursts. This quantitative example demonstrates that tearing and coalescence in the *bursty magnetic reconnection mode* can modulate particle acceleration on time scales that are observed in radio and hard X-rays.

4.3. Dynamics Inferred from Fast Time Structures

The dynamics of magnetic reconnection processes that are responsible for particle acceleration in flares cannot directly be observed with imaging observations, firstly because the reconnecting field lines are not illuminated with over-dense plasma before reconnection, secondly because the currently available cadences (in the order of minutes) of high-resolution imaging (e.g. with *TRACE*) are not sufficiently high to map out the motion of relaxing field lines, and thirdly because the temperature discrimination of broadband filters (e.g. from *Yohkoh*) is insufficient to separate newly-reconnected field lines that are freshly filled with hot plasma from the underlying relaxed field lines that contain older cooling plasma. Therefore, the most direct witnesses of reconnection dynamics are still fast time structures of accelerated particles, as detected in (non-imaging) hard X-rays and radio wavelength time profiles. In the following Section we review the highlights and most relevant results

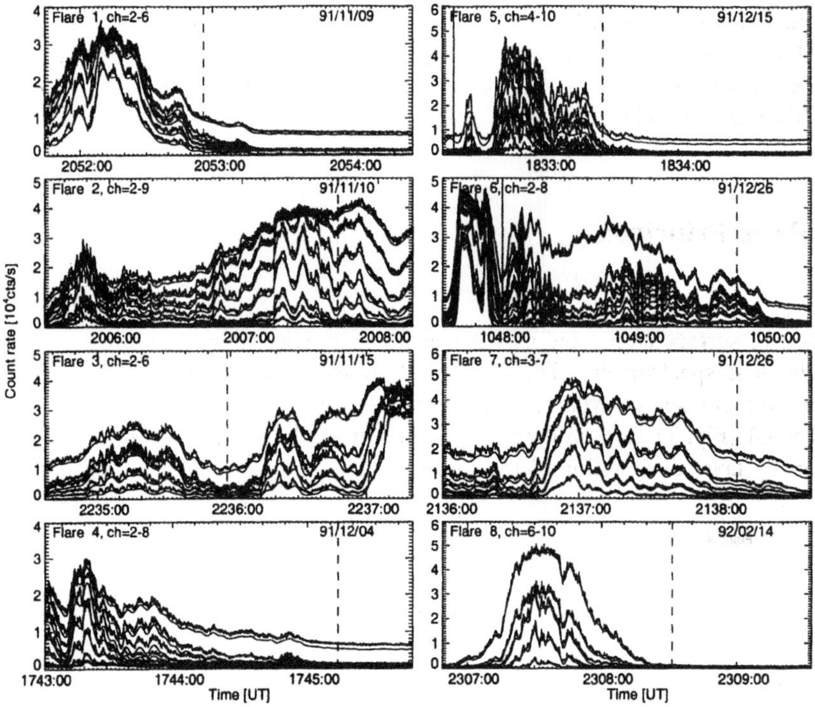

Figure 30. Hard X-ray time profiles of 8 flares commonly observed with *BATSE/CGRO* and *Yohkoh*. The time profiles were recorded in the 16-channel *Medium Energy Resolution (MER)* mode with a time resolution of 64 ms. The frames have a duration of 163.84 s. The lower envelope to the rapid fluctuations was constructed with a FFT lowpass filter with cutoffs of $\tau_{filter} = 1.5 - 3.6$ s [from Aschwanden et al. 1996c].

of fast time structure analysis for the understanding of the dynamics in acceleration regions.

Pulsed (sub-second) time structures observed in nonthermal ($\gtrsim$ 20 keV) hard X-rays are generally interpreted in terms of bremsstrahlung from electrons that have been accelerated in the near-collisionless corona and propagate directly to the highly-collisional chromosphere, where they lose their energy instantly. The duration of such hard X-ray pulses is thus an almost direct measurement of the duration of the injection or acceleration pulses. The fastest time structures discovered with the *Hard X-Ray Burst Spectrometer (HXRBS)* on the *Solar Maximum Mission (SMM)* spacecraft have been found down to rise and decay times of $\approx$ 20 ms and with pulse widths of $\gtrsim$ 45 ms (Kiplinger et al., 1983). Such fast time structures were not very common in *SMM* data, because only about 10% of all flares detected with *HXRBS* have a sufficiently high count rate to detect significant variations on a sub-second time scale (Dennis, 1985).

With the much more sensitive *BATSE/CGRO* detectors, which have a sensitive area of 2025 cm^2 each, sub-second time structures were detected virtually in all flares recorded in flare mode (activated by a burst trigger). A systematic study of

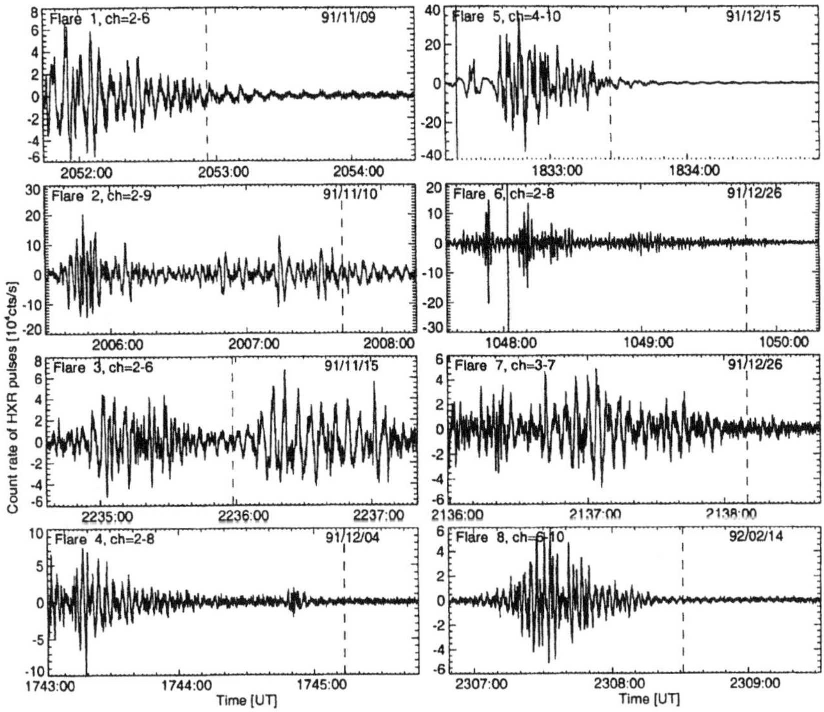

Figure 31. Hard X-ray time profiles of the same 8 flares shown in Fig. 30 with a lower envelope subtracted [from Aschwanden et al. 1996c].

640 *BATSE* flare events with an automated pulse detection algorithm (Aschwanden et al., 1996c) revealed a total of 5430 individual pulses, for which the distribution of pulse widths is shown in Fig. 29. This distribution shows that hard X-ray pulses with durations of $\tau_X \approx 0.3 - 1.0$ s are most typical.

Let us explore how these pulse durations detected in (non-imaging) time series relate to spatial scales of the corresponding flare loops where they originate. In Fig. 30 we show *BATSE* time profiles of the 8 largest flares commonly observed with *BATSE/CGRO* and *Yohkoh*, a sample analyzed in Aschwanden et al. (1996c). In order to separate the fast pulses, which supposedly are produced by directly-precipitating electrons, from the slowly-varying hard X-ray flux, which is supposedly produced by trapped electrons, we subtract a lower envelope. The envelope-subtracted time profiles are shown in Fig. 31, showing significant rapid fluctuations in excess of Poisson statistics. In order to characterize the typical time scale in each flare, we perform an auto-correlation of the envelope-subtracted time profiles. The autocorrelation functions are shown in Fig. 32, yielding full widths at half maximum (FWHM) between $T_p \approx 0.4$ and 1.6 s. This sample of 8 large flares essentially shows double footpoint structures in hard X-rays and dipolar loops in soft X-rays, for which the footpoint separation was measured. The flare loop radius $r = d/2$ measured from this footpoint separations d varies between $r = 5600$

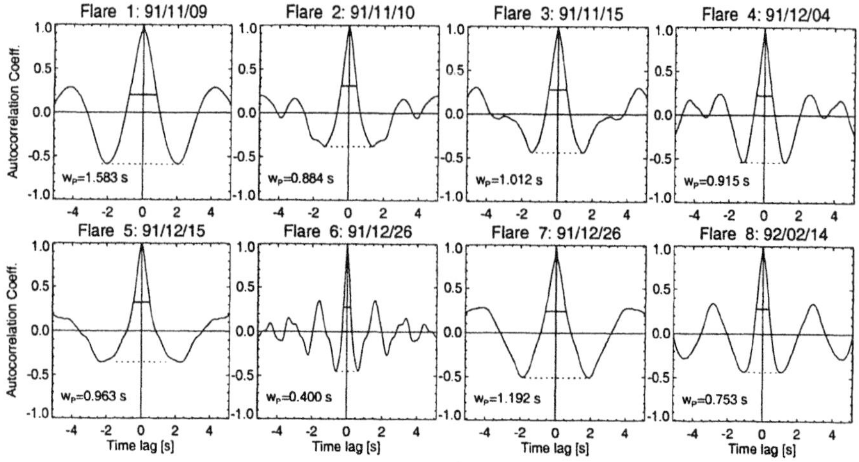

Figure 32. Auto-correlation function of envelope-subtracted hard X-ray time profiles of the same 8 flares shown in Fig. 30 and 31. The pulse widths T_p indicated in each panel (labeled w_p) are measured from the FWHM of the central peak of the autocorrelation function [from Aschwanden et al. 1996c].

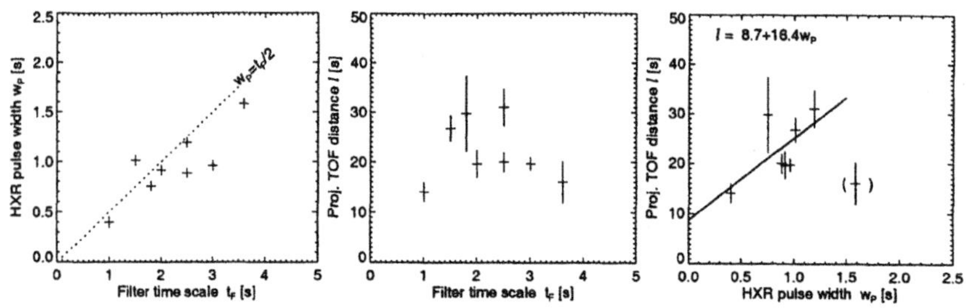

Figure 33. The relations between the hard X-ray pulse widths T_p, the filter time scales t_F, and the projected electron time-of-flight distance l are shown. Excluding one flare with ambiguous time scales (marked with a parentheses in the right panel), a linear regression fit yields the linear relation $l = 8.7 + 16.4T_p$ [from Aschwanden et al. 1996c].

km and 17,000 km. Moreover, the electron time-of-flight distance l'_{TOF} could be measured for all these 8 flares from the energy-dependent hard X-ray time delays, ranging from $l'_{TOF} = 12,400$ km to 25,800 km. Let us now compare the time scales of the pulse widths T_p with the spatial scale l'_{TOF}. Fig. 33 (left panel) shows that the hard X-ray pulse widths correspond roughly to the half filter time scale, as expected for a highpass filter. The time-of-flight distances l_{TOF} seem to be correlated with the pulse widths T_p (Fig. 33 right panel), having a relation of

$$l_{TOF} \approx 8.7 + 16.4 \left(\frac{T_p}{1\ s} \right) [Mm] \approx 2 \times 10^9 \left(\frac{T_p}{1s} \right) cm \qquad (19)$$

but are uncorrelated to the filter time scale t_F (Fig. 33, middle panel), and therefore not an artifact of the filter method. This positive correlation, although obtained with small statistics, may indicate an important physical scaling law between the

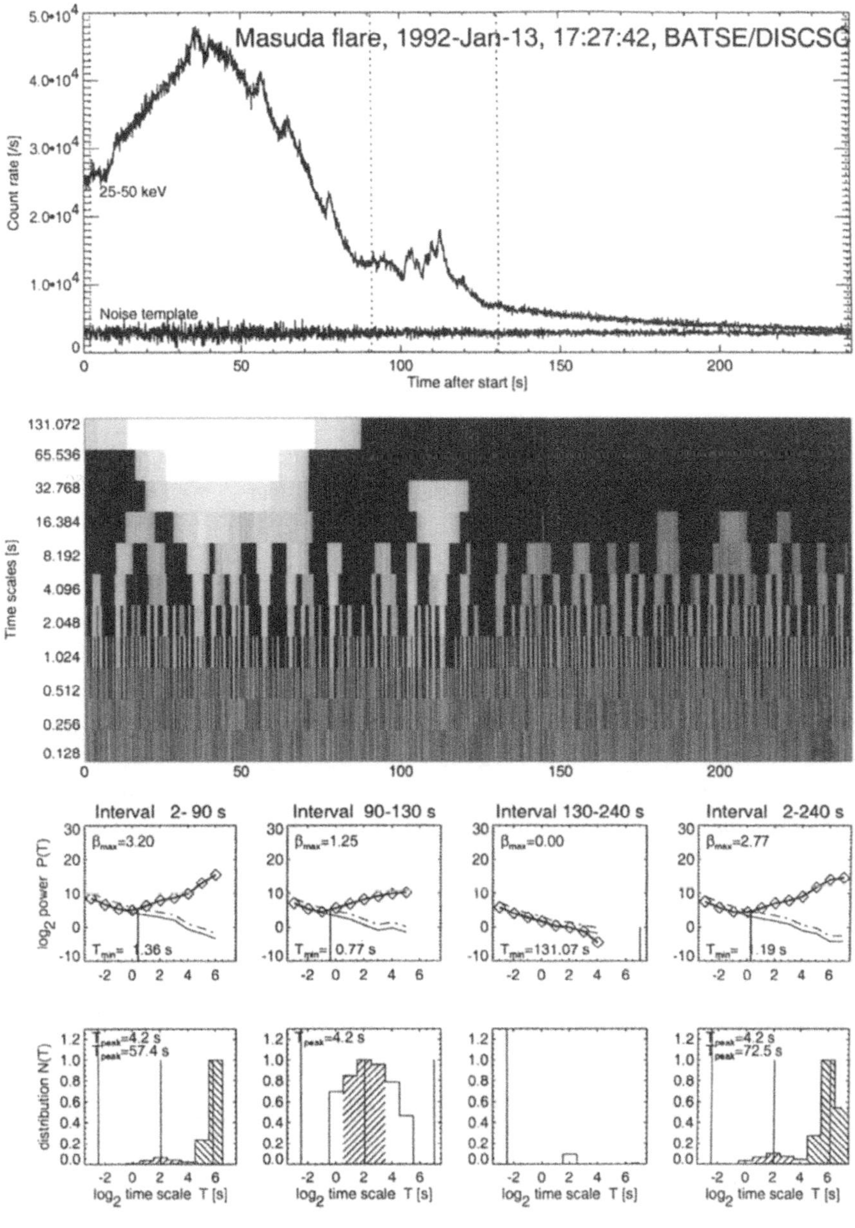

Figure 34. Example of a wavelet analysis of a hard X-ray flare time profile, showing the time profile with noise model (top panel), the wavelet scalogram $P(T, t)$ (middle panel), the time-averaged scalegrams $S(T) = < P(T, t) >$ of the data (diamonds) as well as of the noise model (third row), and the inverted time scale distributions $N(T)$ (forth row) for the Masuda flare, 1992-Jan-13, 1727:42 UT (see also Figs. 8 and 9), observed with *BATSE/CGRO* [from Aschwanden et al. 1998c].

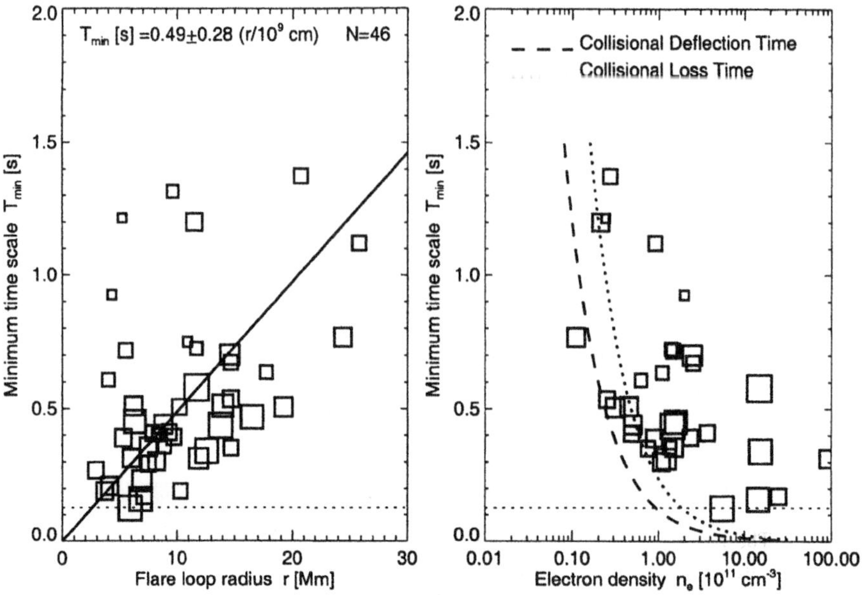

Figure 35. Correlation of the minimum timescale T_{min} with the flare loop radius r (left panel) and trap electron density n_e (right panel) for a set of 46 flares simultaneously observed with *BATSE/CGRO* and *Yohkoh*. The symbol size of the data points is proportional to the logarithm of the count rate. The mean ratio T_{min}/r is indicated (solid line in left panel), and collisional timescales for 25 keV electrons are shown (dashed and dotted line in right panel) [from Aschwanden et al. 1998c].

spatial scale of the acceleration region (e.g. defined by the Alfvénic crossing time, $l_{acc} = v_A T_p$) and its altitude (e.g. defined by the propagation distance l_{TOF}). We will see that a similar spatio-temporal relation is found in another time series study based on wavelet analysis.

The characterization of time scales in solar flares is not straightforward, because subsequent pulses seem to be piled up on top of each other, and they are also convolved with longer time scales, so that the particle injection function can not easily be decomposed from the slower-varying thermal and trapping processes. Fourier power spectra are useless because flare time profiles are relatively short, intermittent, and non-periodic. An objective method to investigate time scales is a multi-resolution time series analysis, a special form of wavelet analysis. Such a wavelet analysis has been applied to 647 events with high time-resolution (64 ms) hard X-ray data by Aschwanden et al. (1998c). In addition to the evaluation of standard scalegrams $S(T)$ (which correspond to power spectra as function of time scales rather than frequencies), a deconvolution method was developed that allowed to extract the approximate distribution $N(T)$ of time scales. The scalegrams $S(T)$ as well as the time scale distributions $N(T)$ show generally a cutoff at a minimum time scale T_{min} (see example in Fig. 34), which corresponds to a cutoff of physical time scales contributing to the modulation of the time profile. So this minimum time scale T_{min} measured in hard X-ray time profiles quantifies the

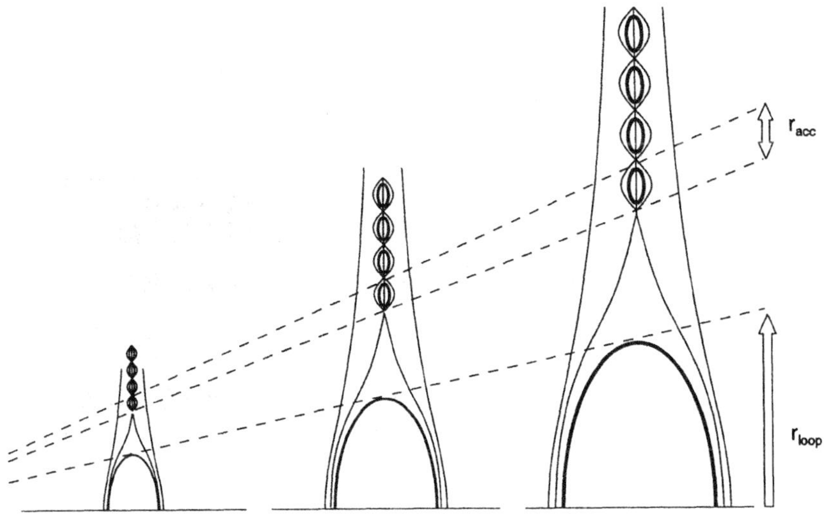

Figure 36. The proportionality between the spatial scale of the acceleration region r_{acc} and the flare loop radius r_{loop}, as implied by the spatio-temporal relation between the hard X-ray minimum time scale T_{min} and loop size r_{loop}, is visualized in the context of a scale-invariant reconnection geometry with magnetic islands in the current sheet region. The scale-invariant geometry appears to be consistent with the tearing mode. For larger loops at the base of the current sheet, the width l_{CS} may be larger. The length scale of the islands, which probably corresponds to the tearing mode with the largest growth rate and scales with l_{CS}, is then also expected to be larger.

fastest significant modulations of the electron acceleration or injection function, while the longer time scales $T > T_{min}$ correspond to clusters of the elementary pulses in these intermittent time series. For 46 events, the spatial scale of flare loops (i.e. the radius r_{loop}) could be measured from *Yohkoh* images. Interestingly, a scatterplot of the wavelet minimum time scales T_{min} versus the flare loop radius r_{loop} reveals a positive correlation, which can be expressed by a linear regression fit (Fig. 35 left),

$$T_{min} = (0.49 \pm 0.28) \left(\frac{r_{loop}}{10^9 cm} \right) \ s \ , \tag{20}$$

found for time scales in the range of $0.15 < T_{min} < 1.5$ s and spatial scales in the range of 3000 km $< r_{loop} <$ 30, 000 km.

Let us now apply these spatio-temporal scaling laws (Eqs. 19, 20) to an acceleration scenario in the *bursty reconnection regime* (Section 4.2), as visualized in Fig. 36. Using the scaling law between time-of-flight distance l'_{TOF} and flare loops radius r_{loop} from Eq. 5, $l'_{TOF} \approx 1.4 \times s = 1.4 \times (\pi/2) r_{loop} \approx 2 \times r_{loop}$, and relating the minimum time scale T_{min} from the wavelet scalegrams to the risetime of the hard X-ray pulses, i.e. $T_{min} \approx T_p/2$, we see that the two observational spatio-temporal relations (Eqs. 19, 20) are fully equivalent,

$$T_p = 2T_{min} = 1.0 \left(\frac{r_{loop}}{10^9 cm} \right) \ s \ \approx 1.0 \left(\frac{l'_{TOF}}{2 \times 10^9 cm} \right) \ s \ , \tag{21}$$

so both the auto-correlation analysis (yielding pulse widths T_p) and the wavelet analysis (yielding minimum time scales T_{min}) measure typical time scales of the *same* rapidly-fluctuating phenomenon. Applying this now to the case of *bursty reconnection*, we can interpret the pulse duration T_p of an elementary particle injection as the coalescence time τ_{coal} (Eq. 18) of a magnetic island, which roughly represents the life span from formation and dissipation of a magnetic island, which may determine the approximate duration T_p of an elementary particle acceleration episode. Thus, setting $\tau_{coal} \approx T_p$ and inserting the observed scaling of Eq. 21, $T_p \approx (r_{loop}/10^9 \text{ cm s}^{-1})$ into Eq. 18, we find the following scaling for the size of a coalescing magnetic island,

$$\lambda_{coal} = \tau_{coal}\zeta v_A \approx 0.2\, r_{loop}\, \zeta \left(\frac{B}{100G}\right)\left(\frac{n_e}{10^{10}cm^{-3}}\right)^{-1/2} . \tag{22}$$

This yields a range of $\lambda_{coal} \approx 60-6000$ km for the sizes of magnetic islands, based on a range of loop sizes $r_{loop} \approx 3000 - 30,000$ km (Fig. 12) and $\zeta = 0.1 - 1$, for $B = 100$ G and $n_e = 10^{10}$ cm^{-3}, using the same parameters estimated in Kliem (1995). Thus, our spatio-temporal scaling law provides us with a relation of the size of coalescing magnetic islands, as function of two measurable parameters (flare loop size r_{loop}, density n_e) and two model parameters (magnetic field B and coalescence speed ratio $\zeta = u_{coal}/v_A$). If the coalescence speed u_{coal} could be determined by other means, the size of the coalescing magnetic island λ_{coal} would directly follow from

$$\lambda_{coal} = \tau_{coal}u_{coal} = T_p u_{coal} \approx r_{loop}\left(\frac{u_{coal}}{10^9 cm\ s^{-1}}\right) . \tag{23}$$

Magnetic islands with sizes of $\lambda_{coal} \approx 60 - 6000$ im in flare loops with radii of $r_{loop} \approx 3000 - 30,000$ km would then correspond to coalescence speeds of $u_{coal} = 200 - 2000$ km s^{-1}.

With the same spatio-temporal scaling law we can estimate the rate R_{acc} of accelerated particles per magnetic island, defined by the product of the volume $V \approx \lambda^3_{coal}$, the electron density n_e, and the acceleration efficiency q_{acc}, which yields with Eq. 22,

$$R_{acc} = \frac{q_{acc}\lambda^3_{coal}n_e}{\tau_{coal}} = 8 \times 10^{34}\, q_{acc}\zeta^3 \left(\frac{r_{loop}}{10^9 cm}\right)^2 \left(\frac{B}{100G}\right)^3 \left(\frac{n_e}{10^{10}cm^{-3}}\right)^{-1/2} s^{-1} . \tag{24}$$

This relation indicates that large flares require fast coalescence speeds $\zeta \lesssim 1$ (or $u_{coal} \lesssim v_A$) and high magnetic fields $B \gtrsim 100$ G, which both scale with the third power to the acceleration rate, while the acceleration efficiency q_{acc} scales only linearly. Also the number of magnetic islands would increase the rate of accelerated particles only with linear scaling. Interestingly, a lower density yields a higher acceleration rate than a higher density, because a lower density speeds up the Alfvén velocity v_A or coalescence speed $u_{coal} = \zeta v_A$.

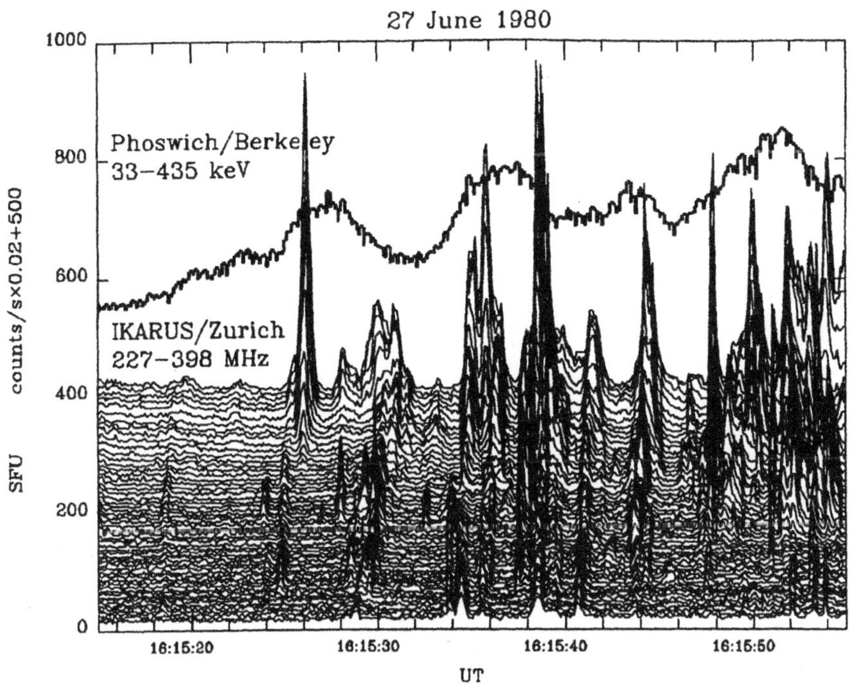

Figure 37. Part of the impulsive phase of the solar flare on 1980 June 27. The top time profile is the $>$ 33 keV hard X-ray flux measured with a balloon-borne phoswich detector, with a time resolution of 128 ms. The time profiles in the lower part displays a section of the dynamic radio spectrum recorded by the digital spectrometer *IKARUS* at *ETH* Zurich, containing 57 frequency channels each with 3 MHz bandwidth and 0.1 s time resolution. Broadband intense metric type IIIs and weak fine structures (blips) occur in clusters. The type III bursts are clustered around the four smooth hard X-ray peaks [from Aschwanden et al. 1990].

4.4. Dynamics Inferred from Spatio-Temporal Fragmentation

In the last Section we reviewed results from the statistics of time structures observed in hard X-rays, which is essentially 1-dimensional (1-D) time series analysis. The number of elementary time structures detected in a 1-D time series represents a lower limit on the true number of elementary energy dissipation processes, because at high rates, when the average time interval is shorter than the pulse duration itself, the pulses pile up in superposition, and we lose the information of their original generation rate. In order to retrieve the original rate, we need 2-dimensional (2-D) information or higher, where an additional parameter can be measured that discriminates near-simultaneous energy dissipation events. A suitable second parameter is the coronal altitude h, besides the first parameter of time t. So, if we could plot some radiative signature of the energy dissipation events as function of these two parameters, say the flux $F(t, h)$, instead of the 1-D time profiles $F(t)$ as we used in hard X-rays, we have a more accurate diagnostic of the true fragmentation rate of the flare energy release. Fortunately, radio dynamic

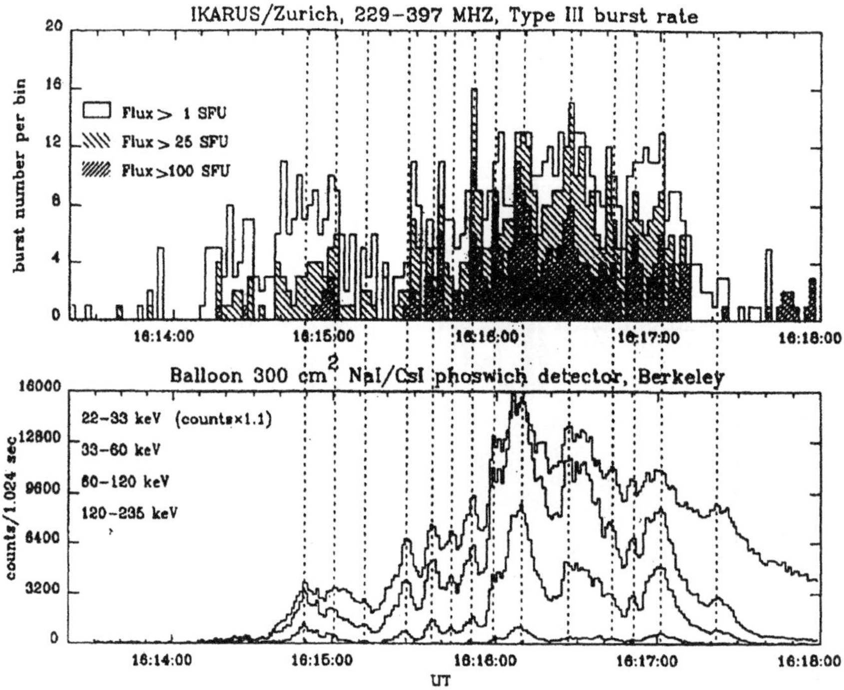

Figure 38. Top: The number of type III bursts, as shown partially in Fig. 37, are counted in 2-s bins, for different threshold (background-subtracted) flux levels (1, 25, and 100 solar flux units [sfu], shown with different shadings). The time profile of the type III burst rate reveals about 9 peaks in all three threshold levels. *Bottom:* Hard X-ray emission is measured with the balloon-borne scintillator of UCB and UCSD. The displayed count rates cover the energy range of 22-235 keV and are integrated in bins of 1.024 s. The hard X-ray count rate shows about 14 peaks (with 4-σ significance), indicated with vertical dashed lines, in intervals of $\approx$ 10 s. Nine of these 13 peaks line up with maxima in the type III burst rate within 3 s [from Aschwanden et al. 1990].

spectra $F(t, f)$ provide almost such a 2-D record, if one relates the frequency f to the coronal altitude h. For many types of radio emission, in particular for type III bursts, which are interpreted as electron beams generating plasma emission, the transformation is straightforward: $f(h) \mapsto f(f_p[n_e(h)])$, where $f(f_p) = sf_p$ means that the frequency f corresponds to the $s - th$ (usually first) harmonic of the plasma frequency f_p, the plasma frequency $f_p(n_e)$ is a function of the electron density (Eq. 14), and $n_e(h)$ can be inferred from a coronal density model, e.g. a hydrostatic model $n_e(h) = n_0 \exp(-h/\lambda_T)$, with $\lambda_T = 46,000$(T/1 MK) km the hydrostatic scale height.

The immense gain of 2-D statistics compared with 1-D statistics can be seen in Fig. 37, which shows a record of a radio dynamic spectra $F_R(t, f)$ and a simultaneous hard X-ray time profile $F_X(t)$ observed during the 1980 June 27 flare. Numerous radio type III bursts show up in the dynamic spectrum, many of them which almost coincide in time and can only be separated in frequency, while the hard X-ray time profiles show a modulation with only about 4 significant peaks.

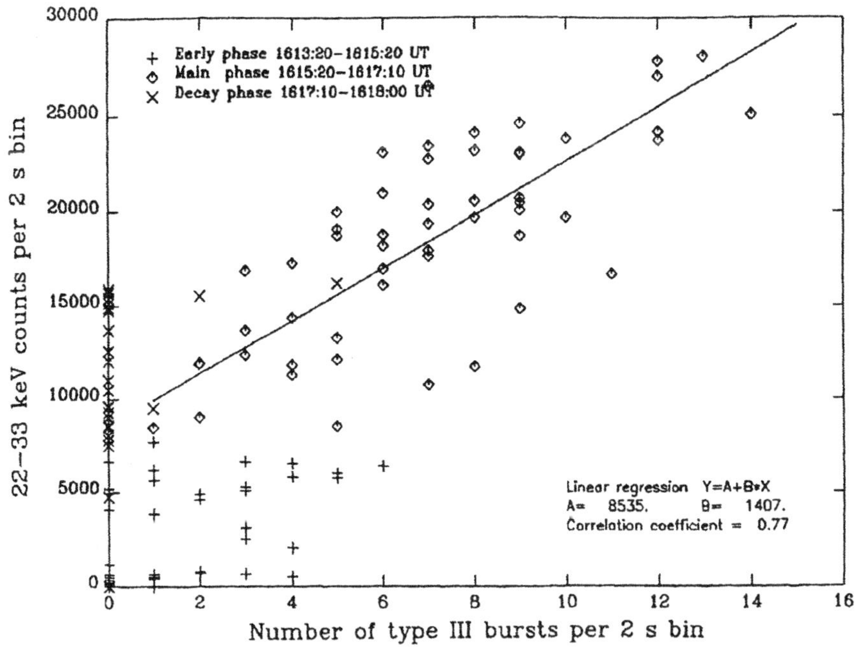

Figure 39. Correlation of hard X-ray flux (per 2 s) versus rate of type III bursts (per 2 s), obtained from the data shown in Fig. 38. Datapoints during the early phase (crosses), the main phase (diamonds) and decay phase (x) are distinguished with symboles. The straight line shows a linear regression fit of the points during the main phase [from Aschwanden et al. 1990].

Obviously, the rate of ratio bursts reveals much more numerous acceleration events than the hard X-ray bursts. A detailed count of type III bursts during the entire flare revealed about 754 individual bursts, while the hard X-ray time profile shows about 13 significant peaks (Fig. 38). If we plot the type III burst rate as function of time, most of the hard X-ray peaks correspond to a peak in the radio burst rate, which is shown in 2-s bins in Fig. 38 (top panel). If we plot the radio burst rate $N_{III}(t)$ versus the hard X-ray flux $F_X(t)$, we find a clear correlation (Fig. 39),

$$F_X(t) \approx 700[N_{III}(t) + 6] \text{ cts s}^{-1} , \qquad (25)$$

with the radio burst rate peaking at $max[N_{III}(t)] \lesssim 7$ bursts s^{-1}. The correlation suggests that a proportional number of electron beams are accelerated in upward (producing radio) and downward direction (producing hard X-rays), but that the degree of spatio-temporal fragmentation can be detected with higher resolution in radio than in hard X-rays. Thus, radio bursts reveal, under favorable circumstances (when emission is not suppressed by free-free absorption), a higher rate of acceleration episodes ($\lesssim 7$ s^{-1}) than we would infer from the fastest hard X-ray time structures ($\lesssim 1.6$ s^{-1}, if we take the shortest significant hard X-ray pulses of $\tau_X \gtrsim 0.3$ s (Fig. 29), which require a minimum separation of $2\tau_X = 0.6$ s). This has important consequences for the number of dissipated magnetic islands in the *bursty reconnection* scenario (Section 4.2).

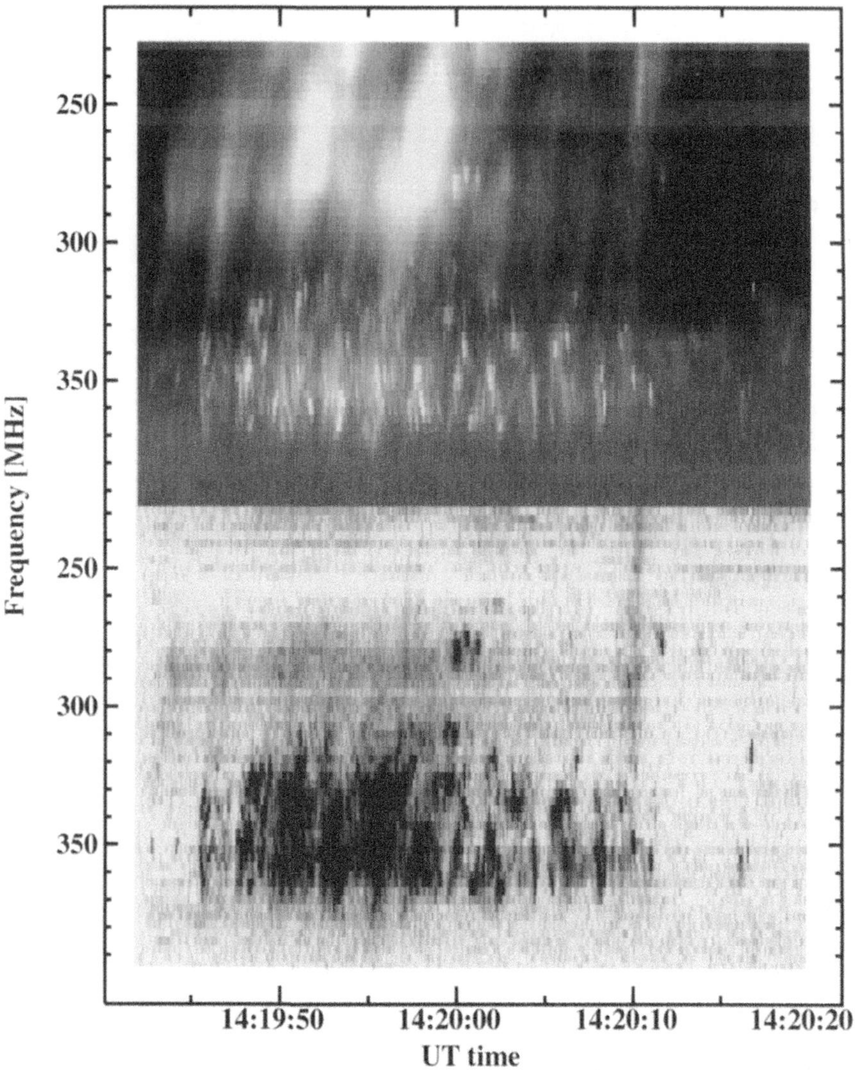

Figure 40. Dynamic spectrum of radio flux (top) and polarization (bottom) of a metric type III group, accompanied by *metric spikes* below their start frequency, recorded by *ETH* Zurich on 1980-Dec-03, 14:20 UT. The degree of polarization is shown in dark for left-hand circular sense. Note that the metric spikes show a strong polarization, while the type III bursts are almost unpolarized, indicating two different emission mechanisms or absorption conditions [from Benz et al. 1996].

The most direct radio diagnostic of the acceleration region probably comes from fast narrowband bursts, termed *metric spikes*, that have been discovered just below the start frequency of metric type III bursts (Benz et al., 1992; 1996). An example of such an observation is shown in Fig. 40. There are strong reasons to believe that they are a direct signature of the acceleration process. First of all, they occur highly coincident with the start of type III burst extrapolated to the same frequency (with an insignificant delay of $< \Delta t >= 0.03 \pm 0.04$ s). Secondly, they occur at a slightly higher frequency $< \Delta\nu_{sep} >= 23 \pm 8$ MHz), as expected for an acceleration source below an upward-accelerated electron beam. Their narrowbandedness ($< \Delta\nu >= 9.9 \pm 0.9$ MHz) indicates a small vertical extent of the accelerator. If we assume plasma emission in a $T = 1.0$ MK degree plasma, their frequency bandwidth (10 MHz) translates into a vertical distance of ≈ 3000 km, a size that falls into the range of coalescing magnetic islands ($\lambda_{coal} \approx 60 - 6000$ km) we estimated with Eq. 23. While this type of *metric spikes* obviously provides a direct diagnostic of the acceleration region, it has not yet been observed during flares (Benz et al., 1996), in association with hard X-rays or microwaves, so their existence may be hidden during flares because of strong absorption of their plasma emission in the denser regions of flare environments. Spatial imaging of simultaneous type III sources and metric spike sources confirmed that the location of the metric spikes coincides with the trajectory of type III electron beams, and thus corroborates their genuine relationship to the acceleration region (Paesold et al., 2001). The altitude of such a metric spike source was estimated from imaging observations with the *VLA* and magnetic field extrapolations from a footpoint radio source, yielding a height of $h \approx 500,000$ km (Krucker, Benz, and Aschwanden, 1997), which is substantially higher than typical flare acceleration sources (see Table II).

A similar type of radio emission has been observed at higher frequencies, termed *decimetric millisecond spikes* (Benz, 1985; 1986). Because (1) they occur only during flares, (2) they show a strong correlation with hard X-rays (Güdel, Aschwanden, and Benz, 1991), (3) they occur in the decimetric frequency band that roughly represents a separatrix between upward-accelerated electron beams (type III bursts) and downward-accelerated electron beams (type RS bursts), they were associated with the acceleration region (Benz, 1985). Because they show a high degree of spatio-temporal fragmentation (up to $\approx 10^4$) per flare, they were interpreted as signature of the fragmented energy release and acceleration region, consisting of elements with sizes of $\lesssim 200$ km, time scales of $\lesssim 50$ ms, and energies of $\approx 10^{26}$ erg (Benz, 1985; 1986). However, there is a big puzzle that questions their possible role as immediat signature of elementary acceleration processes. Although the integrated flux time profile $F_{spikes}(t)$ shows a close co-evolution with the hard X-ray time profile $F_X(t)$, which can be fitted with a convolution function, the best fits show that a convolution delay of $\tau \approx 2-5$ s is necessary to mimic the evolution of the radio spikes emission (Aschwanden and Güdel, 1992). Such a delay suggests that spiky radio emission is not produced by the same acceleration process that generates hard X-ray-emitting electrons, but rather is produced by a secondary radiation mechanism that kicks in after some delay, e.g. at the precipitation site after some trapping delay (of $\approx 2 - 5$ s). Thus we dont pursue further the possible role of *decimetric millisecond spikes* as diagnostic of the spatio-temporal fragmentation of the acceleration region.

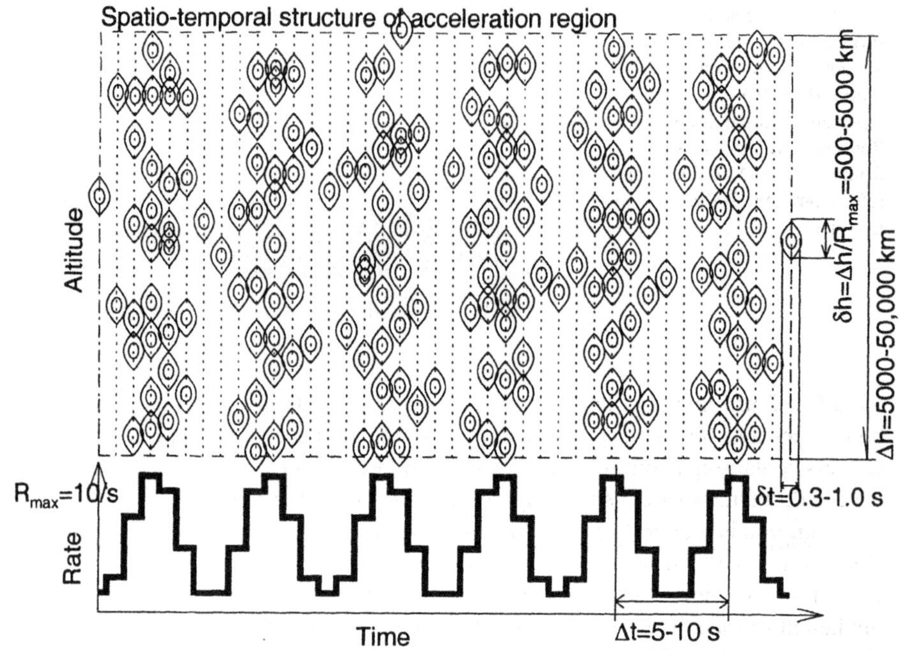

Figure 41. Spatio-temporal evolution of elementary acceleration episodes (e.g. by magnetic island coalescences) occurs in intermittent (quasi-periodic) bursts, which have a spatial scatter in altitude, but are clustered in time. The rate (bottom) of acceleration events can be inferred from radio dynamic spectra, and their altitude distribution from the radio frequency.

Let us summarize, what can we infer from radio dynamic spectra about the spatio-temporal evolution of elementary acceleration events. The example of the 1980-Jun-27 flare (Fig. 38), as well as many other flares, shows that the rate of acceleration episodes fluctuates strongly, exhibiting an intermittent (sometimes quasi-periodic) temporal pattern. The quasi-periodic pattern has typical periods of 5-10 s (dubbed *elementary flare bursts* by De Jager and De Jonge (1978) earlier), but consists of a much higher rate of elementary acceleration episodes (up to $\lesssim$ 10 s^{-1}), totaling up to $\lesssim 10^3$ acceleration episodes per flare. These acceleration episodes have some height distribution (which has not yet been specified in detail, except for its typical vertical extent of $\Delta h \approx 5000 - 50,000$ km, see Table II). If we interpret elementary acceleration episodes in terms of dissipation events of magnetic island coalescences (Fig. 41), the maximum rate $R_{max} \lesssim 10$ s^{-1} constitutes a limit on the maximum height extent of a magnetic island, which is $\delta h \approx \Delta h / R_{max} \approx 500 - 5000$ km. This is comparable with the size of coalescing islands, $\lambda_{coal} \approx 60 - 6000$ km, inferred from the spatio-temporal scaling law (Eq. 23) with *Yohkoh* and *CGRO* data, with the separation distance $\delta h \approx 100 - 1000$ km inferred from type III+RS bursts (Table II), and with the frequency bandwidth of *metric spike* bursts ($\Delta h \approx 3000$ km).

4.5. Dynamics Inferred from Spatial Evolution

After we exploited some dynamical information on the acceleration region from temporal (Section 4.3) and spatio-temporal (Section 4.4) observables, let us turn now to spatial structures observed in solar flare images. There is numerous literature which describes the spatial morphology and evolution of flare emission in radio, hard X-rays, soft X-rays, and EUV, which provides some indirect information on the dynamics of the acceleration region. Radio and soft X-ray images generally show rather diffuse flare sources, mainly because of the instrumental sensitivity over quiet broad temperature ranges. Recent hard X-ray images, which provide the most direct radiation signatures of accelerated particles, have a limited spatial resolution of $\approx$ 5" – 10" (*Yohkoh/HXT*). The highest spatial resolution ($\approx$ 1") available in flare images to-date comes from EUV images of *TRACE*, which has narrowband temperature filters at $T \approx 1 - 2$ MK. So, let us explore the spatial information on the dynamic evolution of the acceleration region from these high-resolution images.

Numerous previous analyses of soft and hard X-ray flare data tell us that plasma temperatures of $T \approx 20 - 40$ MK are obtained in large flares. The *TRACE* 195 Å filter was designed, besides its maximum sensitivity to 1.5 MK plasma, also to have some sensitivity to Fe XXIV plasma ($T \approx 20$ MK) at the λ192 Å line (Golub et al. 1999), which is sometimes detectable at the beginning of large flares. In two large flares, such high-temperature plasmas were detected (Warren et al., 1999; Warren and Reeves, 2001). Both flares were observed at the limb and clearly revealed from this favorable perspective that they had a double-ribbon arcade geometry, with the high-temperature region ($T \approx 20$ MK) located in the cusp above the flare arcade. The *TRACE* localization of these hot regions was found to be co-spatial with those identified by *Yohkoh/SXT*, but the *TRACE* narrowband filters revealed, due to the better temperature discrimination, a more compact high-temperature region above the flare arcade (Fig. 42), than is usually seen in *Yohkoh/SXT* images. The location of the high-temperature region is fairly co-spatial with the X-point at a height of $\approx$ 1.5 times the arcade height, as it is expected in the standard bipolar 2D-reconnection scenario of the Kopp-Pneuman type (Section 2.1, Fig. 7, top right). The evolution is such that the highest temperature is observed at the very beginning of the impulsive flare phase, when magnetic reconnection and particle acceleration happens, while the temperature in the cusp drops in the postflare phase, when the arcade loops become filled with heated chromospheric plasma. Moreover, a strong enrichment of Fe over its photospheric value is observed in the hot region, suggesting in-situ heating of this plasma (Warren and Reeves, 2001). This observation probably represents the best resolved images of plasma heating in the reconnection and particle acceleration region we have available to-date.

Once the flare plasma cools down from the initial 20-40 MK to temperatures of 1-2 MK, it becomes visible in EUV again. The prominent features of the 1-2 MK plasma seen in EUV images during the postflare phase, are relaxed, near-

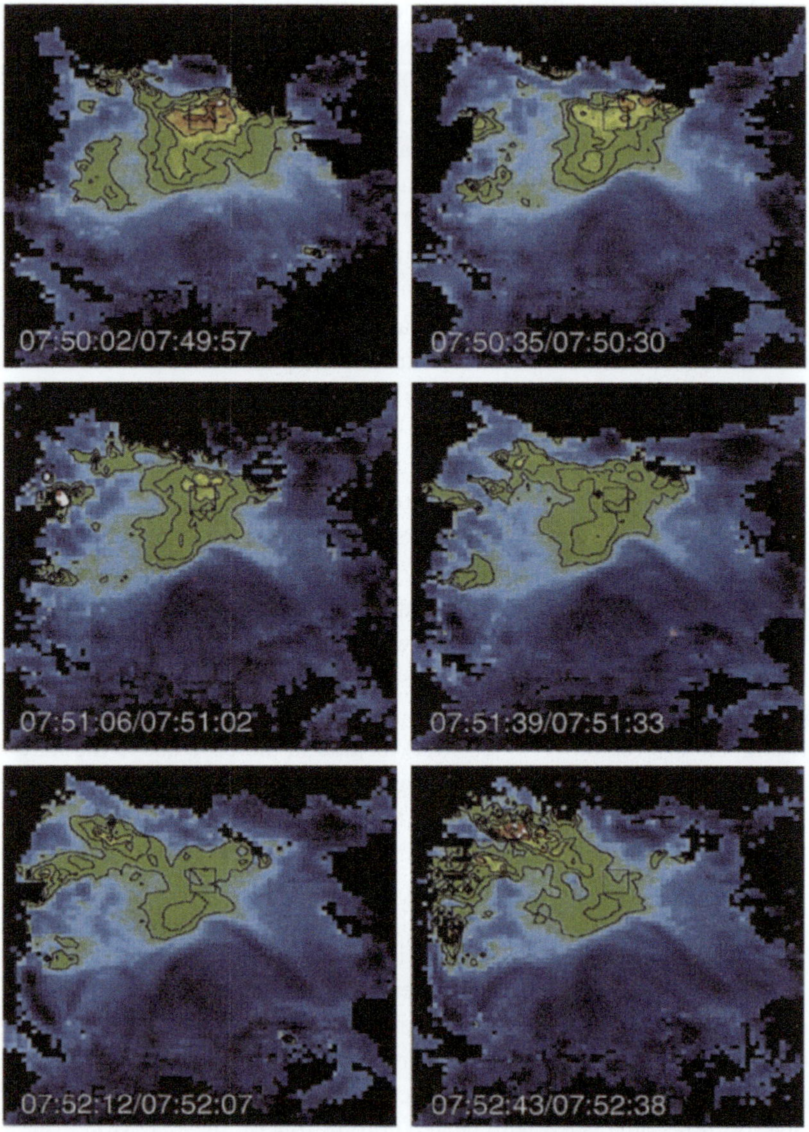

Figure 42. TRACE 195/171 Å ratio images of an above-the-looptop high-temperature region with $T \approx 20$ MK during the 2000-Mar-14, 07:50, X1.8 GOES-class flare. The flare arcade is recognizable as black loop patterns in the blueish background in the lower half of the images. The images are scaled linearly between 0 and 18 and have been rotated counterclockwise. Contours are drawn at ratios of 8, 10, 12, 14, and 16. Red corresponds to ratios above 14. The box centered on the hot region is 2.5" on a side [from Warren and Reeves, 2001].

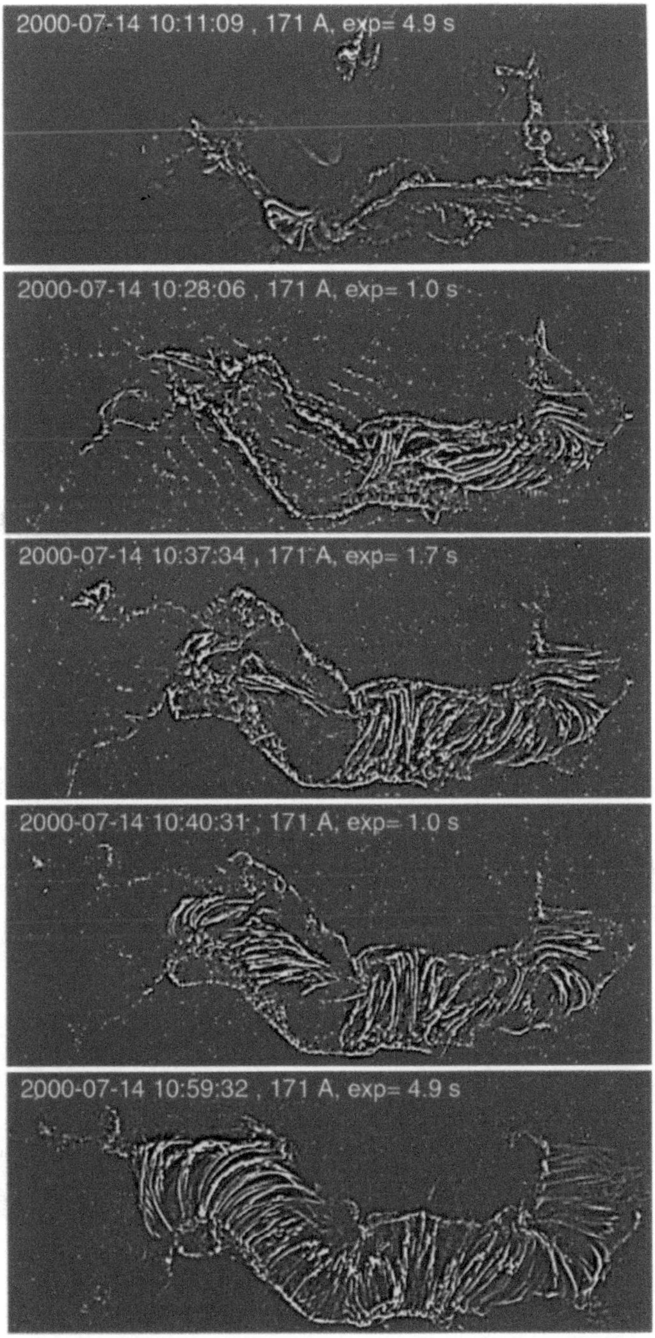

Figure 43. TRACE 171 Å high-pass filtered images of the Bastille-Day flare, 2000-Jul-14, 10:59 UT, flare (see Fig. 3 bottom), recorded at 5 different time periods of the flare between 10:11 and 10:59 UT. The field-of-view is 320" × 128". Note the high degree of spatial fragmentation, which evolves as sequential brightening of highly-sheared to less-sheared loops, first in the western half, and later in the eastern half of the arcade.

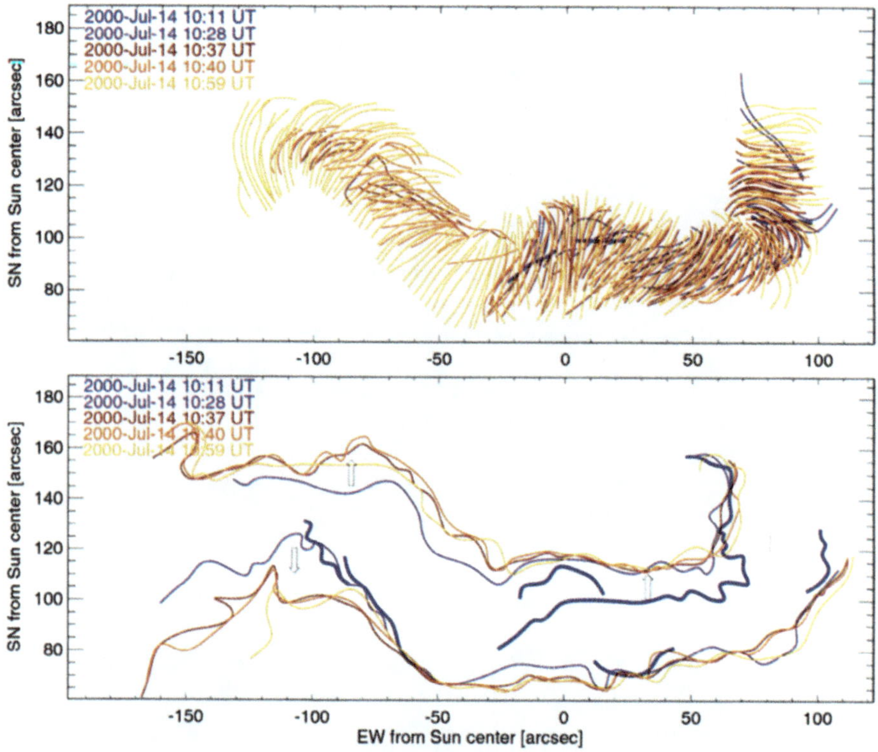

Figure 44. Top: Tracings of individual flare loops from the same 5 *TRACE* 171 Å images shown in Fig. 43. The five sets of loops traced at 5 different times are marked with different colors. Note the evolution from highly-sheared to less-sheared loops. *Bottom:* The position of the two flare ribbons traced from 171 Å images. Note the increasing footpoint separation with time.

dipolar loops that make up the flare arcade, after they have been filled by the chromospheric evaporation process. A detailed study of the flare plasma cooling using combined data with comprehensive temperature coverage in the temperature range from $T \approx 40$ MK down to $T \approx 1$ MK revealed that it takes the plasma a cooling time of about ≈ 13 minutes to cool down over this temperature range during the peak of the Bastille-day 2000-Jul-14, 10:59 UT, flare (Aschwanden and Alexander, 2001). So, whatever spatial structure we observe in *TRACE* 171 Å images, is delayed by about $\gtrsim 10$ minutes with respect to the moment when magnetic reconnection occurred. Nevertheless, the finestructure-rich images we observe in 171 Å suggest that the original fragmentation of the energy release is preserved over this cooling time, otherwise we would see a rather diffuse emission of the entire flare arcade. We show in Fig. 43 high-pass filtered *TRACE* 171 Å images of the Bastille-day flare (see also Fig. 3 bottom for an unfiltered image). There are literally some 100 resolved postflare loops visible, which characterize to some degree the spatial evolution of the energy release process that occurred about $\gtrsim 10$

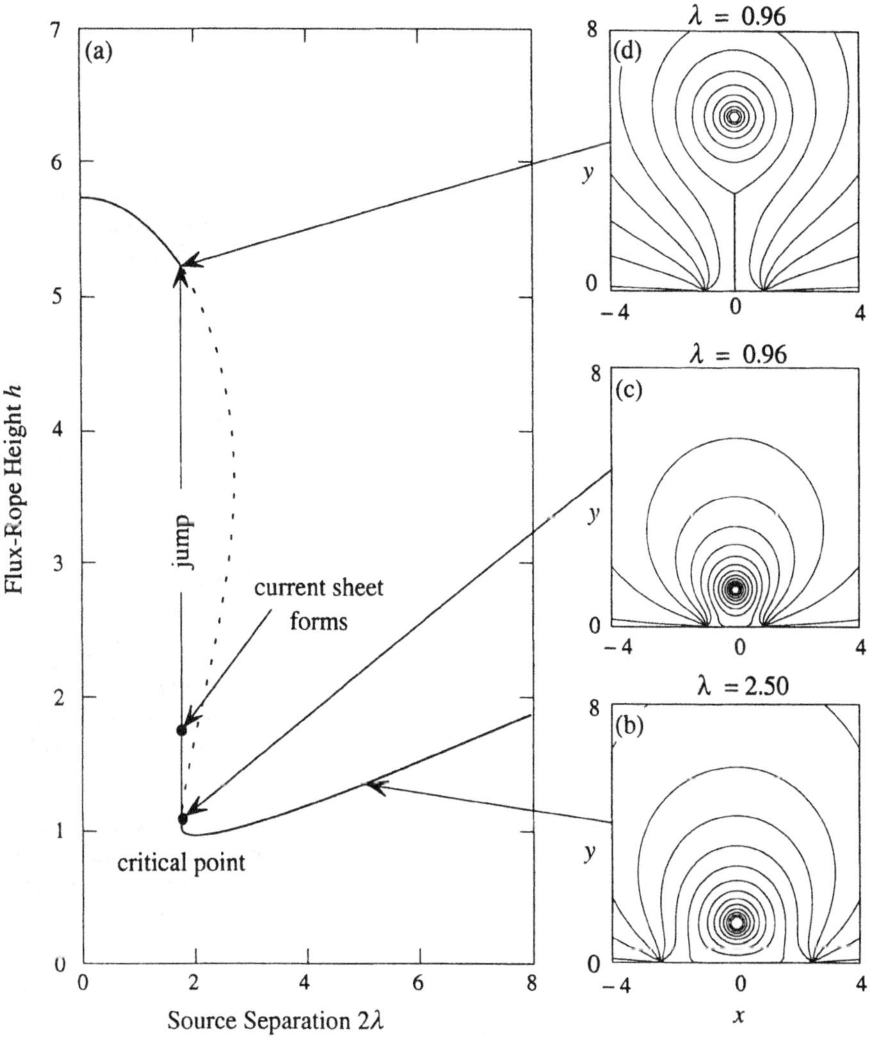

Figure 45. Flare dynamics in the model of Forbes and Priest (1995), inferred from the ideal MHD evolution of a two-dimensional arcade containing an unshielded flux rope (a)-(c). The flux rope and arcade jumps upwards when the two photospheric field sources are pushed too close to one another. (d) The vertical current sheet is subject to magnetic reconnection if enhanced or anomalous resistivity occurs [after Forbes and Priest, 1995].

minutes earlier than the 171 Å images were taken. We traced out all resolved loops and their footpoint ribbons during 5 times, as shown in Fig. 44.

The evolution can be described by two main phases, which repeat in a similar fashion first in the western half, and about 30 minutes later in the eastern half of the flare arcade. First we see some very highly sheared low-lying loops brighten up near the neutral line, while gradually less-sheared and higher-lying loops brighten up later on, until the whole flaring arcade is enclosed with a topmost layer of fully relaxed, dipole-like series of loops, dubbed *slinky* due to its appearance of a "bendable spiral". This evolution is consistent with the basic 2D-dipolar reconnection scenario of Kopp-Pneuman (Section 2.1), which predicts a rise of the reconnection region and an associated separation of the footpoint ribbons [but see Fletcher and Hudson (2001) for discrepancies between observations and model predictions]. The tracings of the field lines in the 171 Å images teach us two more things that are not explicitly predicted or quantified in the 2D reconnection model: (1) the degree of field line shearing decreases with height, and (2) the spatial structuring along the neutral line reveals an intermittent, quasi-periodic pattern. The first characteristics implies an evolution from highly-sheared to less-sheared reconnection, which corresponds to a transition from a highly non-potential to a more potential-like magnetic field in the acceleration region. The height increase moreover implies a systematic density decrease in the acceleration region. The second characteristics implies a spatial fragmentation of the reconnection region, e.g. formation of magnetic islands and coalescences in highly-sheared magnetic field configurations. The length of the arcade is roughly $L_{arcade} \approx 180,000$ km for this Bastille-Day flare and we count roughly $N_{loop} \approx 100$ loops along this arcade. This yields an average loop separation of $\Delta L_{loop} = L_{arcade}/N_{loop} \approx 1,800$ km. If we interpret this as mean separation distance between magnetic islands formed in a sheared arcade, we find good agreement with the estimates of magnetic island sizes from other methods, i.e. with the *bursty reconnection scenario* and the spatial-temporal scaling law from *Yohkoh* and *CGRO* ($\lambda_{coal} \approx 60 - 6000$ km, Eq. 22), with the height extent estimated from the maximum type III burst rate ($\delta_h \approx \Delta h/R_{max} = 500 - 5000$ km, Fig. 41), with the separation distance inferred from combined type III+RS bursts ($\delta h \approx 10 - 1000$ km), and with the frequency bandwidth of *metric spike bursts* ($\Delta h \approx 3000$ km).

The temporal sequence of arcade loop brightenings seems to be consistent with a first filament erupting in the western half, and a second filament erupting in the eastern half arcade later on. The vertical evolution can be understood in terms of a dynamic model by Forbes and Priest (1995), shown in Fig. 45: In this model the footpoint ribbons are moved together until a critical point is reached, where a catastrophic loss of magnetic equilibrium is triggered, which accelerates the filament upward and forms a current sheet underneath. The vertical current becomes stretched and is subject to tearing mode (after the length exceeds about 2π times its width, Furth et al., 1963), which leads to formation of magnetic islands and island coalescences.

We illustrate the spatio-temporal evolution and dynamics in the acceleration region during the Bastille-day flare in Fig. 46, using the constraints of the horizontally projected geometry of magnetic field lines obtained from the tracings of *TRACE* 171 Å images (Figs. 43–44), and combining it with the vertical evolution as suggested in the model by Forbes and Priest (1995). The evolution sketched in Fig. 46 shows magnetic reconnection starts at low-lying highly-sheared field lines above the neutral line and progresses gradually higher to less-sheared field

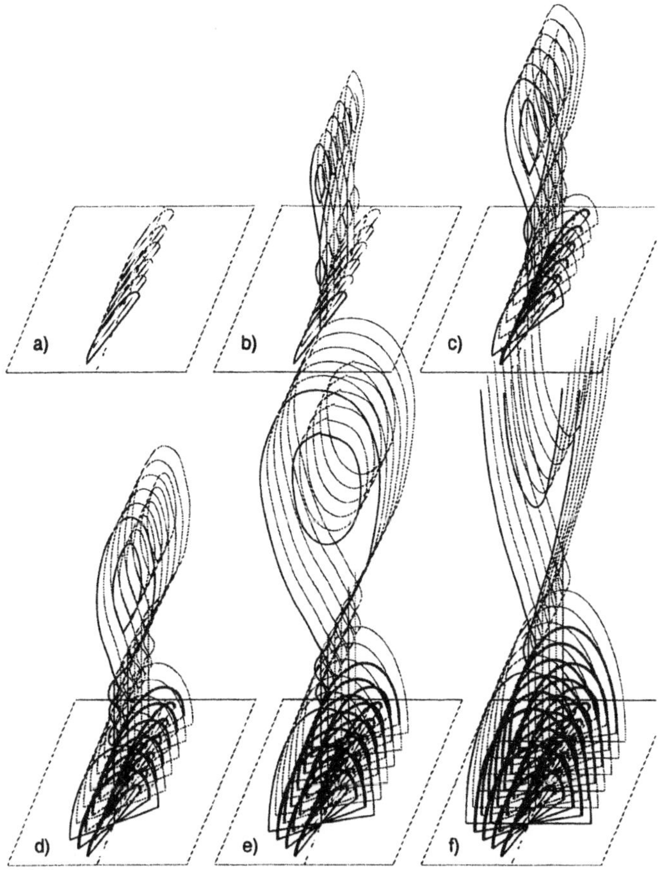

Figure 46. Scenario of the dynamic evolution during the Bastille-day 2000-Jul-14 flare. *a)* Low-lying highly-sheared loops above the neutral line become first unstable. *b)* After loss of magnetic equilibrium the filament jumps upward and forms a current sheet according to the model by Forbes and Priest (1995). When the current sheet becomes stretched, magnetic islands form and coalescence of islands occurs at locations of enhanced resistivity, initiating particle acceleration and plasma heating. *c)* The lowest-lying loops relax after reconnection and become filled due to chromospheric evaporation (loops with thick linestyle). *d)* Reconnection proceeds upward and involves higher-lying less-sheared loops. *e)* The arcade fills gradually up with filled loops. *f)* The last reconnecting loops have no shear and are oriented perpendicular to the neutral line. At some point the filament disconnects completely from the flare arcade and escapes into interplanetary space.

lines. The same height evolution has to be inferred for the acceleration regions, which most likely coincide with magnetic islands in the evolving current sheet. The upward motion of the (fragmented) acceleration regions goes along with a decrease of electron density, magnetic field, and non-potential free energy as function of time. The Bastille-Day flare clearly shows an evolution from high to low shear in the acceleration region, which corresponds to an evolution from high non-potential magnetic fields (with strong currents) to a near-potential configuration (with no currents), and thus implies an evolution from initially strong (current-driven) electric fields to weak electric fields, which modulates the tearing instability and particle acceleration efficiency in a proportional way.

5. Accelerating Electromagnetic Fields and Waves

In order to understand how particle acceleration works in solar flares, we have first to infer what the electromagnetic field configuration is, and its dynamic evolution. Once we have a model of the spatial evolution of the time-dependent electric fields **E**(t) and magnetic fields **B**(t) we can in principle calculate the accelerating forces on charged particles,

$$m\frac{d\mathbf{v}(t)}{dt} = q[\mathbf{E}(t) + \frac{1}{c}\mathbf{v}(t) \times \mathbf{B}(t)] \,, \tag{26}$$

and find their orbits by integrating the equations of motion, at least for the case of adiabatic particle motion in collisionless plasmas.

So, we have to begin with a realistic description of the electromagnetic fields that occur in solar flares. The two major physical processes we consider is magnetic reconnection and shocks. Classical treatments of magnetic reconnection processes involve long current sheets (Sweet-Parker model) or short X-points (Petschek model). These two models have often been treated with a 2-dimensional geometry (with the main magnetic field in the x-y plane), assuming invariance in the third dimension (z-axis). In many models the third dimension coincides with the direction of the accelerating field **E**, so it is important to know its extent, because that determines the electric potential difference over which a particle can be accelerated. Older simplified treatments assume an arbitrary extent of the current sheet in the third dimension, setting it equal to the extent of the flare arcade along the neutral line, or adjusting its length to the maximum electric potential difference needed to explain the highest observed energies of nonthermal particles. Such large-scale electric fields in planar (Sweet-Parker) current sheets (Fig. 47 left) have to be considered to be unrealistic for many reasons: First of all, a stretched current sheet becomes unstable to tearing mode after its length (in x-direction) exceeds about 2π times its width (Furth et al., 1963), and similar instabilities result in the third dimension (e.g. see 3D MHD simulations by Schumacher et al. 2000), so that a fragmented topology of magnetic X- and O-points results (Fig. 47 right), which leads to small-scale electric fields in magnetic islands. Moreover, large-scale electric fields would lead to a charge separation of electrons and ions that sets up a current in excess of the limit set by Ampére's law and the assumed magnetic field. The existence of small-scale electric fields is also corroborated observationally by the spatial structuring of arcade loops (§4.5), by the ubiquitous manifestations of fast (sub-second) time structures (§4.3), and (3) by spatio-temporal scaling laws (§4.4), which all yield typical sizes of $\lesssim$ 1000 km for the extent of elementary acceleration regions. A similar argument can be made for X-point configurations

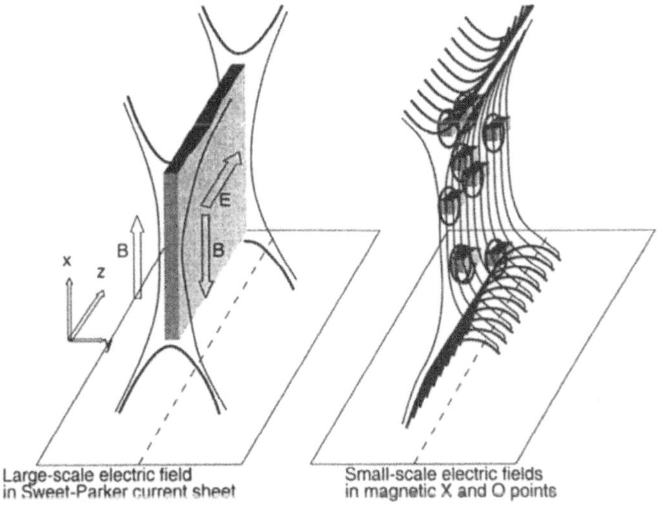

Figure 47. Paradigm shift of current sheet structure: *Left:* Classical models assume large-scale electric fields based on Sweet-Parker magnetic reconnection, which have a much larger extent in x- and z-direction than their width in y-direction. *Right:* Theory and MHD simulations, however, imply small-scale electric fields in magnetic X-points and coalescing islands with magnetic O-points.

in terms of the Petschek mechanism (Fig. 48 left). Since the scales of the diffusion region satisfy $l_x \approx l_y \ll l_z$, any weak external perturbation that is inhomogeneous in z-direction is very likely to fragment the diffusion region and to break the single X-line, so that we end up with a similarly high-fragmented picture (Fig. 48 right). This leaves us essentially with small-scale electric fields around magnetic X- and O-points as most likely acceleration scenarios in solar flares.

5.1. Electric DC Field Acceleration

Strong electric fields are generated in magnetic reconnection regions that can accelerate particles directly, e.g. along the X-line of 2D reconnection models or along separators in 3D reconnection models. Such accelerating electric fields are called *Direct Current (DC)*. These electric DC models can further be subdivided with respect to their orientation to the magnetic field, into *parallel* and *perpendicular* DC fields. The equation of motion (Eq. 26) can then be subdivided into a parallel component

$$m \frac{d\mathrm{v}_{\parallel}}{dt} = q \mathcal{E}_{\parallel} \quad (27)$$

and a perpendicular component,

$$m \frac{d\mathbf{v}_{\perp}}{dt} = q \left[\mathbf{E}_{\perp} + \frac{1}{c} (\mathbf{v}_{\perp} \times \mathbf{B}) \right] . \quad (28)$$

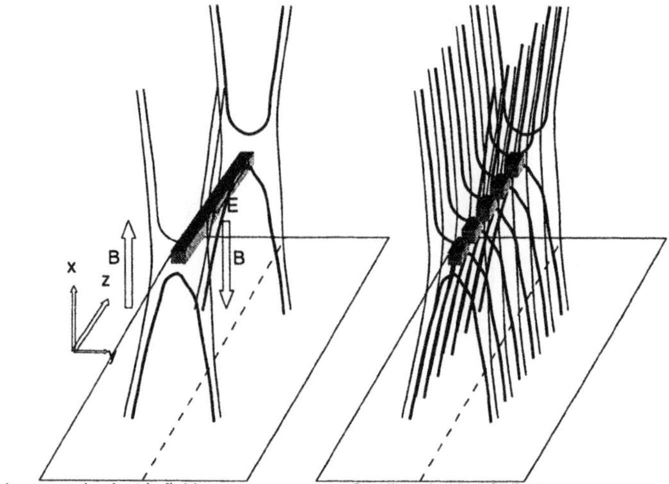

Figure 48. Paradigm shift for Petschek-type reconnection: *Left:* Classical models assume elongated electric fields in 2D Petschek-type magnetic reconnection, which have a much larger extent in z-direction than their width in x and y-direction. *Right:* Theory and MHD simulations, however, imply small-scale electric fields in broken-up X-points along the neutral line.

Acceleration by parallel electric fields $\mathcal{E}_\parallel$ is further distinguished for weak and strong electric fields, depending on whether the electric field strength is below or above the critical Dreicer field $\mathcal{E}_D$ (Dreicer, 1959),

$$\mathcal{E}_D = \frac{q_i \ln \Lambda}{4\pi \epsilon_0 {\lambda_D}^2} \tag{29}$$

where $\ln \Lambda$ is the Coulomb logarithm and $\lambda_D = \sqrt{[\epsilon_0 k_B T/(n_e e^2)]}$ the Debye length. The process of electron acceleration in weak (sub-Dreicer) electric fields is called *runaway acceleration*, which occurs for electrons that overcome the collisional friction between electrons and ions. The value of the runaway speed is

$$\mathrm{v}_r = \mathrm{v}_{Te} \left(\frac{\mathcal{E}_D}{\mathcal{E}_\parallel} \right)^{1/2} \tag{30}$$

with v_{Te} the thermal speed of the electrons.

5.1.1. *DC Field Acceleration in Sub-Dreicer Fields*

Runaway acceleration in sub-Dreicer fields has been applied to solar flare scenarios by Holman (1985), Tsuneta (1985), and Benka and Holman (1994). The electric field in this model is of order $\mathcal{E} \approx 10^{-5}$ V cm^{-1}, extending over spatial scales of $L \approx 30,000$ km, which is compatible with the flare loop size. Holman (1985) finds that the energy gain scales with

$$W - W_c = 7.0 \left(\frac{T}{10^7\, K} \right)^{1/2} \left(\frac{\nu_e}{10\, s^{-1}} \right) \left(\frac{L}{10^9\, cm} \right) \left(\frac{\mathrm{v}_c}{\mathrm{v}_e} \right)^{-2} \text{keV} . \tag{31}$$

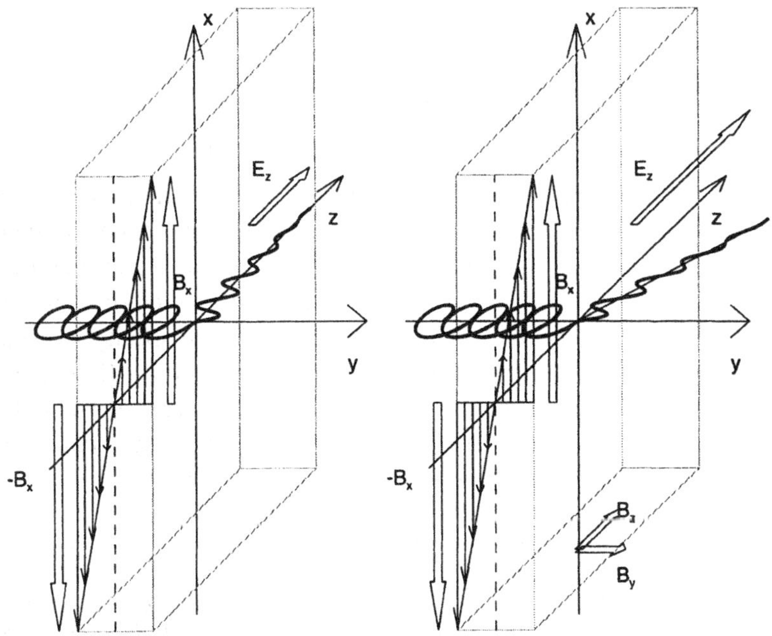

Figure 49. Particle orbit in sub-Dreicer electric DC field (left) and super-Dreicer electric DC field (right). In sub-Dreicer fields, the particle is accelerated the full length of the current sheet in z-direction (e.g. model by Holman 1985), while it is scattered out of the current sheet after a short distance in the super-Dreicer field, due to an additional weak perpendicular magnetic field components $(0, B_y = B_\perp, 0)$ to the longitudinal field $(0, 0, B_z = B_\parallel)$ (e.g. model by Litvinenko 1996).

yielding electron energies of $W \approx 100$ keV for an electron temperature of $T = 10^7$ K, a collision frequency of $\nu_e \approx 2 \times 10^3$ s^{-1}, a ratio $v_c/v_e \approx 4$ of the critical runaway speed v_c to the thermal speed v_e, a critical energy $W_c = 8$ keV, and a length scale of $L \approx 10^9$ cm. This energy gain is sufficient for the bulk of electrons observed in most hard X-ray flares, supposed that such large-scale electric fields over distances of $L = 10,000$ km exist. Higher energies could be achieved by assuming anomalous resistivity, which enhances the value of the effective Dreicer field, and thus the maximum values of sub-Dreicer electric fields. Regarding the transverse extent of current channels, Holman (1985) concludes that a fragmentation of about $\approx 10^4$ current channels is needed due to the maximum magnetic field limit imposed by Ampére's law. If we take the additional fragmentation in longitudinal direction into account, which yields, as suggested by observations, typical sizes of $L \approx 1000 - 3000$ km for elementary acceleration regions (Table II), the energy gain in sub-Dreicer fields is substantially reduced, to $W - W_c \approx 10$–30 keV. DC electric fields may exist in X-points between coalescing islands, but runaway acceleration in sub-Dreicer fields would only produce $\lesssim 10 - 30$ keV electrons in those regions.

5.1.2. *DC Field Acceleration in Super-Dreicer Fields*

Applications of DC electric fields in super-Dreicer fields, $\mathcal{E} \gg \mathcal{E}_D$, to solar flares have been demonstrated by Litvinenko (1996). He calculated particle orbits in standard current sheet geometries (Fig. 47 left), but assumed besides the main magnetic field $\mathbf{B}_\parallel = (0, 0, B_\parallel)$ and the parallel electric field $\mathbf{E} = (0, 0, \mathcal{E}_\parallel)$, also a weaker perpendicular magnetic component, $\mathbf{B}_\perp = (0, B_\perp, 0)$, which serves to scatter electrons accelerated along the $\mathcal{E}$-field direction out of the current sheet before they reach the end of the current sheet. The magnetic field inside the current sheet is thus $\mathbf{B} = (-y/\Delta w_y, B_\perp, B_\parallel)$. In this configuration, the maximum particle energy W is determined not only by the electric field $\mathcal{E}_\parallel$, but also by the ratio of the parallel to the perpendicular magnetic field,

$$W = \frac{B_\parallel}{B_\perp} e \Delta w_y \mathcal{E}_\parallel , \tag{32}$$

with Δw_y the width of the current sheet. The acceleration time is

$$\Delta t = \sqrt{\frac{B_\parallel}{B_\perp} \frac{2\Delta w_y m_e}{e\mathcal{E}_\parallel}} . \tag{33}$$

For typical values, $B_\parallel = 100$ G, $B_\perp = 1$ G, $\mathcal{E}_\parallel = 10$ V cm^{-1}, $\Delta w = 10^2$ cm, one obtains electron energies of $W \approx 100$ keV. A typical particle orbit for such a configuration is shown in Fig. 49. Essentially the particle spirals around the guiding magnetic field during acceleration, until the $\mathcal{E} \times B_\perp$ drift scatters it out of the thin current sheet. The typical acceleration length in direction of the electric field is only

$$\Delta l = \Delta w_y \frac{B_\parallel}{B_\perp} \tag{34}$$

which is about $\Delta l \approx 10^4$ cm for the parameters above. Thus, the electric field or current sheet does not need a large distance as required in sub-Dreicer field acceleration ($L \approx 10^9$ cm). Therefore, the required spatial extents of super-Dreicer electric fields fit everywhere into a fragmented acceleration region scenario, as suggested in the *bursty reconnection model* (Fig. 47 right). Numerical simulations of proton acceleration near an X-type reconnection point exhibited a powerlaw spectrum of $N(E) \propto E^{-2}$ (Mori et al. 1998).

5.1.3. *DC Field Acceleration During Current Loop Coalescence*

While we considered static current sheets in the previous two sections, we turn now to dynamic current sheets. The dynamics of accelerating electric fields has been studied for the case of two coalescing current-carrying loops during an *explosive magnetic reconnection process* in great detail by Tajima and Sakai (1986), Tajima et al. (1987), Sakai and Ohsawa (1987), and Sakai and De Jager (1991). For the initial configuration of the current sheet they assume a 2D geometry with width w_y

and length l_x (as illustrated in Fig. 47 left and Fig. 49). The reconnection process is driven by the lateral inflow v_y, producing a reconnection outflow v_x with the local Alfvénic velocity, $v_x = v_A \approx B_x/\sqrt{n_e}$, which are related to the geometric dimensions of the current sheet by the mass conservation law (in an incompressible plasma),

$$l_x v_y = w_y v_x \ . \tag{35}$$

If the lateral magnetic influx, $v_y B_x$, is constant, the reconnection rate would be just the Sweet-Parker reconnection rate in the case of long current sheets, $l_x \gg w_y$, or the Petschek reconnection rate, in the case of X-type short current sheets ($l_x \approx w_y$). The essential feature of non-steady reconnection processes is the dynamics of the driver. In the treatment of Sakai and Ohsawa (1987) it is assumed that the lateral magnetic influx increases explosively,

$$v_y(t) \propto -\frac{y}{(t - t_0)} \ , \tag{36}$$

dynamically driving the dimensions of the current sheet as

$$w_y(t) \propto w_{y0}\eta(t_0 - t) \tag{37}$$

$$l_x(t) \propto l_{x0}\eta(t_0 - t)^2 \ . \tag{38}$$

Once the time t approaches t_0, the length of the current sheet $l_x(t)$ decreases faster than the width $w_y(t)$, so that the Sweet-Parker current sheet ($l_x \gg w_y$) makes a transition to a Petschek type ($l_x \approx w_y$). The associated change in magnetic flux, $\Delta\Psi(t)$ becomes then explosive,

$$\Delta\Psi(t) \propto \Delta\Psi_0(t_0 - t)^{-4/3} \tag{39}$$

which evolves into a very rapid reconnection rate, independently of the plasma resistivity η. The associated electric and magnetic field components show the following nonlinear time dependences,

$$\mathcal{E}_y(t) \propto -\frac{y}{(t_0 - t)^2} \tag{40}$$

$$B_x(t) \propto \frac{x}{(t_0 - t)^{4/3}} \tag{41}$$

$$\mathcal{E}_z(t) \propto c_1\frac{x^2}{(t_0 - t)^{7/3}} + c_2\frac{1}{(t_0 - t)^{5/3}} \tag{42}$$

producing complicated particle orbits that require numerical computations. Sakai and Ohsawa (1987) calculate the evolution of this dynamic reconnection process with analytical approximations, as well as perform numerical simulations, using the standard MHD equations for a two-fluid (electrons and ions) plasma. They find a variety of dynamical evolutions: (1) explosive collapse, (2) nonlinear oscillations between magnetic collapses and recoveries, and (3) double-peak structures during nonlinear oscillations (Fig. 50), depending on the particular value of the plasma-beta parameter $\beta = c_s^2/v_A^2$. Oscillatory evolutions result from the counter-acting forces of the $\mathbf{j} \times \mathbf{B}$-term, which drives the magnetic collapse, and the pressure gradient term ∇p inside the current sheet. These results received support by two particular flare observations, which showed indeed quasi-periodic sequences of double-peak structures, i.e. 1980-Jun-7, and 1982-Nov-26 (Sakai and Ohsawa, 1987). In hindsight, given the high-resolution observations of *TRACE*, which show a high degree of spatially-fragmented loop arcades, we think that most of the (generally aperiodic) fast hard X-ray time structures result from spatially-separated magnetic collapses, rather than from a quasi-periodic oscillation of a single collapse region. Moreover, multiple, spatially separated magnetic collapses ease the number problem of accelerated particles considerably, compared with the operation of a single collapse region which has to be driven by a fast inflow to constantly replenish particles. Nevertheless, the work of Sakai and Ohsawa (1987) yields valuable quantitative physical insights into the nonlinear evolution of the 3D electric and magnetic field components that accelerate electrons and ions in a collapsing current sheet.

5.1.4. *DC Field Acceleration in Filamentary Current Sheets*

While Sakai and Ohsawa's (1987) treatment of particle acceleration essentially involves a single X-point in a reconnection region, a more general treatment of a filamentary current sheet includes multiple X-points, which also imply intervening magnetic islands with O-points, e.g. as illustrated in the concept by Furth et al. (1963) (see Fig. 23). The presence of magnetic islands add an interesting additional feature to the dynamics of accelerated particles, namely that it allows for temporary trapping of accelerated particles inside the magnetic islands, which increases the total acceleration time and thus requires less demanding DC electric fields to achieve the same final kinetic energy, compared with electric DC fields in X-point geometries. Particle orbits, trapping, and acceleration in such a filamentary current sheet has been explored by Kliem (1994). Consider a chain of current filaments (magnetic islands) that are naturally formed by the tearing instability in a reconnection region with enhanced resistivity (§4.2). There are two different mechanisms of particle acceleration in a reconnecting current sheet, given by the parallel ($\mathcal{E}_\parallel$) and perpendicular ($\mathcal{E}_\perp$) electric field components, respectively. The maximum parallel electric field $\mathcal{E}_\parallel$ is the Dreicer field (Eq. 29). The convective electric field $\mathbf{E}_{conv} = -\mathbf{u} \times \mathbf{B}/c$) is typically orders of magnitude larger for fast reconnection, $u \approx (0.01 - 0.1)v_A$. It is often considered as irrelevant for particle acceleration, but in the inhomogeneous vicinity of magnetic X-lines and O-lines, the guiding center drift of the particles points into the direction of the convective electric field ($\mathcal{E}_\perp$), which, therefore, contributes to the particle acceleration. One perpendicular electric field component results from the lateral inflow motion. In addition, another perpendicular electric field component is induced from the approaching motion of the coalescing magnetic islands. The directions of the electric field components at

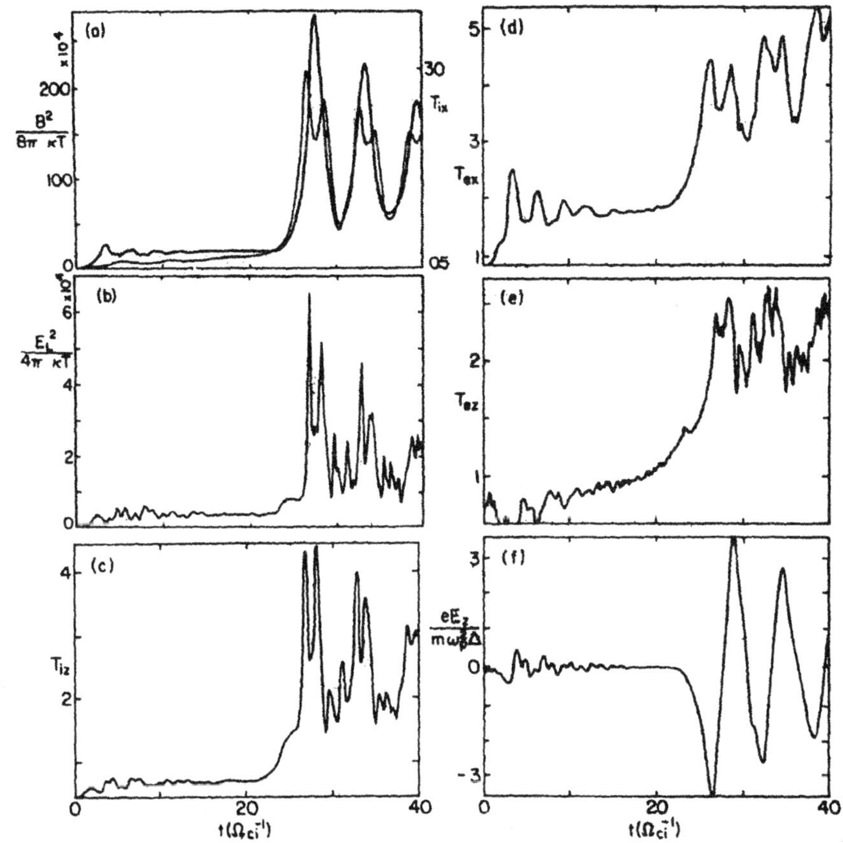

Figure 50. The dynamical evolutions of magnetic current loop collapses: (a) The effective potential describing the magnetic collapse. The schematic pattern of the nonlinear oscillations after the explosive phase for (b) $B^2 \approx B_y^2$, (c) $E_L \approx E_x^2$, and (d) T_{ix}. The oscillation period T is of the order of the explosion time t_0 (from Tajima et al. 1987).

various locations inside and outside the magnetic islands is indicated in Fig. 51 for a pair of approaching islands.

The particle motion for this configuration of two approaching magnetic islands have been numerically computed in a Fadeev equilibrium by Kliem (1994). Three types of net motion in directions across the magnetic fields are found: meander orbits at magnetic X- and O-lines, the magnetic gradient and curvature drift, and the $\mathbf{E} \times \mathbf{B}$ drift. The drift motion in inhomogeneous magnetic fields near the X-lines can be characterized by

$$\mathbf{v}_\nabla = \frac{mc}{2q}(v_\perp^2 + 2v_\parallel^2)\frac{\mathbf{B} \times \nabla \mathbf{B}}{B^3} \tag{43}$$

which is exactly parallel to $\pm\mathbf{E}$ in the two-dimensional field case. The drift and meander orbit of a test particle is shown in Fig. 52, obtained from numeric calculations by Kliem (1994). It shows a particle that enters the separatrix of a magnetic island

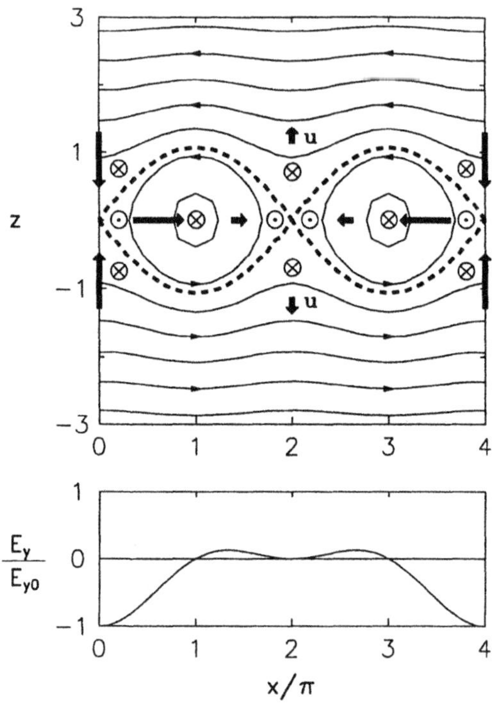

Figure 51. Electric field components (arrow heads ⊙ and tails ⊗), magnetic field $\mathbf{B}(x, y)$ (solid lines), separatrices (dashed lines), and mass flows (thick arrows, $\mathbf{u}$) in a pair of approaching magnetic islands (from Kliem 1994).

and becomes trapped inside. The particle experiences the largest acceleration kick during the first part of the orbit after entering the magnetic island near the X-point, and *meanders* then around the magnetic O-point in a stable orbit, slowly drifting along the y-axis, dragged by the ∇B and $\mathbf{E} \times \mathbf{B}$ force. The stability of meander orbits essentially traps particles sufficiently long so that they can experience the full energy gain provided by the strength and extent of the convective electric field near the O-lines. On the other hand, only particles with a sufficient high threshold energy (probably requiring a pre-accelerated seed population) enter a meander-like orbit within one characteristic coalescence time, but those gain the highest kinetic energies. Also, only particles that come close to an X-point experience the largest acceleration kick (e.g. in Fig. 52).

The maximum energy particles can obtain in coalescing magnetic islands is essentially given by the convective electric field, which can be as high as

$$\mathcal{E} \approx u_{inflow} B_0 / c \approx 0.03 \mathrm{v}_{A0} B_0 / c \approx 3 \times 10^3 \ \mathrm{V\ m^{-1}} \tag{44}$$

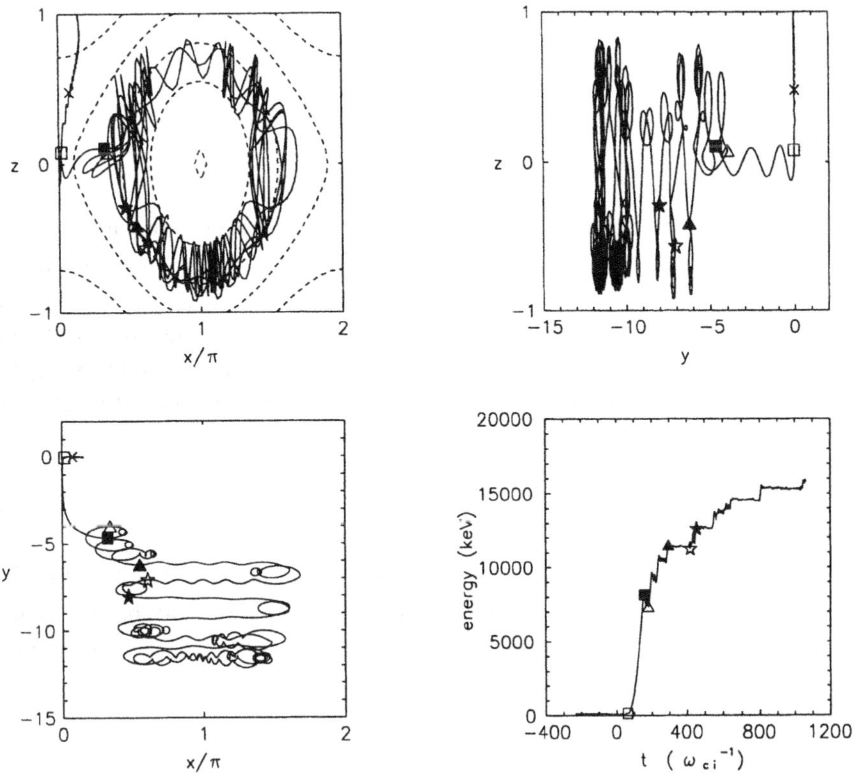

Figure 52. Test particle orbit near a magnetic O-point in a magnetic island configuration as shown in Fig. 51. The particle is carried into the vicinity of the X line by the **E** × **B** drift (equivalent to the reconnection inflow). There it experiences a ∇B drift acceleration with a meander component (top right), which leads to the largest acceleration kick (bottom right). The acceleration continues due to further ∇B drifts in the trapped orbit around the O line. The symbols mark characteristic points of the orbit (from Kliem 1994).

for a main magnetic field of $B_0 = 200$ G and a density of $n_0 = 10^{10}$ cm^{-3}. The half width of a current sheet l_{CS} was estimated using the minimum critical current j_{cr} for onset of anomalous resitivity:

$$l_{CS} \approx \frac{cB}{4\pi j_{cr}} \approx 7 \times 10^3 \text{ cm} , \tag{45}$$

and the typical acceleration length was numerically found to be of order of the magnetic island (or coalescence) length, $\approx L_{CI}$. The maximum energy a particle can obtain in such a static, convective electric DC field is then $W_{max} \approx 200$ keV, which is significantly higher than the particles would gain by the parallel field ($W_{max} \approx eL_z\mathcal{E}_\parallel \lesssim 20$ keV). Acceleration to higher energies would require multiple magnetic islands or dynamic reconnection.

In summary, we can say that the parallel electric fields could be important for the bulk acceleration, while the convective electric fields provide acceleration from a superthermal seed population to higher energies for a subset of particles that pass close to an X-line and become trapped inside the magnetic islands around O-lines. We should also not forget that the tearing instability that drives the entire process of magnetic island coalescence requires anomalous resistivity, e.g. provided by turbulence from current-driven instabilities at X-lines. Recent work describes the topology of reconnection in multiple magnetic islands with a fractal geometry (Shibata and Tanuma, 2001).

5.1.5. *Field-Aligned Potential Drop Acceleration*

Field-aligned potential drops have been widely used to explain particle acceleration in magnetospheric auroras, but were also applied to solar flares (Haerendel 1994). In contrast to DC electric field acceleration, as discussed in the previous Sections, field-aligned potential drops can be generated by unstable field-aligned currents, which convert energy stored in magnetic shear stresses into kinetic energy. The key difference to DC electric fields is that the field-aligned potential drops are not static, but rather originate in a highly dynamic way during rapid dissipation of a multitude of small-scale current sheets, similarly as discussed above during the *bursty reconnection mode*. The shear stresses between the coronal reconnection sites and the chromospheric footpoints would then build-up short-lived field-aligned potential drops, in which electron can be accelerated to γ-ray energies (Haerendel 1994). Another advantage of this model is that high-energy particles do not require an initial low-energy bulk acceleration, unlike runaway or stochastic acceleration mechanisms.

5.2. Stochastic Acceleration

In the last Section we saw a gradual increase in complexity, starting from continuous energy gain in simple-structured DC electric fields to (meander-like) cyclic orbits with interchanging episodes of energy gain and loss in multiple electric field components. The great majority of the particle orbits are still deterministic in such prescribed fields, but a few particles that approach an X-line very closely execute chaotic orbits. For even more complex accelerating forces, as provided by turbulence or wave resonances, which generate electromagnetic fields modulated with high frequencies, particle acceleration becomes even less continuous, where energy gain and loss changes during each gyroperiod or waveperiod. However, if there is a systematic trend that particles gain energy over the long term, although they gain and loss seemingly at random in the short term (like a casino owner in Las Vegas), we call this process *stochastic acceleration* (for a recent review see Miller et al., 1997).

The fundamental coupling between electromagnetic waves (characterized by the dispersion relation $\omega(k)$ between the wave frequency ω and wave number $k = 2\pi/\lambda$) and a particle (with speed v, relativistic Lorentz factor $\gamma = 1/\sqrt{1-(v/c)^2}$,

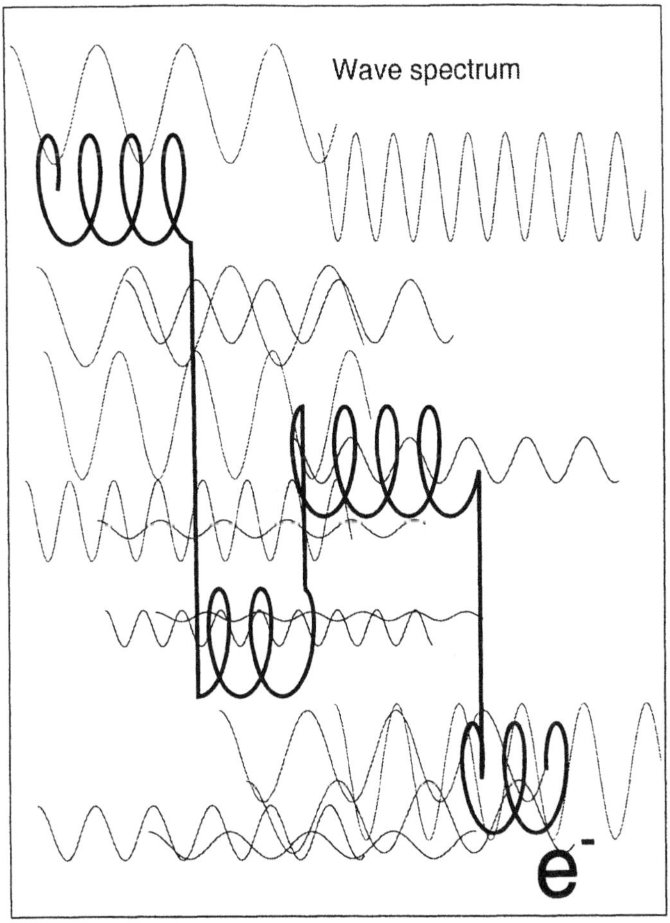

Figure 53. Cartoon of a stochastic acceleration process. A charged particle (e.g. an electron here) executes its gyromotion around the guiding magnetic field. If the Doppler-shifted gyrofrequency is in phase with the frequency of waves, the electron experiences the electromagnetic field of the resonant waves and is accelerated until it becomes detuned. If a broad wave spectrum is present, the electron suffers many random acceleration and deceleration episodes, which add up to a net gain in energy for suitable wave spectra.

gyrating in a guiding magnetic field with gyroperiod $\Omega = eB/m_e c$) is the *Doppler resonance condition*,

$$\omega - s\Omega/\gamma = k_{\parallel} v_{\parallel} \tag{46}$$

which specifies when the gyromotion of the particle is in phase with the Doppler-shifted wave vector, where maximum energy exchange can be obtained. If the harmonic number is positive ($s > 0$), we have *normal Doppler resonance*, for negative integer numbers ($s < 0$) we have *anomalous Doppler resonance*, and the resonance of $s = 0$ is referred to as *Landau* or *Cerenkov resonance*. The Doppler resonance condition can be written in more explicit form, to show the dependence on the wave dispersion relation $\omega(k)$,

$$\omega - s\Omega/\gamma = kc\beta \cos\vartheta \cos\alpha \tag{47}$$

where β =v/c is the relativistic speed, $\alpha = \arccos(\mathrm{v}_\parallel/\mathrm{v})$ the pitch angle of the particle, and $\vartheta = \arccos(k_\parallel/k)$ the angle of the wave vector to the guiding magnetic field. The *index of refraction*, $N = kc/\omega$, characterizes the dispersion relation $\omega(k)$ for different wave modes, which are tabulated in Table IV.

TABLE IV

Summary of dispersion relations of elementary waves for parallel and perpendicular directions in cold plasma (after Chen 1974)

Wave direction	Dispersion relation	wave mode
Electron waves (electrostatic)		
$\mathbf{B}_0 = 0$ or $\mathbf{k} \parallel \mathbf{B}_0$	$\omega^2 = \omega_p^2 + \frac{3}{2}k^2\mathrm{v}_{th}^2$	Langmuir waves (P)
		(Plasma oscillations)
$\mathbf{k} \perp \mathbf{B}_0$	$\omega^2 = \omega_p^2 + \omega_c^2$	Upper hybrid waves
Ion waves (electrostatic)		
$\mathbf{B}_0 = 0$ or $\mathbf{k} \parallel \mathbf{B}_0$	$\omega^2 = k^2\mathrm{v}_s^2$	Ion sound waves
	$\omega^2 = k^2\left(\frac{\gamma_e k_B T_e + \gamma_i k_B T_i}{M}\right)$	(Acoustic waves)
$\mathbf{k} \perp \mathbf{B}_0$	$\omega^2 = \Omega_p^2 + k^2\mathrm{v}_s^2$	Electrostatic ion
		cyclotron waves
	$\omega^2 = \omega_l^2 = \Omega_c\omega_c$	Lower hybrid waves
Electron waves (electromagnetic)		
$\mathbf{B}_0 = 0$	$\omega^2 = \omega_p^2 + k^2c^2$	Light waves
$\mathbf{k} \perp \mathbf{B}_0$, $\mathbf{E}_1 \parallel \mathbf{B}_0$	$\frac{c^2k^2}{\omega^2} = 1 - \frac{\omega_p^2}{\omega^2}$	Ordinary waves (O)
$\mathbf{k} \perp \mathbf{B}_0$, $\mathbf{E}_1 \perp \mathbf{B}_0$	$\frac{c^2k^2}{\omega^2} = 1 - \frac{\omega_p^2}{\omega^2}\frac{\omega^2-\omega_p^2}{\omega^2-\omega_h^2}$	Extraordinary waves (X)
$\mathbf{k} \parallel \mathbf{B}_0$	$\frac{c^2k^2}{\omega^2} = 1 - \frac{\omega_p^2/\omega^2}{1-\omega_c^2/\omega^2}$	Right hand waves (R)
		(whistler mode)
	$\frac{c^2k^2}{\omega^2} = 1 - \frac{\omega_p^2/\omega^2}{1+\omega_c^2/\omega^2}$	Left hand waves (L)
Ion waves (electromagnetic)		
$\mathbf{B}_0 = 0$	none	
$\mathbf{k} \parallel \mathbf{B}_0$	$\omega^2 = k^2\mathrm{v}_A^2$	Alfvén waves
$\mathbf{k} \perp \mathbf{B}_0$	$\frac{\omega^2}{k^2} = c^2\left(\frac{\mathrm{v}_s^2+\mathrm{v}_A^2}{c^2+\mathrm{v}_A^2}\right)$	Magnetosonic waves

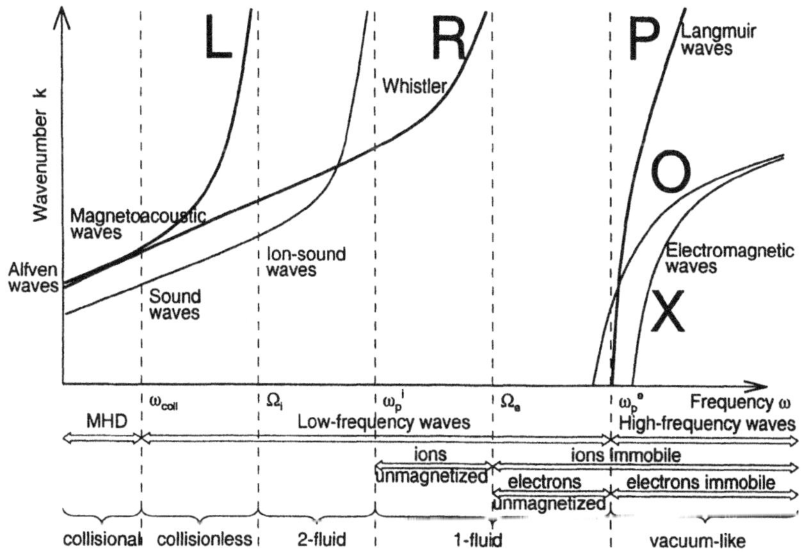

Figure 54. Overview on wave modes in cold plasma that play a role in stochastic wave-resonant particle acceleration. Ions can resonate with low-frequency waves, while electrons resonate mostly with high frequency waves. Indicated are the frequency regimes separated by the collision frequency ω_{coll}, the ion gyrofrequency Ω_i, the ion plasma frequency ω_p^i, the electron gyrofrequency Ω_e, and the electron plasma frequency ω_p^e. L and R are left and right-hand circular polarized low-frequency electromagnetic waves, and O and X are the ordinary and extraordinary mode of the high-frequency electromagnetic waves.

Inserting the dispersion relations $\omega(k)$ for different wave modes (Table IV) into the Doppler resonance condition (Eq. 47) yields a mathematical solution for the wave frequency ω that is resonant with the gyromotion of a particle with velocity v. An overview of the different frequency ranges is given in Fig. 54 for typical conditions in the solar corona. Typical magnetic field strengths (which determine the gyrofrequencies Ω_i and Ω_e of ions and electrons) and densities (which determine the plasma frequencies ω_p^i and ω_p^e) yield the following order of frequencies in the solar corona: $\omega_{coll} < \Omega_i < \omega_p^i < \Omega_e < \omega_p^e$. Fig. 54 shows at one glance which waves are important for the acceleration of ions and electrons: essentially Alfvén waves, magnetosonic waves, and ion sound waves can accelerate ions, while whistler waves, Langmuir waves, and electromagnetic waves are efficient in acceleration of electrons. Whether a net gain of acceleration results for a given particle distribution $f(\mathbf{v})$ (in velocity space), depends on the ambient wave spectrum $N(\omega, k)$. Essentially, if a particle resonates with a particular wave frequency ω, it can gain energy and becomes detuned from the wave, but if there is an adjacent wave with a slightly higher frequency $\omega + \Delta\omega$ present, the particle can be accelerated further. So, stochastic acceleration requires the presence of a broadband wave spectrum.

The wave-particle interaction is generally characterized by a coupled differential equation system that entails the energy change between the wave energy spectrum $W(\omega, k, t)$ in Fourier space $(\omega, \mathbf{k})$ and the particle distribution $f(\mathbf{x}, \mathbf{v}, t)$ in velocity space $(\mathbf{x}, \mathbf{v})$, the so-called *quasi-linear diffusion* equations (given here in generic form),

$$\frac{\partial f}{\partial t} = \frac{1}{p_i^2} \frac{\partial}{\partial p_i} \left[p_j^2 D_{ij} \frac{\partial f}{\partial p_j} \right] - \frac{f}{t_L} \tag{48}$$

$$\frac{\partial W}{\partial t} = \frac{\partial}{\partial k_i} \left[k_i^2 D_{ij} \frac{\partial}{\partial k_j} (k_j^{-2} W) \right] - \gamma W + S \tag{49}$$

where D_{ij} represents the quasi-linear diffusion term of the particles, t_L a particle leakage time out of the system, γ a damping term of the waves due to the acceleration of energetic electrons, and S a source term of wave injection. The quasi-linear equation system allows for energy exchange between waves and particles in both directions. If particles have an anisotropic distribution $f(\mathbf{v})$ in velocity space, e.g. a beam or losscone, they possess free energy that can be converted into waves (e.g. Langmuir waves or electron cyclotron maser emission). On the other hand, if electromagnetic waves are present, e.g. generated by turbulence or convective motion, part of the wave energy can be transferred into kinetic particle energy, at frequencies where the Doppler resonance condition (Eq. 46) is fulfilled. The latter process constitutes the prime mechanism of *stochastic particle acceleration.*

The mutual wave-particle interaction is characterized by the diffusion terms D_{ij} in the *quasi-linear diffusion equations* (Eqs. 48 and 49). If energy is transferred from particles into waves, e.g. by gyroresonance emission, the diffusion term is defined as function of the particle momentum vector $\mathbf{p}$, i.e. $D_{ij}(\mathbf{p})$. The wave energy $W(\mathbf{k})$ can be defined (e.g. Melrose 1980a, p.146) by a *photon occupation number* $N(\mathbf{k})$ at wave number $\mathbf{k}$, multiplied with the photon energy $\hbar\omega$ that is carried by every photon,

$$W(\mathbf{k}) = N(\mathbf{k}) \hbar\omega \, . \tag{50}$$

The diffusion coefficient $D_{ij}(\mathbf{p})$ can then be defined (e.g. Melrose 1980a, p.162) by the transition probability $w^\sigma(\mathbf{p}, \mathbf{k}, s)$ of a photon with wave vector $\mathbf{k}$, stimulated by a particle with momentum $\mathbf{p}$ in wave mode σ and at harmonic number s, integrated over the entire wave vector space,

$$D_{ij}(\mathbf{p}) = \int \frac{d\mathbf{k}^3}{(2\pi)^3} N^\sigma(\mathbf{k}) w^\sigma(\mathbf{p}, \mathbf{k}, s) \cdot (\hbar k_i)(\hbar k_j) \, . \tag{51}$$

The transition probability contains a delta-function with the resonance condition (Eq. 46),

$$w^\sigma(\mathbf{p}, \mathbf{k}, s) = \frac{1}{\hbar} [\eta_s^\sigma(\mathbf{k}) \delta(\omega - s\Omega_e/\gamma - k_\parallel v_\parallel)] \tag{52}$$

where $\eta_s^\sigma(\mathbf{k})$ is the emissivity of gyroemission at the s-th harmonic and wave vector $\mathbf{k} = (\omega, \vartheta)$ (for a derivation see, e.g. Melrose 1980a, p.103; Melrose 1980b, p.275).

For stochastic acceleration, energy is transferred from wave photons to particle momentum p_{ij}, so the diffusion term D_{ij} in the quasi-linear diffusion equations is defined as function of wave energy, i.e. $D_{ij}(\mathbf{k})$. The diffusion coefficients $D_{ij}(\mathbf{k})$ are calculated as scattering coefficients of resonant particles interacting with a wave spectrum $W(\mathbf{k}) = W(k, \vartheta)$ (e.g. see Melrose 1980b, p.20). Such diffusion coefficients D can also be obtained from turbulence theory. In solar flares, it is assumed that a MHD-turbulent cascade generates a broad wave spectrum (Miller, LaRosa, and Moore, 1996). For wave turbulence there are two different types of spectra depending on the cascade phenomenology. In the Kolmogorov treatment, the spectral energy transfer time at a wavelength λ is the eddy turnover time $\lambda/\delta\mathrm{v}$, where $\delta\mathrm{v}$ is the velocity fluctuation of the wave. In the Kraichnan treatment, the transfer time is longer by a factor of $\mathrm{v}_A/\delta\mathrm{v}$. The two types of spectra are defined in Zhou and Matthaeus (1990) and applied to solar flares in Miller et al. (1996),

$$D(k) = \begin{cases} C^2 \mathrm{v}_A k^{7/2} \left[\frac{W(k)}{2U_B}\right]^{1/2}, & (Kolmogorov) \\ C^2 \mathrm{v}_A k^4 \left[\frac{W(k)}{2U_B}\right], & (Kraichnan) \end{cases} \quad (53)$$

where C^2 is the Kolmogorov constant, $U_B = B_0^2/8\pi$ is the background magnetic field energy density. The Kolmogorov phenomenology is appropriate for strong turbulence, and it appears also to be better than the Kraichnan phenomenology for weak turbulence (Miller and Roberts, 1995).

After this cursory review of the theory underlying stochastic acceleration, we turn now to the results of numerical calculations, where we largely follow the excellent summary of Miller et al. (1997), for electromagnetic waves that accelerate electrons (§5.2.1), and for ions (§5.2.2).

5.2.1 *Electromagnetic Waves - Electrons*

From the wave mode diagram in Fig. 54 we see that the electron gyrofrequency Ω_e is slightly above the wave frequencies of whistler (R) waves, so electrons can easily be accelerated by gyroresonant interaction with whistler waves. Whistlers, which have frequencies in the range of $\Omega_p \ll \omega < \Omega_e$, yield (with Eq. 46) a resonance requirement of $\gamma \mathrm{v}_\parallel \gg (m_p/m_e)^{1/2}\, \mathrm{v}_A$. Thus, the threshold condition for whistlers is $\gtrsim$ 20 keV, so whistler waves would accelerate mildly relativistic, hard X-ray-emitting electrons to higher energies (Melrose, 1974; Miller and Ramaty, 1987).

However, the electrons have first to be accelerated from their thermal distribution ($T \approx 1 - 10$ MK, i.e. $E \approx 0.04 - 0.4$ keV) to mildly-relativistic energies. Because gyroresonant stochastic acceleration seems not to be efficient for such small energies, the *Landau* or *Cerenkov* resonance ($s = 0$) was considered, i.e. $\omega = k_\parallel \mathrm{v}_\parallel$ in Eq. 46, using the compressive magnetic field component of (magnetosonic) fast mode waves. The resonance condition can be written as $\mathrm{v}_\parallel = \mathrm{v}_A/\eta$, where $\eta = k_\parallel/k$, which shows that magnetosonic waves (which have similar speeds as Alfvénic waves, i.e. $\mathrm{v}_A \approx 1000$ km/s), can resonate with thermal electron

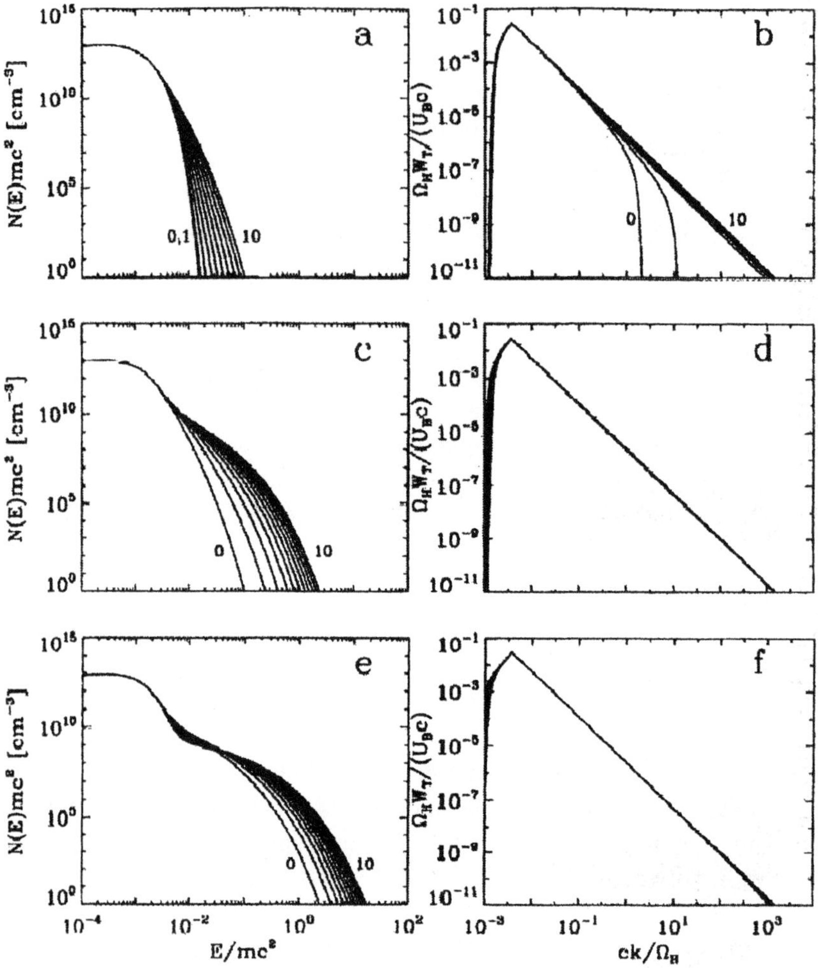

Figure 55. Electron energy spectrum $N(E)$ (left) and wave spectral densities W_T (right) resulting from cascading and transit time damping of fast mode waves. The waves were injected at a wavelength of $\approx 10^7$ cm, at a rate of about 19 ergs cm^{-3} s^{-1}, and over a time of 0.6 s. The ambient electron density was 10^{10} cm^{-3}. (a) and (b) Evolution from $t = 4 \times 10^5 T_H$ to $5 \times 10^5 T_H$. N and W_T are shown at times $t_n = (4 \times 10^5 + 10^4 n) T_H$, for n=0,...,10. (c) and (d) Evolution form $t = 10^6 T_H$ to $3 \times 10^6 T_H$. N and W_T are shown at times $t_n = (10^6 + 2 \times 10^5 n) T_H$, for n=0,...,10. Here $T_H = \Omega_H^{-1} \approx 2.1 \times 10^{-7}$ s and $U_B = B_0^2/8\pi$ is the ambient magnetic field energy density [From Miller et al. 1996].

speeds. This process is the magnetic equivalent to Landau damping and is called *transit-time damping* (Lee and Völk, 1975; Fisk, 1976; Achterberg, 1979; Stix, 1992), because the transit time of a particle across a wavelength is equal to the period of the wave. The only drawback of this mechanism is that electrons are accelerated only in parallel direction, so that a highly beamed distribution would result ($v_\parallel \gg v_\perp \approx v_{th}$), and thus some additional (unknown) pitch-angle scattering is required, which transfers momentum to the perpendicular component ($v_\perp \approx v_\parallel$).

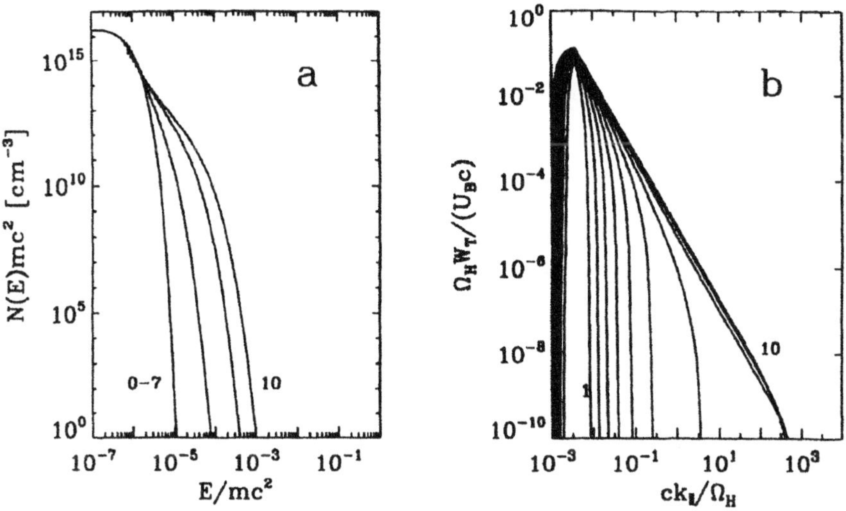

Figure 56. Proton energy spectrum $N(E)$ (left) and wave spectral density W_T (right) resulting from cascading and cyclotron damping of Alfvén waves. The waves were injected at a wavelength of $\approx 10^7$ cm, at a rate of about 100 ergs cm^{-3} s^{-1}, and over a time of 2 s. The ambient proton density was 10^{10} cm^{-3}. (a) N at times $t_n = n(5 \times 10^4 T_H)$, for n=0,...,10. The leftmost curve is the spectrum for $n \leq 7$, and the remaining curves, from left to right, are the spectra for n=8, 9 and 10. (b) Spectral densities at the same times. [From Miller and Roberts 1995]

Miller et al. (1996) have conducted numerical simulations of transit-time damping, starting from an MHD-turbulent cascade wave spectrum, and could demonstrate electron acceleration out of the thermal distribution up to relativistic energies during subsecond time intervals (see example in Fig. 55). Similar results have been obtained by Hamilton and Petrosian (1992) for the evolution of an initially thermal electron distribution, subjected to stochastic acceleration by whistler waves and energy loss via Coulomb collisions.

While electron transit-time damping represents a stochastic acceleration process that works for weak turbulence, the strong-turbulence case with large amplitudes of MHD waves ($\delta B/B \approx 1$) corresponds to the classic *Fermi mechanism* (Fermi, 1949) of collisions between electrons and magnetic scattering centers, which was also applied to solar flares (Ramaty, 1979; LaRosa and Moore, 1993).

5.2.2. *Electromagnetic Waves - Ions*

If the wave frequency of resonant waves is assumed to be much smaller than the ion gyrofrequency, $\omega \ll \Omega_H$, the Doppler resonance condition (Eq. 46) simplifies to $\Omega_H \approx k_\parallel v_\parallel$, which implies $|v_\parallel| \gg v_A$ (with the dispersion relation for Alfvén waves, Table IV). This yields a threshold of $E \approx (1/2) m_p v_A^2 \approx 20$ keV for typical coronal Alfvén speeds, $v_A \approx 2000$ km s^{-1}. Above this suprathermal threshold, Alfvén waves can easily accelerate protons to GeV nucleon^{-1} energies on time scales of $\approx 1 - 10$ s (Barbosa 1979; Miller, Guessoum, and Ramaty 1990; Steinacker and Miller, 1992)

Again, similar to the electrons, there is an injection problem in the sense that a suitable acceleration mechanism needs to be found to accelerate the ions from their thermal energy ($\lesssim 1$ keV) to the Alfvénic acceleration threshold (≈ 20 keV). It was proposed that nonlinear Landau damping of Alfvén waves can lead to rapid proton heating and energization above the Alfvénic acceleration threshold (Lee and Völk 1973; Miller 1991; Miller and Ramaty 1992; Smith and Brecht 1993). Alternatively, higher-frequency waves of the Alfvénic branch through gyroresonant interaction were also shown to be able to accelerate protons directly from thermal energies (Eichler 1979; Zhou and Matthaeus 1990; Miller and Roberts 1995). An example of numerical simulations of proton acceleration employing an MHD-turbulent cascade to generate the initial wave spectrum with a quasi-linear code is shown in Fig. 56. Acceleration of protons to energies > 30 MeV on timescales of ≈ 1 s is found, consistent with gamma-ray observations. However, the detailed shape of the spectrum depends on the balance between the acceleration rate and escape rate, which requires a detailed trap-plus-precipitation model.

An attractive feature of gyroresonant stochastic acceleration is also its ability to explain the enhanced ion abundances, which is not easy to accomplish with DC electric field acceleration. Essentially, electron beams excite Alfvén waves around the cyclotron frequencies Ω_i of Ne, Mg, Si, and Fe, and this way lead to preferential acceleration of those, creating an abundance of those ions at gamma-ray energies over C, N, O, and ^{4}He. Some problems still remain to reproduce the observed $^3He/^4He$ abundance. Table V shows the enhanced ion abundances observed in impulsive and gradual flares (e.g., Reames, Meyer, and VonRosevinge, 1994), that are largely reproduced with a quasi-linear code, by simultaneously solving all ion diffusion equations and the wave equations (Miller and Reames 1996).

TABLE V

Enhanced Elemental Abundances During Flares[1]

Element ratio	Impulsive flares	Gradual flares (corona)
^{3}He/^{4}He	≈ 1 ($\times 2000$ increase)	≈ 0.0005
^{4}He/O	≈ 46	≈ 55
C/O	≈ 0.436	≈ 0.471
N/O	≈ 0.153	≈ 0.128
Ne/O	≈ 0.414 ($\times 2.8$ increase)	≈ 0.151
Mg/O	≈ 0.413 ($\times 2.0$ increase)	≈ 0.203
Si/O	≈ 0.405 ($\times 2.6$ increase)	≈ 0.155
Fe/O	≈ 1.234 ($\times 8.0$ increase)	≈ 0.155
H/He	≈ 10	≈ 100

[1]) from Miller and Reames (1996).

Let us conclude this Section with a brief summary of the *pros* and *cons* of the stochastic acceleration model. The *pros* are: (1) No direct current of accelerated particles is injected, and thus no such strong filamentation of the acceleration region as in the DC electric field models is required (e.g. $\approx 10^4$ current channels, see Sect.5.1.1); (2) The acceleration volume is much larger than the vicinity of X- and O-point regions in a current sheet, which eases the electron number problem; Some *cons* are: (1) The input turbulent wave energy, as well as the efficiency of the turbulent cascade, are ad hoc assumption tuned to reproduce the observations (e.g. Miller et al. 1996); (2) Support for the assumed amount of turbulent wave energy input has yet to come from reconnection physics.

5.3. SHOCK ACCELERATION

Shock waves are believed to play an important role in particle acceleration during solar flares. The most conspicuous flare phenomena associated with shock waves are reconnection outflows, radio type II bursts, filament eruptions, and coronal mass ejections. In this Section we focus on particle acceleration models that are particularly tailored to the solar flare situation, while there exists a large body of alternative literature on shock acceleration in other astrophysical plasmas, such as interplanetary shocks, bow shocks between planetary magnetospheres and the solar wind, supernova remnants, or other sources of cosmic rays.

Basic introductions into shock waves can be found in Priest (1982, p.189), Benz (1993, p.234), Kirk, Melrose, and Priest (1994, p.226), Kivelson and Russell (1995, p.129). Shocks are waves with nonlinear amplitudes that propagate faster than the sound speed of the ambient medium. Shocks can be classified either by (1) the change in magnetic field direction from upstream to downstream (*slow-mode, intermediate-mode, fast-mode*), (2) the particle velocity distribution (*collisional* or *collisionless shocks*), (3) the ion acceleration (*subcritical* or *supercritical shocks*), or (4) the driving agent (*blast wave* or *piston-driven wave*). From the point of view of particle acceleration, one distinguishes between *shock-drift acceleration* and *diffusive-shock acceleration*, where the first type is closer to the case of *DC-electric field acceleration*, while the second one is closer to the case of *stochastic acceleration*.

5.3.1. *Shock-Drift Acceleration*

Most shocks in the solar corona have a sufficiently low density so that they are essentially collisionless during the passage of a particle, and thus adiabatic particle orbit theory can be applied. The normal component of the magnetic field is continuous across the shock front ($B_{n1} = B_{n2}$), while the tangential component varies, most strongly for fast shocks ($B_{t1} \gg B_{t2}$). Therefore the total magnetic

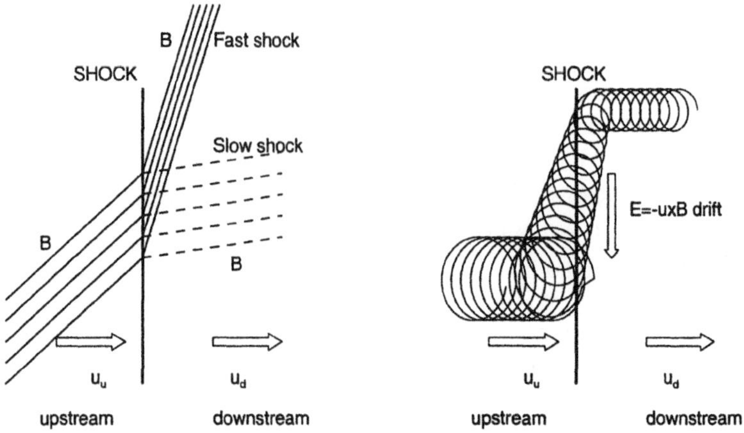

Figure 57. Definition of fast and slow shocks (left) and adiabatic orbit of a particle during shock passage (right). The magnetic field lines are closer together for a fast shock, indicating that the field strength increases when the field is bent away from the shock normal. During the passage of a shock front, a charged particle experiences a drift along the shock front due to the inhomogeneity of the magnetic field. This drift is parallel to the electric field $\mathcal{E}_{drift} = -u \times B$. The particle orbit is shown in the rest frame of the moving shock front.

field strength increases across the shock front, $B_1 \gg B_2$, and the particle gains perpendicular velocity due to conservation of the magnetic moment (in first order),

$$\mu = \frac{\frac{1}{2}m\mathrm{v}_{\perp,1}^2}{B_1} = \frac{\frac{1}{2}m\mathrm{v}_{\perp,2}^2}{B_2} . \tag{54}$$

The particle experiences a $\mathbf{E} \times \mathbf{B}$ drift, which is the same thing in the test particle picture as the fluid velocity u that produces the same electric field,

$$\mathcal{E}_{drift} = \frac{1}{c}\mathbf{u} \times \mathbf{B} , \tag{55}$$

which can be transformed away in a coordinate system co-moving with the drift speed perpendicular to the shock vertical, so that the electric field $\mathcal{E}_{\mathrm{drift}}$ due to the fluid motion **u** vanishes (de Hoffman-Teller frame). If electrons pass the shock front only in a single encounter, the energy gain is limited to the downstream/upstream ratio B_1/B_2 of the magnetic field strengths, which is typically a factor of 4. Higher energies can be achieved if the magnetic field has a trapping region upstream of the shock, so that particles are mirrored multiple times at the shock front and gain energy each time. When the shock propagation becomes near-perpendicular ($1^0 - 2^0$) to the upstream magnetic field (fast shock), acceleration becomes most efficient (Wu 1984; Krauss-Varban et al. 1989), but on the other side, this small angle restriction limits the high acceleration efficiently only to $\approx 1\%$ of the electrons. Given these restrictions, shock-drift acceleration was mainly applied to the Earth' bow shock (Jokipii 1966; Krauss-Varban and Burgess, 1991) and to radio type II bursts (Holman and Pesses 1983).

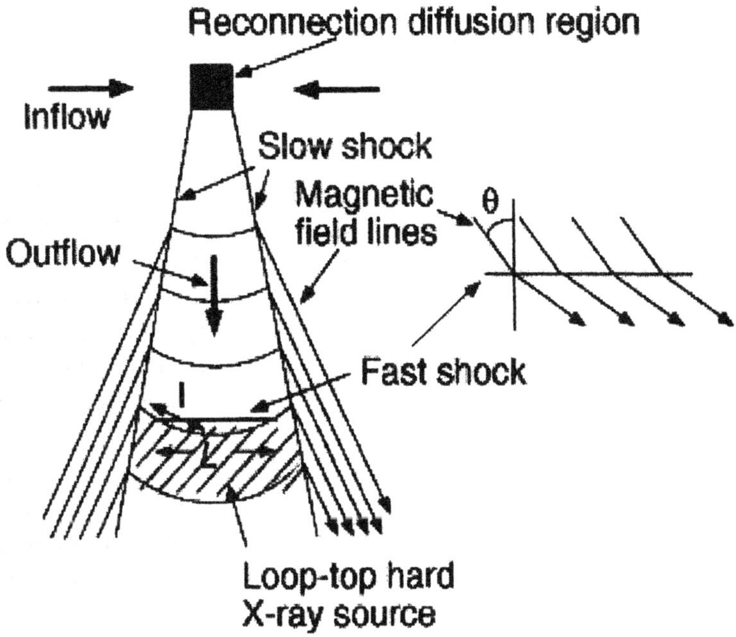

Figure 58. Magnetic field line configuration of the reconnection region. An Alfvénic downward outflow is sandwiched by the two steady slow shocks. A fast shock with length L forms between the slow shocks. Magnetic disturbances both upstream and downstream of the fast shock scatter the electrons being accelerated. The total length of the diffusion region along the field lines is l. [Tsuneta and Naito 1998].

5.3.2. *First-Order Fermi Acceleration at Fast Shock*

Recently, a special application of shock-drift acceleration, namely first-order Fermi acceleration at the fast shock, was applied to solar flares, after the geometry of fast shocks in the outflows from magnetic reconnection regions became clearer with the discovery of loop-top hard X-ray sources by Masuda et al. (1994a). Tsuneta and Naito (1998) proposed that nonthermal electrons can be efficiently accelerated by the first-order Fermi process at the fast shock that is expected to form where the downward-directed reconnection outflows from the cusp regions hit the soft X-ray bright flare loops underneath, filled by upflowing plasma into relaxed magnetic field lines that reconnected earlier on. The situation is sketched in Fig. 58, showing a fast shock at the downward front of the outflows, sandwiched by two slow shocks on either side of the outflows.

The model of first-order Fermi acceleration is viable for 10-100 keV electrons if the following four observational requirements are met (Tsuneta and Naito, 1998): (1) The net acceleration rate has to overcome the (relatively high) collisional loss rate,

$$\frac{dE}{dt} = \left(\frac{dE}{dt}\right)_{acc} - \left(\frac{dE}{dt}\right)_{coll} > 0 \,, \tag{56}$$

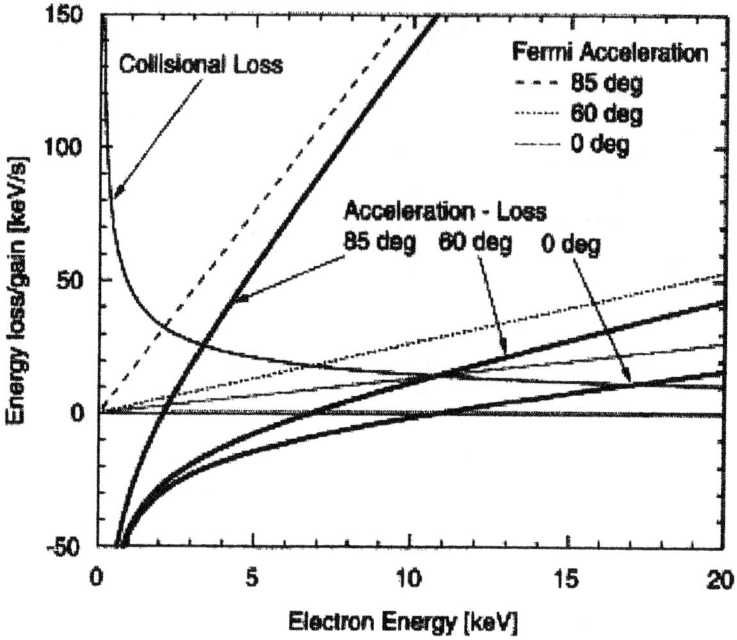

Figure 59. Collisional energy loss rate and Fermi acceleration rate for three different shock angles. The net energy gain rate (thick lines) is the energy gain rate (dotted and dashed lines) minus the loss rate (thin line). [Tsuneta and Naito 1998].

(2) the energy gain has to be sufficiently high to explain the $\approx$ 50 keV hard X-ray emission of Masuda's above-the-loop-top sources, (3) the acceleration time has to be sufficiently fast ($\lesssim$ 1 s) to explain the impulsive hard X-ray bursts, and (4) the number of accelerated electrons has to meet the hard X-ray inferred electron injection rates of $\approx 10^{34} - 10^{35}$ electrons s^{-1}. The acceleration rate in the fast shock was estimated to

$$\left(\frac{dE}{dt}\right)_{acc} = \frac{\Delta E}{\Delta t} = \frac{2}{3} E \frac{u}{l \cos \vartheta} , \tag{57}$$

where $u \approx v_A \approx 1000$ km s^{-1} (Tsuneta 1996) is the the speed of the outflow from the reconnection region, $l \approx 500$ km is the estimated diffusion length, and ϑ the angle between the fast-shock normal and the magnetic field line crossing the shock (Fig. 58). The net acceleration rate (Eq. 56), after subtraction of the collisional losses,

$$\left(\frac{dE}{dt}\right)_{coll} \approx 47 \frac{n_{10}}{\sqrt{E_{keV}}} , \tag{58}$$

is shown in Fig. 59, for shock angles of $\vartheta = 0^0, 60^0$, and 85^0. The diagram shows that the net acceleration exceeds the collisional loss at energies of $E \gtrsim 4$ keV for shock angles of $\vartheta = 85^0$. So, if a bulk energization mechanism exists that pre-accelerates electrons out of the thermal distribution ($T \approx 10-20$ MK, $E \approx 0.5-1$

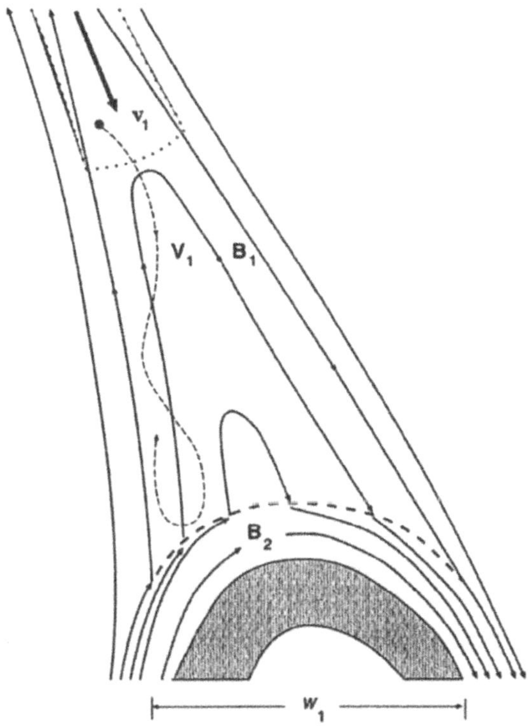

Figure 60. Magnetic trap between the *High-Temperature Turbulent-Current Sheet (HTTCS)* and the shock front in the downward-directed outflow of a reconnection region. Accelerated particles move with velocity v_1 along the field lines, are reflected at the fast shock above the soft X-ray flare loop (grey), and experience first-order Fermi acceleration. [Somov and Kosugi, 1997].

keV) to $E_0 \gtrsim 4$ keV, first-order Fermi acceleration will accelerate them up to ≈ 1 MeV energies at the fast shock. Tsuneta and Naito (1998) estimate that the pre-acceleration could be provided by the slow shocks. A diffusion length of $l \approx 500$ km is needed at a fast shock angle of $\vartheta \approx 85^0$ to satisfy the maximum energies ($E \lesssim 1$ MeV), for acceleration time scales of $t_{acc} \approx 0.3 - 0.6$ s and a number of accelerated electrons of $N_{acc} = n_e u L^2 \approx 5 \times 10^{35}$ electrons s^{-1}. They point out that this scenario considerably ameliorates the injection problem of earlier first- and second-order Fermi acceleration scenarios, where relatively high initial energies ($E_0 \approx 20 - 100$ keV) were required (Bai et al. 1983; Ramaty 1979).

5.3.3. *First-Order Fermi Acceleration in Mirror-Trap of Reconnection Outflows*

Somov and Kosugi (1997) developed a more detailed scenario of particle acceleration in the reconnection outflow region derived by Tsuneta (1996), by including the aspects of particle trapping in the cusp region. They assume that initial heating and bulk acceleration is accomplished by a *high-temperature turbulent-current sheet (HTTCS)*, which injects pre-accelerated particles into the cusp region between the HTTCS reconnection point and the *fast oblique collisionless shock (FOCS)*

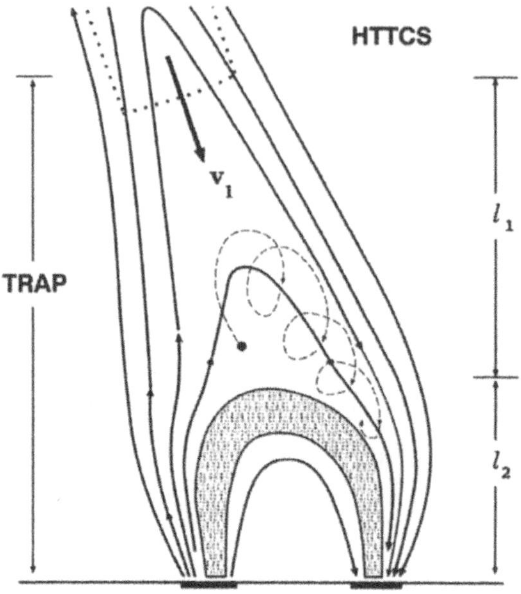

Figure 61. Magnetic trap without a shock. [Somov and Kosugi, 1997].

situated above the soft-X-ray emitting flare loop (Fig. 60). Because the magnetic field is increasing in downward direction, it produces a magnetic mirror for most of the electrons that have not too small pitch angles. In addition, each newly-reconnected magnetic field line that leaves the reconnection point, relaxes from an initial cusp-shaped line to a dipole-like line at a lower altitude (Fig. 61), carrying with it the guiding center of trapped electrons. This relaxation is expected to happen on Alfvénic time scale, but recent observations indicate a slower velocity field of $v \approx 100 - 200$ km s^{-1} in the reconnection outflow region (McKenzie and Hudson 1999). There are two major effects that contribute to particle acceleration: (1) First-order Fermi acceleration at the fast collisionless shocks, and (2) collapse of the mirror trap driven by relaxing field lines which increases the magnetic moment of the trapped particles. So, the particles gain energy (1) from the increase in parallel momentum (due to spatial length decrease of the collapsing magnetic field lines), (2) from the increase in perpendicular momentum (due to the increasing magnetic field for downward collapsing magnetic field lines), and (3) from first-order Fermi acceleration at each mirroring at the fast shock front. A logical next step would be to numerically simulate the total energy gain and loss of particles trapped in such a collapsing cusp region, using a Fokker-Planck code.

5.3.4. *Diffusive Shock Acceleration*

The basic problem of first-order Fermi acceleration, the limited energy gain during a single shock encounter, can be overcome in inhomogeneous plasmas, where particles are scattered many times forth and back across the shock front, so that

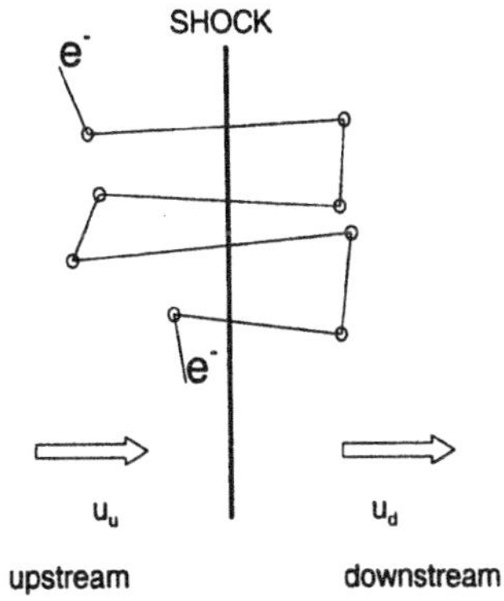

Figure 62. Particle orbit during diffusive shock acceleration. The electron crosses the shock front multiple times, scattered by wave turbulence, entering it from the upstream direction.

they experience a cumulative acceleration effect during multiple shock encounters (Fig. 62). The original concept of Fermi is based on particles that encounter a cloud of scattering centers moving in random directions, where colliding particles gain more energy than they lose in the average, a second-order effect. In solar flares, wave turbulence is mainly employed to provide particle scattering. However, diffusive shock acceleration requires efficient particle scattering in the regime of *strong wave turbulence*, while *stochastic acceleration* based on gyro-resonant wave-particle interactions (described in the previous Section) works in the regime of *weak wave turbulence*.

Diffusive-shock acceleration is generally described by a *diffusion convection equation* (e.g. Parker, 1965; Priest and Forbes, 2000, p.504),

$$\frac{\partial f}{\partial t} + \mathbf{u} \cdot \nabla f = \nabla \cdot (\kappa \nabla f) + \frac{1}{3} \nabla \cdot \mathbf{u} p \frac{\partial f}{\partial p} + \frac{1}{p_i^2} \frac{\partial}{\partial p_i} \left(p_j^2 D_{ij} \frac{\partial f}{\partial p_j} \right) + I - L \,, \quad (59)$$

where the terms describe the time dependence of the particle distribution f, spatial advection, spatial diffusion (κ), adiabatic expansion or compression (at the shock front), momentum diffusion (D_{ij}), particle injection (I), and escape or loss (L).

Some first energy spectra of ions accelerated in diffusive shocks were calculated by Ramaty (1979), which could reproduce observed gamma ray spectra and interplanetary particle spectra, and was found to be a viable mechanism to accelerate ions up to $\approx$ 100 MeV within $\lesssim$ 1 s. Wave-particle interactions in diffusive shocks, e.g. scattering by whistlers and Alfvén waves, was included by Achterberg and Norman (1980) and Decker and Vlahos (1986). However, an important problem that was identified was the pre-acceleration to $E \approx 20$ keV for electrons, requiring another mechanism for *first-phase acceleration*, while protons and ions can directly be accelerated out of the Maxwellian distribution. Diffusive-shock acceleration has been simulated numerically, and it confirmed that ions can promptly be accelerated to gamma-ray energies (Cargill, Goodrich, and Vlahos, 1988), and that shock waves can form electron streams capable of radio type II emission (Lemberge 1995).

There is a large body of literature on shock acceleration that we dont cover in this review. Diffusive shock acceleration is considered to operate in a wide number of transient phenomena in and around flares, including reconnection outflows, flare-initiated *EIT* and Moreton waves, coronal mass ejections, filament eruptions, radio type II bursts, interplanetary shocks, etc. In this review, however, we focus on reconstructing the physics of acceleration processes in solar flares from high-precision measurements of the particle kinematics (involving propagation times of less than a few 100 ms, which restricts propagation distances to the lower corona), and thus pay less attention to propagation of particles and shock waves in the outer corona, heliosphere, or interplanetary space, where the timing between acceleration and radiative signatures becomes gradually more blurred with increasing distance.

6. Particle Kinematics

In this Section we describe the kinematics of relativistic particles in solar flares. This is a fairly new research area in solar physics, which only became possible by means of the high-sensitivity and high-precision energy-dependent timing measurements with the *Compton Gamma Ray Observatory (CGRO)* during 1991-1999. Previously, flare researchers attempted to obtain information on particle acceleration and propagation by studying spectral changes of hard X-ray emission as function of time, say spectral variations $N(\epsilon, t)$ (e.g. Kiplinger et al. 1984). The new approach that was taken with *CGRO* observations was to measure energy-dependent time delays $\Delta t(\epsilon)$ between time profiles $N(t, \epsilon_i)$ of different energies, rather than measure spectral changes $\Delta N(\epsilon, t_i)$ at different times t_i. In principle, both approaches are mathematically equivalent, but energy-dependent time scales $\tau(\epsilon)$ can be extracted from time delays $\Delta t(\epsilon)$ with higher accuracy than from spectral changes in $\Delta N(\epsilon, t_i)$, because radiative processes occur instantaneously, only delayed by the speed-of-light propagation time, and thus the time history $t(\epsilon)$ of the original signal is preserved with high accuracy, while energy spectra represent a broadband convolution of time scales $\tau(\epsilon)$ with the radiation process and instrumental response function, so that the inversion of the energy-dependent timing $t(\epsilon)$ is far less accurate. The basic physics of particle kinematics in flares can then be broken down into a sequence of time intervals t_i, each related to a distinct dynamic process, and the combined timing, $t(\epsilon) = \sum_{i=1}^{n} t_i(\epsilon)$, can be used for modeling and fitting of the observed energy-dependent time delays.

The type of particle kinematics we introduce here for solar flare applications has long been practiced in laboratory experiments of elementary particle physics. In a typical particle collision experiment, a beam of high-energy particles is deflected from an accelerator exit at time $t = t^{acc}$, and propagates then on a well-defined path with length L to a collisional target during the time-of-flight interval t^{TOF},

TABLE VI

Overview of Particle Acceleration Mechansims in Solar Flares

Acceleration Mechanisms	Electromagnetic fields
DC Electric Field Acceleration:	
– Sub-Dreicer fields, runaway acceleration[1]	$\mathcal{E} < \mathcal{E}_D$
– Super-Dreicer fields[2]	$\mathcal{E} > \mathcal{E}_D$
– Current sheet (X-point) collapse[3]	$\mathcal{E} = -u_{inflow} \times B$
– Magnetic island (O-point) coalescence[4]	$\mathcal{E}_{\text{conv}} = -u_{coal} \times B$
– (filamentary current sheet: X- and O-points)	
– Double layers[5]	$\mathcal{E} = -\nabla V$
– Betatron acceleration (magnetic pumping)[6]	$\nabla \times \mathcal{E} = -(1/c)(dB/dt)$
Stochastic Acceleration:	
Gyroresonant wave-particle interactions (weak turbulence) with:	
– whistler (R-) and L-waves[7]	$k \parallel B$
– O- and X-waves[8]	$k \perp B$
– Alfvénic waves (transit time damping)[9]	$k \parallel B$
– magnetoacoustic waves[10]	$k \perp B$
– Langmuir waves[11]	$k \parallel B$
– Lower hybrid waves[12]	$k \perp B$
Shock Acceleration:	
First-order Fermi acceleration, Shock-drift acceleration[13]	
– Fast shocks in reconnection outflow[14]	
– Mirror-trap in reconnection outflow[15]	
– Diffusive-shock acceleration[17]	
Second-order Fermi acceleration[16]	
– Strong turbulence scattering[18]	

References: [1]*) Holman (1985), Tsuneta (1985), Benka and Holman (1994);* [2]*) Litvinenko and Somov (1995), Litvinenko (1996);* [3]*) Tajima and Sakai (1986), Sakai and Ohsawa (1987), Sakai and De Jager (1991);* [4]*) Furth et al. (1963), Pritchett and Wu 1979; Biskamp and Welter 1979, Kliem (1994), Kliem et al. (2000);* [5]*) Block (1978), Volwerk and Kuijpers (1994), Volwerk (1993);* [6]*) Brown and Hoyng (1975), Karpen (1982);* [7]*) Melrose (1974), Miller and Ramaty (1987), Steinacker and Miller (1992), Hamilton and Petrosian (1992);* [8]*) Karimabadi et al. (1987);* [9]*) Lee and Völk (1975), Fisk (1976), Achterberg (1979), Barbosa (1979), Stix (1992), Miller et al. (1990), Hamilton and Petrosian (1992), Steinacker and Miller (1992), Miller et al. (1997);* [10]*) Zhou and Matthaeus (1990); Eichler (1979); Miller and Roberts (1995);* [11]*) Melrose (1980a,b);* [12]*) Papadopoulos (1979), Lampe and Papadopoulos (1977), Benz and Smith (1987), McClements et al. (1990);* [13]*) Fermi (1949), Jokipii (1966), Bai et al. (1983), Ellison and Ramaty (1985);* [14]*) Tsuneta and Naito (1998);* [15]*) Somov and Kosugi (1997);* [16]*) Parker (1965), Jones (1994);* [17]*) Ramaty (1979), Achterberg and Norman (1980), Decker and Vlahos (1986); Cargill, Goodrich, and Vlahos, (1988), Lemberge (1995);* [18]*) LaRosa, Moore, and Shore (1994).*

where secondary particles and photons are produced and recorded by detectors. The identity of a secondary particle can then be determined from the measured velocity $v = L/t^{TOF}$, which leads to the kinetic energy and rest mass of the secondary particle. The only major difference of such a laboratory experiment to a solar flare observation is that we do the "experiment" the other way round: we do not know the propagation distance L, but infer it from the energy-dependent timing $t^{TOF}(E)$ and the kinetic energy relation $v(E)$ of a-priori known particles, namely electrons. In essence, this allows us to localize the unknown particle accelerator in solar flares. Although L is only a one-dimensional distance, the knowledge of the location of the collisional target in the chromosphere (from hard X-ray maps) combined with the likely propagation path along connecting magnetic field lines, allows us in principle to determine the 3-dimensional location of the particle accelerator in the solar corona.

We start with the kinematics of particles that are detected in hard X-rays, which are mainly produced by electrons, although it can also be applied to protons and ions, but their radiative signatures in gamma-rays cannot be measured equally accurate, due to the poorer photon statistics. We distinguish at least 5 different physical processes (Aschwanden 1998a), which occur basically in sequential order, some near-simultaneously. These five processes include:

(1) particle acceleration,

(2) particle injection onto a magnetic field line that escapes from the acceleration region,

(3a) free-streaming particle propagation,

(3b) particle mirroring in a magnetic trap (plus precipitation or not),

(4) energy loss by radiation.

Step (3a) and (3b) occur in parallel, there is generally a bifurcation in the particle path, some particles propagate free-streaming directly to the energy loss site, while other particles become stored in a magnetic trap and mirror many times before they lose their energy or escape by precipitation and propagate to their final energy loss site. These conceptual five steps are illustrated in Fig. 63. Each process has its own energy-dependent timing,

$$t(E) = t^{acc}(E) + t^{inj}(E) + t^{prop}(E) + t^{trap}(E) + t^{loss}(E) \tag{60}$$

which has to be included in a model of the observed timing of a hard X-ray or radio spectrum. In the following subsections we describe first the basic relations of relativistic kinematics, and then the energy-dependent timing of these five processes (Eq. 60), applied to the flare models and geometries we discussed in the foregoing Sections.

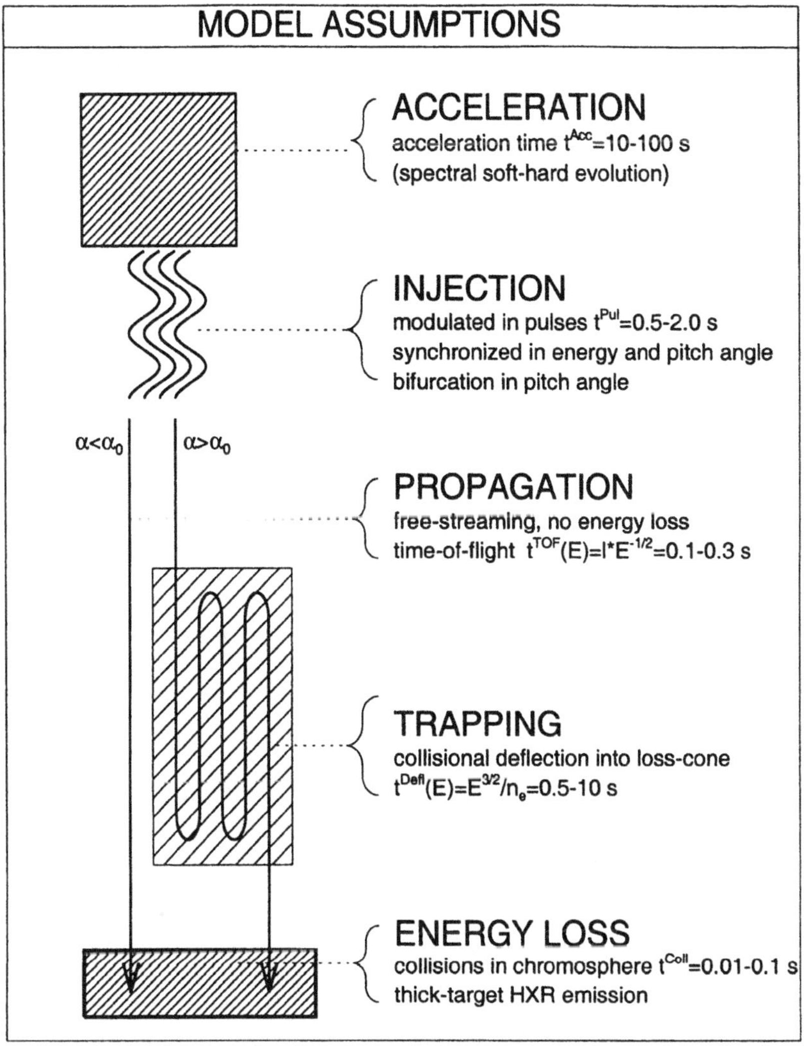

Figure 63. Conceptual breakdown of the flare kinematics into five different physical processes, described in this Section.

6.1. Kinematics of Relativistic Particles

The rest mass of an electron is

$$m_e c^2 = 511 \text{ keV} . \tag{61}$$

The total energy E^{total} of a relativistic electron is composed of the rest mass $m_e c^2$ and the kinetic energy E,

$$E^{total} = m_e c^2 + E = m_e c^2 \gamma \ , \tag{62}$$

also specified in terms of the relativistic Lorentz factor γ,

$$\gamma = \frac{1}{\sqrt{1-\beta^2}} = \frac{1}{\sqrt{1-\frac{v}{c}^2}} \tag{63}$$

where the velocity v is often expressed by the dimensionless variable β =v/c (not to be confused with the plasma β-parameter). So, the kinetic energy E is related to the Lorentz factor γ or velocity v by

$$E = m_e c^2(\gamma - 1) = m_e c^2 \left(\frac{1}{\sqrt{1-(\frac{v}{c})^2}} - 1\right) = \frac{1}{2} m_e v^2 + ... \tag{64}$$

Often we want to know the electron speed v for a given kinetic electron energy E, which follows from Eqs. 61–64,

$$v(E) = c\sqrt{1 - \frac{1}{\gamma^2}} = c\sqrt{1 - \frac{1}{(1+\frac{E}{m_e c^2})^2}} \ . \tag{65}$$

The way we detect electrons in the solar corona is by hard X-ray bremsstrahlung or by radio emission, both produced by nonthermal electrons. An electron with kinetic energy E loses its energy by bremsstrahlung in a collisional plasma, from which we can detect photons with energies $\epsilon \le E$ that are lower than the kinetic energy E of the incident electron. The observed hard X-ray spectrum $N(\epsilon)$ represents then a convolution of the incident electron energy E with the bremsstrahlung cross-section $\chi(E, \epsilon)$ and the instrumental response function $R(\epsilon)$,

$$N(\epsilon)d\epsilon = R(\epsilon)d\epsilon \int_{\epsilon}^{\infty} \chi(E, \epsilon) n(E) dE \tag{66}$$

where $n(E)$ represents the distribution of the electron injection spectrum. If we assume a mono-energetic electron injection spectrum with a δ-function in time, $n(E = E_0, t = t_0)$ that peaks at time t_0, there will be a velocity dispersion during propagation, so that the electron spectrum at the hard X-ray emission site peaks at different times for different energies, according to the time-of-flight propagation time interval t^{TOF}=L/v(E),

$$n(E, t, x = L) = n(E, t - \frac{L}{v(E)}, x = 0) \tag{67}$$

and the resulting hard X-ray spectrum as function of time is

$$N(\epsilon, t)d\epsilon = R(\epsilon)d\epsilon \int_{\epsilon}^{\infty} \chi(E, \epsilon) n(E, t - \frac{L}{v(E)}) dE \ . \tag{68}$$

Hard X-ray time profiles $N(\epsilon, t)$ at a particular detected energy ϵ will show a sharp pulse too, if the electron injection pulse was a short pulse, but the pulse shape

TABLE VII

Mean photon and electron energies detected with the *CGRO/BATSE/DISCLA* energy channels.

Nominal Hard X-ray range	Mean photon energy ϵ	Mean electron energy E	Lorentz factor γ	Relativistic speed β=v/c
25-50 keV	39 keV	44 keV	1.086	0.389
50-100 keV	70 keV	79 keV	1.154	0.499
100-300 keV	153 keV	172 keV	1.336	0.663
300-1000 keV	483 keV	543 keV	2.063	0.875

will be convolved with the velocity dispersion, the bremsstrahlung cross-section, and the instrumental response function. Moreover, the peak time will be energy-dependent, peaking earlier at high energies and later at low energies. A simple approach is to define a hard X-ray energy ϵ_E that peaks at the same time as the arrival time of (mono-energetic) electrons with kinetic energy E at the hard X-ray emission site. This energy ϵ_E of hard X-ray photons is lower than the electron energy E, because the bremsstrahlung cross-section is zero for $\epsilon > E$. Thus we define an *electron-to-photon energy conversion factor* $q_E(E, \gamma, E_0)$, which generally depends on the spectral shape of the injection spectrum, which can be characterized by a single powerlaw in the simplest case (with powerlaw slope γ and high-energy cutoff E_0),

$$E(t = t_{peak}) = \epsilon(t = t_{peak}) \times q_E(E, \gamma, E_0) . \tag{69}$$

This conversion factor q_E is implicitly defined by Eq. 68 and its calculation (Aschwanden and Schwartz 1996a) is described in Section 8 along with the bremsstrahlung cross-section. The non-relativistic Born approximation is used for the Bethe-Heitler cross-section here (following Brown 1971), while relativistic treatments can be found elsewhere (Elwert 1939; Koch and Motz 1959; Holt and Cline 1968). For most hard X-ray spectra, the photon-to-electron conversion factor has values in the range of $q_E \approx 1 - 2$. In Table VII we provide some typical values for the *Burst and Transient Source Experiment (BATSE) DISCLA* channels on *CGRO*, which detect hard X-ray photons in the four ranges ϵ=25-50 keV, 50-100 keV, 100-300 keV, and 300-1000 keV, computed for a typical spectral slope of $\gamma = 4$ and $E_0 \gg \epsilon$, which yields $q_E \approx 1.124$.

6.2. Kinematics of Particle Acceleration

The relativistic equation of motion for charged particles (with charge q) in a static, uniform magnetic field $\mathbf{B}$ and electric field $\mathbf{E}$ is

$$\frac{d}{dt}(\gamma m\mathbf{v}) = q\mathbf{E} + \frac{q}{c}(\mathbf{v}\times\mathbf{B}) \ . \tag{70}$$

which can be separated for a unidirectional magnetic field into

$$\frac{d}{dt}(\gamma m\mathbf{v}_\parallel) = q\mathcal{E}_\parallel \ , \tag{71}$$

$$\frac{d}{dt}(\gamma m\mathbf{v}_\perp) = q\mathcal{E}_\perp + \frac{q}{c}(\mathbf{v}_\perp\times\mathbf{B}) \ . \tag{72}$$

For electron time-of-flight measurements, which involve the propagation time along a segment of a magnetic field line, the only acceleration that matters is parallel to the field line (Eq. 71), while the perpendicular component contains perpendicular drifts (caused by $\mathcal{E}_\perp$) and the Larmor motion of the electron (Eq. 72).

Let us consider first the kinematic parameters for an electron that is accelerated by a uniform electric field $\mathcal{E}_\parallel$ over a distance l. The maximum energy E_{max} that an electron acquires over the distance l is then

$$E_{max} = m_e c^2(\gamma_{max} - 1) = \int_0^l e\mathcal{E}_\parallel(s)ds = e\mathcal{E}_\parallel l \ . \tag{73}$$

It is most convenient to express all kinematic parameters as function of the Lorentz factor γ, which has a simple relation to the electron energy E, i.e. $\gamma = 1 + E/m_e c^2$ (Eq. 64). Because the gained kinetic energy E is proportional to the distance x in a constant electric field, we can express the acceleration path length x_{acc} as function of the energy E by the parameters E_{max} and l,

$$x_{acc}(\gamma) = l\frac{E}{E_{max}} = l\frac{(\gamma-1)}{(\gamma_{max}-1)} \ . \tag{74}$$

The velocity $v_{acc}(\gamma)$ at distance $x_{acc}(\gamma)$ is simply obtained from the definition of the Lorentz factor (Eq. 65),

$$v_{acc}(\gamma) = c\sqrt{1-\frac{1}{\gamma^2}} \ . \tag{75}$$

In parallel electric fields, the change of energy is only transferred into parallel direction, so that the final parallel velocity of the accelerated electron is much larger than the perpendicular velocity ($v_\parallel \gg v_\perp$). We can therefore use the approximation $v_\parallel \approx v$ in the parallel equation of motion (Eq. 71). The acceleration time $t_{acc}(\gamma)$ can then simply be evaluated by integrating Eq. 71 ($\gamma m_e v \approx e\mathcal{E}t$) and by substituting Eqs. 73 and 75),

$$t_{acc}(\gamma) = \frac{l}{c}\frac{\sqrt{\gamma^2-1}}{(\gamma_{max}-1)} \ . \tag{76}$$

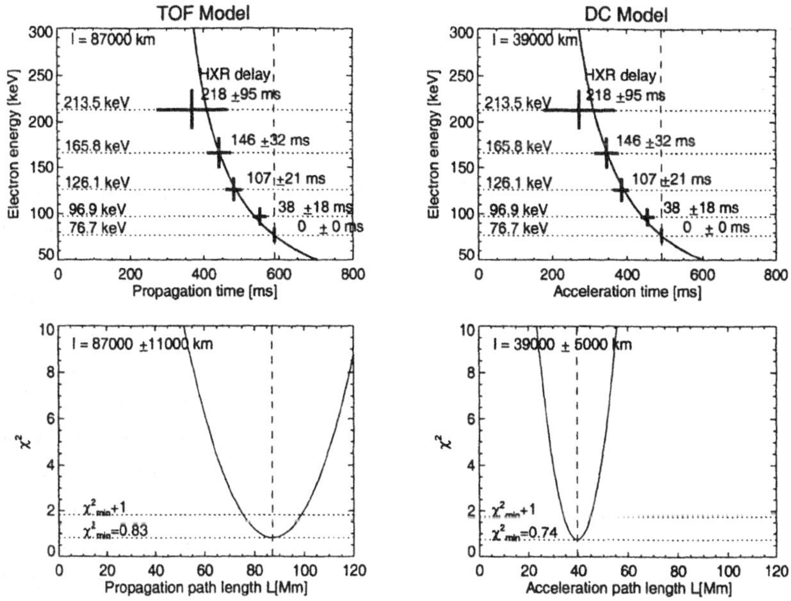

Figure 64. Observed hard X-ray pulse delays (top: thick crosses), observed between the five *CGRO/BATSE/MER* channels (with electron energies indicated), are fitted with a TOF model (propagation path length $l = 87,000$ km; left side), and a DC field model (propagation path length $l = 39,000$ km; right side). The χ^2-square values of the fits are shown for both models (bottom panels), with the 68% confidence range, $\chi^2 + 1$ [from Aschwanden et al. 1996b].

Because the perpendicular velocity $v_\perp$ remains constant in first order (within a factor of γ according to Eq. 71), the change of the pitch angle can be expressed by

$$\alpha_{acc}(\gamma) = \arcsin\left(\frac{v_\perp}{v_{acc}(\gamma)}\right) . \tag{77}$$

The pitch angle therefore decreases monotonically during acceleration in a parallel electric field.

Let us now apply this pure acceleration kinematics to solar flares. Considering the general timing equation (Eq. 60) we need a scenario without free-streaming propagation and trapping. The only applicable scenario are large-scale parallel electric fields which extend from the coronal reconnection site all the way down the chromospheric hard X-ray emission site. Such a model was considered in DC electric field scenarios (Holman, 1985; Tsuneta, 1985; Benka and Holman 1994; Holman 1996). The simplest model version would be to assume a single uniform electric field $\mathcal{E}$ and acceleration of electrons all over the same distance from the reconnection point to the footpoint. The total acceleration time would then be (using Eq. 76)

$$t_{acc}(\gamma_{max}) = \frac{l}{c}\sqrt{\frac{\gamma_{max}-1}{\gamma_{max}+1}} , \tag{78}$$

but this produces a mono-energetic electron injection spectrum at the chromospheric hard X-ray emission site, which produces a very flat bremsstrahlung spectrum with no energy-dependent delays, and is not consistent with the observed hard X-ray delays (e.g. Fig. 64; Aschwanden et al. 1996d).

A second variant would be a scenario with highly filamented current channels, as Holman (1985) concluded from Ampére's limit, where each current channel has a different electric field strength $\mathcal{E}_{\parallel}$, yielding a spectrum of particle energies

$$N(E_{max}) = elN(\mathcal{E}_{\parallel}) \tag{79}$$

This yields energy-dependent time delays according to Eq. 78, because the Lorentz factors $\gamma_{max} = 1 + (E/m_e c^2)$ have now a distribution according to the variable kinetic energies $E = el\mathcal{E}_{\parallel}$ acquired in different electric fields $\mathcal{E}_{\parallel}$. In Fig. 64 (right panels) we show a fit of this acceleration time-delay model to observed hard X-ray delays during the Masuda flare (Aschwanden et al. 1996b). Apparently the model fits the observed energy-dependent delays closely, for a propagation distance of $l = 39,000 \pm 5000$ km. The highest energies of $E_{max} = 200$ keV require a maximum electric field strength of $\mathcal{E}_{\parallel} \lesssim E_{max}/l = 5 \times 10^{-5}$ V cm^{-1}, which is in the sub-Dreicer regime as proposed by Holman (1985). However, we have to consider this model as unrealistic, because we considered only those electrons starting at the upper end of the electric field. In reality, if such a large-scale field is turned on, all electrons along the field would be accelerated, so we have moreover a uniform distribution $N(l)$ of acceleration distances. In this scenario, those electrons starting near the footpoints arrive first, which have been accelerated only to low energies, while the high-energy electrons, which need to be accelerated over long loop segments, will arrive later, just the opposite to the first case where all electrons start at the top. So, this second more realistic scenario would not be consistent with the observed delays of hard X-ray pulses, which always peak at the high energies first. Moreover, there are a number of fundamental problems with large-scale electric field models that have never been addressed: (1) How can large-scale electric fields of order $l \gtrsim 10^9$ cm be generated sufficiently fast in the first place? The electric induction time for such large current systems is much longer than the sub-second hard X-ray pulses we observe during the impulsive flare phase. (2) How can such highly-fragmented bundles of current channels with different electric fields coexist without coalescing? (3) How long can a large-scale electric field stably exist at all? Once a large-scale electric field is turned on, the accelerated electrons generate an oppositely-directed return current of positive ions and protons, which leads to charge separation and ambipolar diffusion. Furthermore, the DC electric field acceleration model has also been criticized for its inability to produce high energies (a continuation of the γ-ray spectrum to many MeV's) and to reproduce the observed heavy ion enhancements (Miller et al. 1997). In summary, given all these difficulties and inconsistencies with the observed hard X-ray delays we conclude that large-scale DC electric fields cannot explain particle acceleration in flares.

6.3. Kinematics of Particle Propagation

We turn now to propagation delays of free-streaming electrons between the acceleration site and the energy loss site. The energy-dependent propagation delays are simply given by the velocity dispersion, i.e. the time-of-flight difference between to different (constant) velocities over a common distance l, (see Eq. 65),

$$\Delta t^{prop} = \frac{l}{v_1} - \frac{l}{v_2} = \frac{l}{c}\left(\sqrt{1 - \frac{1}{\gamma_1}} - \sqrt{1 - \frac{1}{\gamma_2}}\right) , \tag{80}$$

Obviously the faster particles (with higher kinetic energy) arrive first, preceding the slower ones. Thus, if propagation delays dominate the total energy-dependent timing (Eq. 60), hard X-ray pulses should peak first at the highest energies. A fit of this time-of-flight (TOF)-model to the Masuda flare is shown in Fig. 64 (left panels), yielding a propagation distance of $l = 87,000 \pm 11,000$ km. This distance, which measures the path length of an electron spiraling along the magnetic field trajectory with some pitch angle, however, needs to be corrected for the pitch angle and the helical twist of the magnetic field line, in order to obtain a projected length in the loop plane, which then can be compared with other geometric parameters measured in the loop plane (Aschwanden et al. 1996b).

The pitch angle correction can simply be calculated from the ratio of the *loop crossing time* $t_0 = 4L/v$ of a particle moving parallel to the magnetic field (with pitch angle $\alpha = 0$) to the *loop bounce time* t_B of a particle with pitch angle α in a mirror field. This ratio is derived in Section 6.6 on particle trapping to $q_\alpha = t_0/t_B \leq 0.64$. Because the velocity v is conserved during propagation, so that v$= l''/t_0 = l/t_B$, we can use this ratio to determine the projected length l'' of the magnetic field line from the trajectory length l measured with the time-of-flight method, i.e.

$$l'' = l q_\alpha \leq l \times 0.64 \tag{81}$$

A second correction that needs to be applied is the helical twist of the magnetic field line. If we straighten the flare loop to a cylindrical geometry, the projected length l' of the cylinder is related to the length l'' of a helix with radius r by

$$l' = l'' q_h \tag{82}$$

$$q_h = \frac{1}{\sqrt{1 + (2\pi n r / l')^2}} , \tag{83}$$

with n being the number of complete twists by 2π radians. The helicity of coronal loops cannot exceed a few radians before they become magnetically unstable. For instance, erupting prominences show a twist consistent with theoretical models that predict a kink instability when the twist exceeds $\approx 2.5\pi$ (Vršnak, Rudjak, and Rompolt, 1991). Mikić, Schnack, and Van Hoven (1990) predict kink instabilities

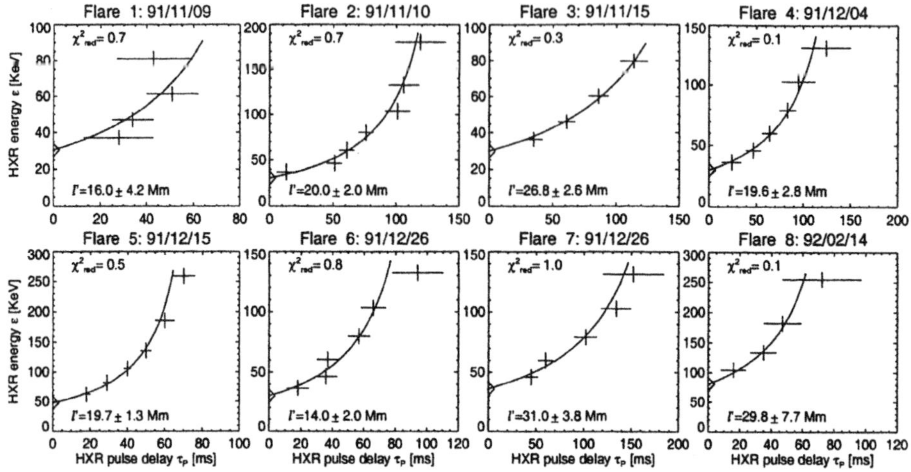

Figure 65. Energy-dependent time delays $\tau_{ij} = t(\epsilon_i) - t(\epsilon_j)$ of the filtered hard X-ray pulses $F^P(t)$ of 8 flares shown in Figs. 30 and 31, measured from the cross-correlation between the energy channels ϵ_i and ϵ_j (with $i = j_{min}$) during the selected time segments shown in Fig. 30. The horizontal bars represent the uncertainties of the delay measurement caused by Poisson noise. The curve represents the best fit of the TOF model $\tau_{ij} = (l/c)(1/\beta_i - 1/\beta_j)$. The projected TOF distance $l' = l \times 0.54$ and the χ^2_{red} of the best fit are indicated in each panel [from Aschwanden et al. 1996c].

for twists in excess of $\approx 4.8\pi$. We use $n = 2.5\pi/2\pi \approx 1$ for the estimated number of twists in our flare models. Based on a loop aspect ratio of $2r/l' \approx 0.2$ (observed for the Masuda flare), we find a correction factor of $q_H = 0.85$ for the average field line. Taking these two correction factors together, the measured electron time-of-flight distances yield a projected field line length l' that is about half as long as the time-of-flight distance l,

$$l' = l q_\alpha q_H \lesssim l \times 0.64 \times 0.85 \approx l \times 0.54 \,. \tag{84}$$

Applying this correction to the Masuda flare, the fitted TOF distance $l = 87,000$ km yields a projected field line length of $l' = l \times 0.54 = 47,000$ km, which corresponds to about the double height of the soft X-ray-bright flare loop ($h \approx 22,000$ km, see Fig. 9).

The pure TOF model, where propagation of free-streaming electrons dominate the energy-dependent timing [$t(E) \approx t^{prop}(E)$ in Eq. 60, with all other terms negligible], seems to work quite well for all fast time structures seen in hard X-ray time profiles. The smoothly-varying background flux has usually a different energy-dependent timing and is attributed to trapped electrons (as discussed later in Section 6.6). However, if the background is subtracted, as shown in the filtering procedure of hard X-ray time profiles shown in Figs. 30 and 31, the residual, rapidly-fluctuating hard X-ray spikes seem all to exhibit energy-dependent time delays that are consistent with the TOF model. Eight examples are shown in Fig. 65 (Aschwanden et al. 1996c), with energy delays of $\tau_{ij} \lesssim 100$ ms in the energy

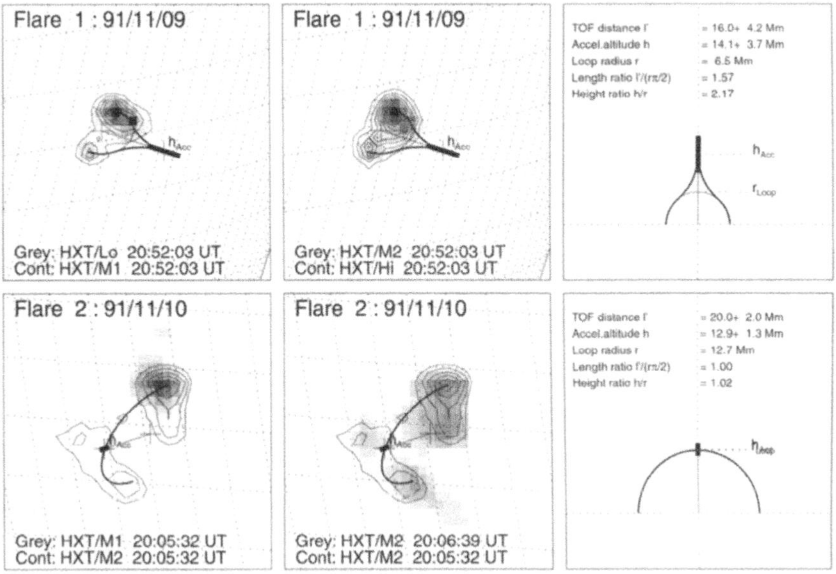

Figure 66. Overlays of coregistered *HXT/Lo,M1,M2,Hi* and *SXT* images (left and middle columns), identified with date (top), start time (bottom), and instrument (bottom) for greyscale (Grey) and contour plots (Cont). The contour levels are always linear, starting above the noise level. The hard X-ray double footpoints are connected with a semi-circular loop geometry in a vertical plane to the solar surface (thin curve). Symmetric open magnetic field lines joining the vertical in the cusp above the loop midpoints are projected in the same loop plane (thick curves). The length of the open field line corresponds to the TOF distance l', with the uncertainties $\pm\sigma_l$ indicated with a thick bar. A projection of the vertical loop plane is shown in the right columns. All panels in Figs. 66-69 have the same field of view (55,000 km), the spacing of the heliographic grid is 1^0 (12,150 km). [from Aschwanden et al. 1996c]

range of $\epsilon \approx 25 - 300$ keV. The inferred time-of-flight distances have a range of $l' \approx 14,000 - 31,000$ km. These measurements have been performed in virtually all available *CGRO* data (Aschwanden et al. 1996d) and confirmed the same result, namely that the fast time structures in hard X-rays show energy-dependent time delays that are consistent with propagation delays (or velocity dispersion).

In the Carmichael-Sturrock-Hirayama-Kopp-Pneuman (CSHKP) model (see Section 2.1) the reconnection site is located in a cusp above the soft X-ray bright flare loops. If we associate this cusp position with an acceleration site, we can estimate the propagation distance of electrons accelerated in the cusp down to the hard X-ray emission site in the chromosphere by modeling a cusp-shaped magnetic field line. If we model the geometry of such a cusp-shaped field line with two circular segments, with the lower segment having the same radius r as the flare loop and the upper segment with a radius constrained by smooth tangential connections at the

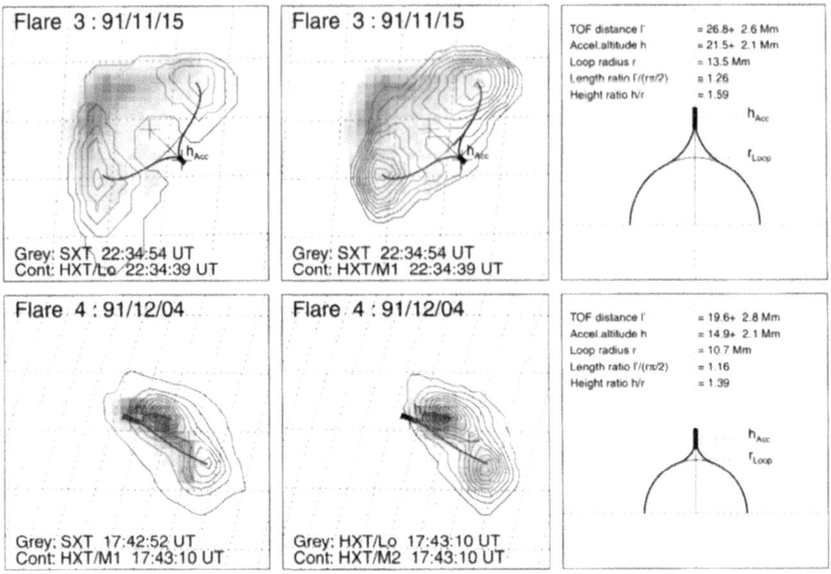

Figure 67.

semicircular flare loop and the vertical in the cusp, we obtain the following length l' for this cusp-shaped field line,

$$l' = \frac{r}{2}\left(1 + \frac{h^2}{r^2}\right) \arctan\left(\frac{2hr}{h^2 - r^2}\right), \tag{85}$$

where h is the height of the cusp above the footpoints. Using now the measured time-of-flight distances $l' = l q_\alpha q_H$ obtained from *CGRO* timing measurements and the radius r of the flare loops determined from the footpoint separation of the flare loops simultaneously observed with *Yohkoh/SXT*, we can measure the height h of the cusp. This type of geometric reconstruction is shown in Figs. 66–69, for the same 8 flares with data shown in Figs. 30, 31, 65). A summary of the measurements of the flare loop radii r, the cusp height h inferred from the TOF distance l' (with Eq. 85), the height ratios h/r, and the cusp field line length ratios l'/L (where $L = r\pi/2$ is the half length of the flare loop) is given in Table VIII (adapted from Table 3 in Aschwanden et al. 1996c). The most interesting result is that we find always a cusp location *above* the flare loop, with height ratios of $h/r \approx 1.0 - 2.0$, or cusp field line lengths of $l'/L \approx 1.0 - 1.6$ (times the flare loop half length). This correlation, which is shown for a larger dataset in Fig. 12, led to the conclusion that the acceleration site is closely associated with the reconnection point in the cusp region.

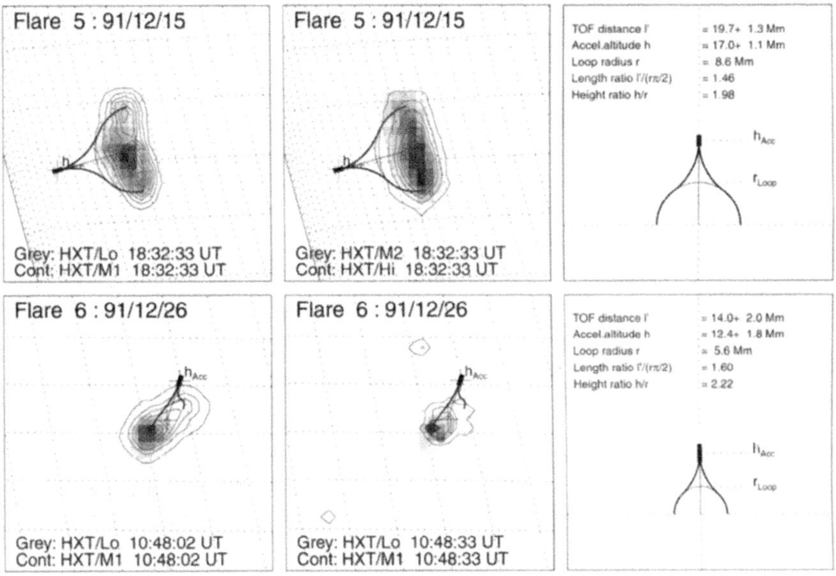

Figure 68.

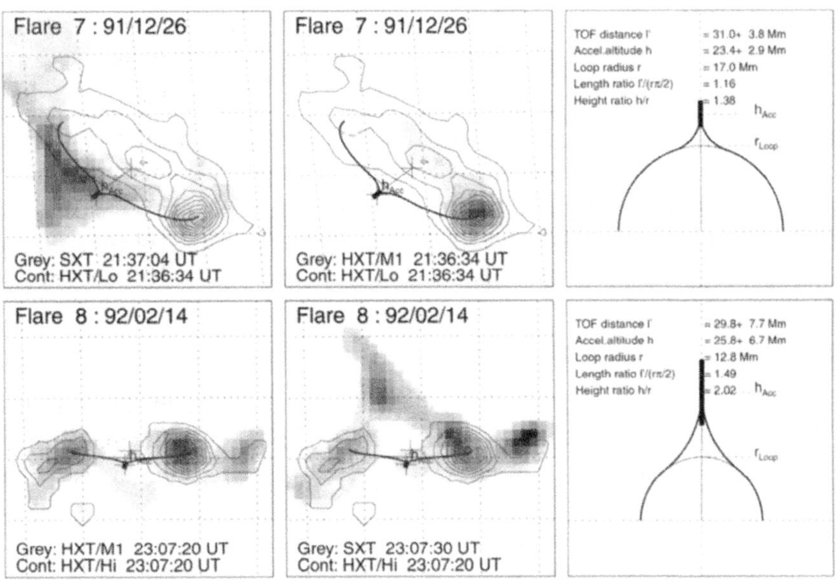

Figure 69.

TABLE VIII

Measurements of flare loop geometries from *Yohkoh/SXT* and TOF distances from *CGRO* in 8 flares.

Flare	Loop radius r [Mm]	TOF distance l' [Mm]	TOF height h [Mm]	Height ratio h/r	Length ratio l'/L
91/11/09 2052:03	6.5	16.0±4.2	14.1±3.7	2.17	1.57
91/11/10 2005:32	12.7	20.0±2.0	12.9±1.3	1.02	1.00
91/11/15 2234:39	13.5	26.8±2.6	21.5±2.1	1.59	1.26
91/12/04 1743:10	10.7	19.6±2.8	14.9±2.1	1.39	1.16
91/12/15 1832:33	8.6	19.7±1.3	17.0±1.1	1.98	1.46
91/12/26 1048:02	5.6	14.0±2.0	12.4±1.8	2.22	1.60
91/12/26 2136:34	17.0	31.0±3.8	23.4±2.9	1.38	1.16
92/02/14 2307:20	12.8	29.8±7.7	25.8±6.7	2.02	1.49

6.4. Kinematics of Combined Acceleration and Propagation

While we considered two extreme cases of electron kinematics so far, one where acceleration dominates the energy-dependent timing (Section 6.2), and one where propagation (or velocity dispersion) dominates (Section 6.3), we study now cases with a combination of the two, i.e. we approximate the general timing equation (60) with the two leading terms,

$$t(E) \approx t^{acc}(E) + t^{prop}(E) , \tag{86}$$

but still neglect the other three terms (t^{inj}, t^{trap}, $t^{loss} \approx 0$). We discuss five different scenarios (Aschwanden 1996f), cases with small-scale and large-scale acceleration regions, which are depicted in Fig. 70. Case 1 and 2 mimic small-scale acceleration regions, one for stochastic and one for electric DC-field acceleration. Case 3 is a large-scale version of electric DC-field accelerations, as we described in Section 6.2. Case 4 and 5 represent variants of large-scale DC-fields, one with a spatial distribution of acceleration starting points, mimicking electrons that enter the accelerating current channels at various places along the field lines, while Case 5 contains a spatial distribution of exit points, where electrons are accelerated initially and exit the fields at various points, and propagate free-streaming thereafter to the footpoints.

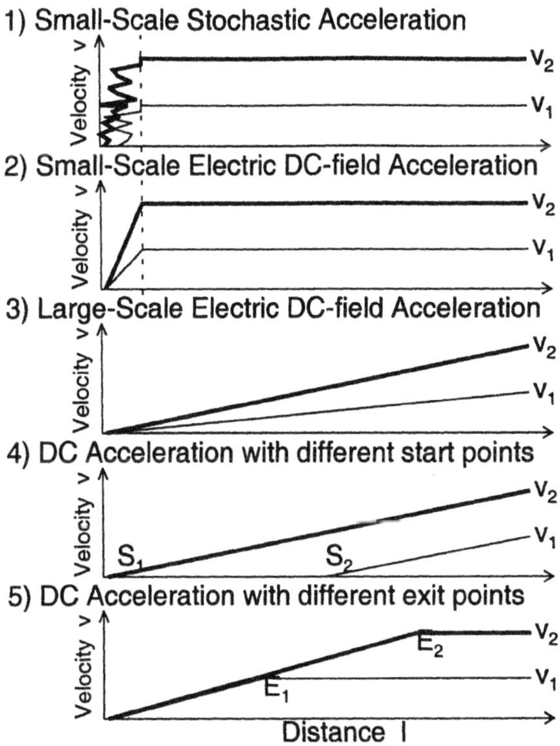

Figure 70. Five different models for the timing of electron acceleration and propagation. The velocity changes of a low-energy (v_1) and a high-energy electron (v_2) are shown along a 1-dimensional path from the beginning of acceleration (left side) to the thick-target site (right side). Model 1 and 2 characterize small-scale acceleration processes, while Model 3-5 depict large-scale acceleration situations. Model 4 and 5 illustrate different start (S_1, S_2) and exit positions (E_1, E_2) for the accelerated electrons [from Aschwanden 1996f].

6.4.1. *Case 1) - Small-Scale Stochastic Acceleration:*

Model 1 illustrates a stochastic acceleration process as it can occur in coronal regions with enhanced wave turbulence (or similarly in shock fronts). However, a small spatial scale of the acceleration region (compared with the TOF propagation distance) does not necessarily imply that the acceleration time is also much smaller than the propagation time. In the case of diffusive stochastic acceleration the particles can be bounced around in a turbulence region significantly longer than the travel time through this region. For instance, LaRosa et al. (1994) estimate the bulk energization time of electrons in a reconnection-driven MHD-turbulent cascade to t^{acc}(E=20 keV)$\approx$ 300 ms, which is comparable with TOF propagation times observed in flares. For a specific example we consider the 1991-Dec-15, 1832 UT flare (the 5th example in Table VIII, and Figs. 30–31 and 65–69), for which t^{prop}(E=20 keV) =l/v =29,000 km /(0.27 c)=360 ms. More specifically, Miller et

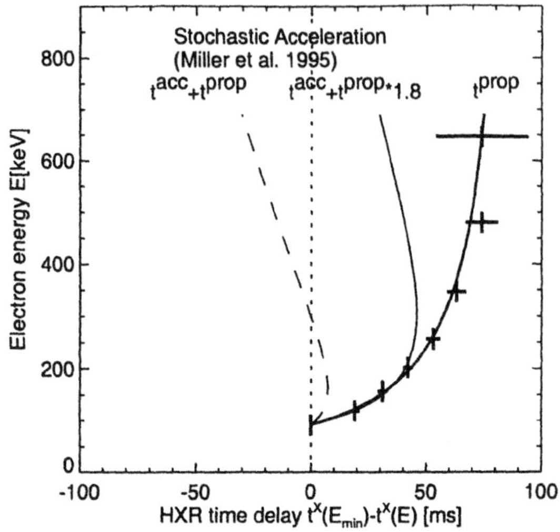

Figure 71. Fit of the stochastic acceleration model of Miller et al. (1996) to the flare of 1991-Dec-15, 1832 UT (crosses represent the measured energy-dependent hard X-ray time delays). Adding the predicted acceleration time to the propagation time (thick line) yields negative delays (dashed line). Adjusting the propagation time by a factor of 1.8 (thin line) yields an acceptable fit in the 100-200 keV range, but not at higher energies.

al. (1996) estimate acceleration times of 70 ms to energize electrons from ≈ 5 to 50 keV, or about 180 ms to 511 keV. They specify an energy dependence of

$$t^{acc}(E) = [(\frac{E}{m_e c^2})^{1/6} - 0.48] \cdot 350 \text{ ms}, \quad (\text{E} > 5 \text{ keV}) \tag{87}$$

to energize electrons by gyroresonant interactions with fast mode waves in an MHD-turbulent cascade. We fit this model to the 1991 Dec 15 flare and show the expected hard X-ray timing in Fig. 71. First we add the acceleration time to the same propagation time inferred from the TOF distance of $l = 29,000$ km. The expected hard X-ray delay becomes negative above 200 keV, meaning that the high-energy electrons arrive later than the low-energy electrons at the chromosphere due to the longer acceleration time. If we perform a fit of the combined expression $t^X = t^{acc} + t^{prop}$ (using Eqs. 80 and 87) we find that the data can be reasonably fitted in the 100-200 keV range with a 1.8 times larger TOF distance (to compensate for the acceleration time), but the hard X-ray delay decreases above 200 keV significantly below the measured values. Thus, the acceleration time scale specified in Miller et al. (1996) cannot fit the observed delays over the entire energy range of 80-800 keV for this specific flare. Thus we conclude that the observed timing is not dominated by acceleration delays, but rather by free-streaming propagation delays, implying that the acceleration times are much smaller than the propagation times ($t^{acc} \ll t^{prop}$).

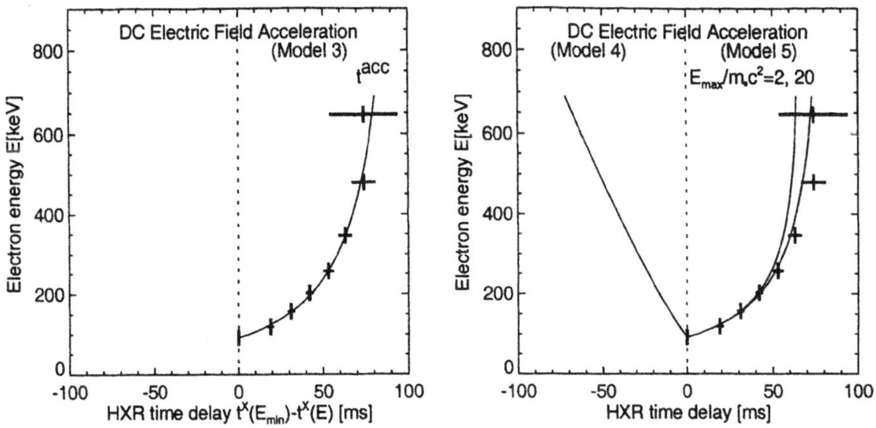

Figure 72. Fit of the DC electric field Model 3 (left) and Model 4 and 5 (right) for the flare of 1991-Dec-15, 1832 UT. With respect to the TOF distance $l^{prop} = 29,000$ km fitted in Fig. 65, the best fit of Model 3 (left) yields an acceleration distance of $l = l^{prop} \cdot 0.44$, Model 4 is shown for $l = l^{prop}$, and Model 5 yields $l = l^{prop} \cdot 1.1$ (for $E_{max}/m_e c^2 = 2$) and $l = l^{prop} \cdot 1.0$ (for $E_{max}/m_e c^2 = 20$).

6.4.2. *Case 2) - Small-Scale Electric DC-field Acceleration:*

The dynamics of magnetic reconnection (described in Section 4.2) suggests that there exist electric fields near X-points in the reconnection region, which are confined to localized regions of the size of magnetic islands (see Figs. 47 and 48 and discussion of filamentary current sheets in Section 5.1). For particles near a magnetic island, the timing very much depends on whether the particles become trapped in the magnetic islands (in meander-like orbits) or are accelerated outside the magnetic island and are able to move away immediately from the X-points. In Case 2 depicted in Fig. 70 we assume direct acceleration over a short distance $x^{acc} \ll L$ and subsequent free-streaming over the distance $L - x^{acc}$. For direct electric DC field acceleration, the acceleration time is indeed proportional to the acceleration path $t^{acc} \propto x^{acc}$, and thus implies $x^{acc} \ll L \mapsto t^{acc} \ll t^{prop}$, in contrast to stochastic or trapped acceleration, where the acceleration and free-streaming propagation times could be compatible, although $x^{acc} \ll L$. Anyway, for small-scale electric DC-field acceleration, where $t^{acc}(E) \ll t^{prop}(E)$ we can neglect the acceleration time difference and the observed hard X-ray delays should be consistent with the time-of-flight differences $t(E) \approx t^{prop}(E)$ (i.e. as the TOF-model fit t^{prop} in Fig. 71 shows).

6.4.3. *Case 3) - Large-Scale Electric DC-field Acceleration*

The case of large-scale electric DC-field acceleration, where the acceleration path comprises the entire distance from the reconnection site to the footpoints, has been described in Section 6.2. The velocity is steadily increasing along the path (Fig. 70 middle), and the variation of different electric fields along different paths produces a spectrum of final energies at the footpoints. The time delay fit of this model

$t(E) = t^{acc}(E)$ to the 1991-Dec-15, 1832 UT, flare are shown in Fig. 72 left. Interestingly, this model shows a very similar energy dependence as the TOF propagation model (shown in Fig. 71) and thus fits the data equally well. The inferred acceleration path length is a factor of 0.44 shorter than the path length in the TOF propagation model, because the average electron speed is about half of the final speed applied in the propagation model (being exactly half in the nonrelativistic limit). Thus, the two models cannot be distinguished from the timing alone, but the inferred distance scale is a factor of ≈ 2 different. However, we do not consider this model as realistic, because only electrons that are injected at the reconnection point are accelerated. More realistic variants are considered in Cases 4 and 5.

6.4.4. *Case 4) - DC Acceleration with Different Start Points:*

In Model 4 and 5 (Fig. 70) we investigate two further variants of electric DC field acceleration, where the accelerated electrons are allowed to enter into (Model 4) or exit from (Model 5) an electric field channel at different locations. In both models the resulting electron energy is proportional to the acceleration path length, assuming a constant (mean) electric field $\mathcal{E}$ in all current channels. Model 4 is a natural situation in the sense that all electrons in a current channel experience acceleration once the electric field is turned on. In this scenario, the acceleration path length of each electron is defined by the distance between its start position and the loop footpoint. The timing follows according to the expression given in Eq. 76, except that we have variable distances $x \leq l$ that are smaller than the total path length l,

$$t^{acc}(\gamma) = \frac{x}{c} \frac{\sqrt{\gamma^2 - 1}}{(\gamma_{max} - 1)} . \quad (88)$$

The resulting timing of Model 4 is shown in Fig. 72 (right) for the same TOF distance $l = 29,000$ km obtained in the 1991-Dec-15, 1832 UT flare, and for $\gamma_{max} - 1 = 2$ (or $E_{max} \approx 1$ MeV). The high-energy electrons arrive later at the thick-target site than the low-energy electrons for every parameter combination, and thus, cannot fit the data. Therefore, Model 4 can clearly be rejected, since all flares with rapid time structures show an opposite sign of the energy-dependent time delays, consistent with electron time-of-flight differences.

6.4.5. *Case 5) - DC Acceleration with Different Exit Points*

In Model 5 the electrons are allowed to exit a current channel with an accelerating electric field at an arbitrary location. Because electron spectra have always negative slopes, this means that more electrons leave the current channel after a short distance than after longer distances. This model may mimic a realistic situation when the current channels are relatively thin (for most extreme aspect ratios see Emslie and Henoux, 1995) so that electrons exit a current channel by cross-field drifts. The length of the acceleration path $x^{acc}(\gamma)$ can then be determined from the final electron energy E using Eq. 74. The resulting timing of electrons arriving at the thick-target site is then composed of the sum of the acceleration time t^{acc} over the acceleration path length x^{acc} and the free-flight propagation time t^{prop} over the remaining path length $l - x^{acc}$,

$$t^X(\gamma) = t^{acc}(\gamma) + t^{prop}(\gamma) = t^{acc}(\gamma) + \frac{l - x^{acc}(\gamma)}{v(\gamma)}$$

$$= \frac{l}{c}[\frac{\sqrt{\gamma^2 - 1}}{(\gamma_{max} - 1)} + \frac{1 - \frac{(\gamma-1)}{(\gamma_{max}-1)}}{\sqrt{1 - \gamma^{-2}}}] . \quad (89)$$

The fit of the timing Model 5 is also shown in Fig. 5 (right) for two different parameter combinations (E_{max} = 1 MeV and 10 MeV). The essential result is that Model 5 fits the data the better the smaller the acceleration time is relative to the propagation time, a situation that approaches asymptotically Model 2 for high electric field strengths. Consequently, the best fit is consistent with a small-scale acceleration region like in Model 2.

6.5. Kinematics of Particle Injection

We proceed to the next term in our energy-timing equation (Eq. 60), the injection time $t^{inj}(E)$. This is the least explored timing parameter and generally neglected, so that we can only discuss it qualitatively at this point. We define the injection time t^{inj} simply as time interval that occurs between acceleration and propagation to the hard X-ray footpoints.

For parallel DC electric field acceleration, the particle leaves the acceleration path directly and continues propagation (e.g. Model 5 in the previous Section) without any intervening delay, so the injection time is zero, $t^{inj} = 0$. Also for perpendicular DC electric field acceleration (Fig. 49; e.g. Litvinenko 1996), there is no time delay between the acceleration inside the current sheet and subsequent propagation once the particle exits the current sheet sideward. For DC field acceleration in magnetic islands (Figs. 51 and 52; e.g. Kliem 1994), for stochastic acceleration (e.g. Miller et al. 1996), for shock acceleration (Fig. 58; e.g. Tsuneta and Naito 1998), in particular in collapsing traps (Figs. 60 and 61; e.g. Somov and Kosugi, 1997), particles are temporarily trapped during acceleration and do not escape onto a magnetic field line (leading to the chromospheric footpoints) before an injection mechanism deflects the particle out of the acceleration trap. This injection mechanism could be the dynamics of a macroscopic structure, e.g. the coalescence of two magnetic islands into a single one, which changes the locations of the separatrix surfaces that divide trapped from free-streaming electrons, or it could be an internal microscopic change, such as a change of particle pitch angles. Generally, when particles are accelerated in parallel direction to the magnetic field, their pitch angles become smaller and the probability to be mirrored in converging magnetic bottles becomes smaller. So every parallel acceleration mechanism controls also the pitch-angle evolution of particles in such a way that they automatically escape after a finite time from the magnetic trap.

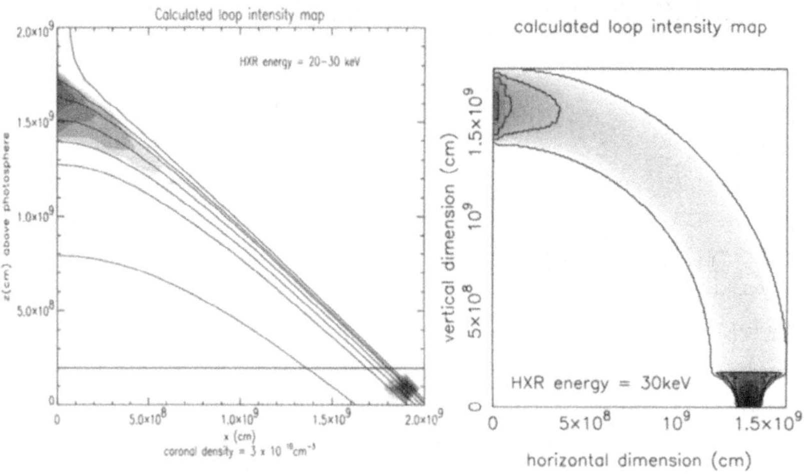

Figure 73. Hard X-ray loop-top model of Fletcher and Martens (1998). Calculated hard X-ray maps in Syrovatskii-type current sheet (left) and in cylindric flux tube, from Fletcher, 1995 (right). The contours are relative to the highest counts in the image, at 12.5%, 17.5%, 25% , 34%, 50%, and 70% of the maximum [from Fletcher and Martens, 1998].

Quantitative models on the particle injection process or escape mechanism from the acceleration region have not been developed yet, but we like to touch on two recent studies that are relevant. The trapping of electrons in the acceleration region has been modeled by a Syrovatskii-type current sheet geometry and using a Fokker-Planck numerical code by Fletcher and Martens (1998). In this study it was simulated how the combined effects of magnetic field convergence (which enlarges the pitch angles for particles approaching the mirror points) and collisional scattering (which scatters electrons into the losscone) form a trapped electron distribution in the cusp region where acceleration takes place. The trapping efficiency was found to be sufficiently high to reproduce the above-the-loop-top sources discovered by Masuda et al. (1994a). Fig. 73 shows an example of the simulations by Fletcher and Martens (1998). At this point it is not clear whether the trap region in the cusp is different from other traditional trap-plus-precipitation regions previously observed by gyrosynchrotron emission in radio wavelengths. However, because previously known trap regions were observed to have the form of large extended post-flare loops, it appears that the cusp regions, which are much smaller than flare loops, indeed represent a separate trap. We should therefore distinguish between a first trapping region that encapsulates the acceleration source (Fig. 74), and a secondary trapping region that is located further down near the converging magnetic fields near the flare loop footpoints. In such a concept, the injection time corresponds to the escape time from the first acceleration trap. In the model of Fletcher and Martens (1998), the injection time could then be computed from the net pitch angle scattering rate between acceleration and collisions in a cusp geometry.

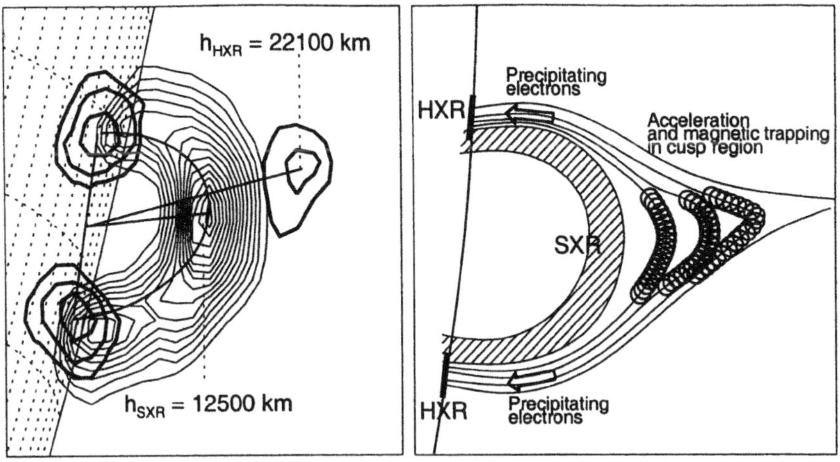

Figure 74. Concept of primary trap in the acceleration region located in the cusp beneath the reconnection point, according to the model of Fletcher and Martens (1998) (right side), which can explain the coronal hard X-ray emission observed during the Masuda flare (left side). The observations in the left panel show a *Yohkoh/HXT* 23-33 keV image (thick contours) and Be119 *SXT* image (thin contours) of the 92-Jan-13, 1728 UT flare (Masuda 1994a et al.) [from Aschwanden 1997c].

In the interest of future developments we would like to point out another dynamic process that may act as an *injection mechanism* for accelerated particles into the underlying flare loop system. This is illustrated in a study that deals with 3D reconnection (Section 2.4). The magnetic topology of the *TRACE*-observed flare on 1993-May-3, 23:05 UT, was inferred to consist of a 3D-reconnection configuration, containing a separatrix dome over an isolated magnetic polarity, with a spine field line above (Fig. 75). Particle acceleration is likely to happen near the reconnection locations (both in the spine field line as well as around the separatrix dome surface) where the magnetic field changes are fastest and induces an electric field, $\nabla \times \mathcal{E} = -(1/c)(dB/dt)$. For spine reconnection, the reconnecting field lines move through the spine, and thus reconfigure a trap on one side to a different connectivity, which essentially acts as a switch or injection mechanism to release accelerated particles onto other field lines. This area of topological changes in magnetic reconnection processes and their consequences on redistributing the paths of accelerated particles has not been explored much, but will undoubtly be an important ingredient in future solar flare models.

6.6. Kinematics of Particle Trapping

Magnetic reconnection reconfigures the connectivity of magnetic field lines in a solar flare environment, but the new-configured magnetic field lines include always *closed field lines* (see Fig. 2). In the classical Kopp-Pneuman model (and in many other reconnection models too), the new-configured field lines have initially a cusp-shaped form and relax then into a dipole-like postflare loop. This means, because

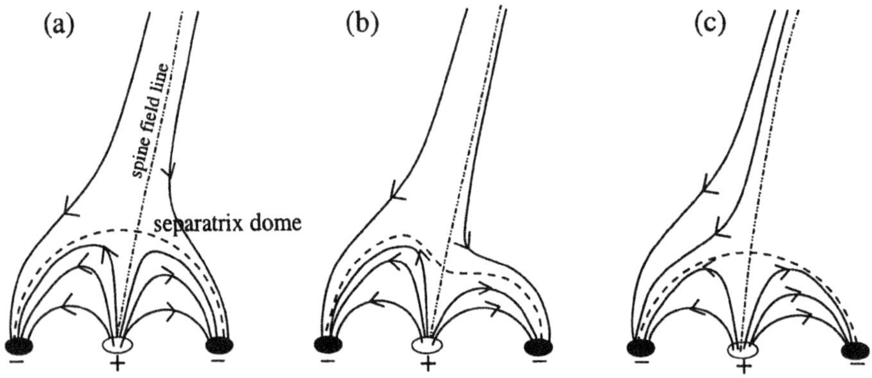

Figure 75. Magnetic field topology inferred in the 1993-May-3, 23:05 UT, flare by Fletcher et al. (2001). The sequence shows a two-dimensional representation of the reconnection process via a separator dome [from Fletcher et al. 2001].

the photospheric magnetic field at the flare loop footpoints is always stronger than the coronal segments of the field lines inbetween, that each closed magnetic field line forms a magnetic trap. In the collisionless limit, the particle motion is adiabatic and the magnetic moment is conserved along the loop coordinate s,

$$\mu = \frac{\frac{1}{2} m_e v_\perp^2(s)}{B(s)} = \frac{\frac{1}{2} m_e v^2 \sin^2 \alpha(s)}{B(s)} = const\ . \tag{90}$$

The pitch angle $\alpha(s)$ changes as function of the magnetic field $B(s)$ along the field line, while the velocity v is constant for adiabatic motion. In the solar corona, there are essentially three different types of flare loop traps, depending on the degree of plasma filling by chromospheric evaporation: (1) low-density or collisionless traps, (2) high-density, fully collisional traps, and (3) intermediate cases, where parts of the loops are filled with dense plasma (Fig. 76). In a low-density trap, the magnetic mirror points are essentially in the transition region. In half-filled loops the mirror points are at the interface between the collisional (high-density footpoints) and collisionless (low-density loop top) regions. In a high-density loop, such as the soft X-ray-bright flare loops and postflare loops, nonthermal particles become quickly thermalized and their mean free path length is less than a loop transit time. Interestingly, because chromospheric evaporation eventually fills all postflare loops, particle trapping is only a temporary phenomenon for the small-scale flare loops below the reconnection point, but longer-living traps are possible for large-scale loops that start in upward direction from the reconnection point to a remote footpoint.

To understand the basic kinematics in magnetic traps we assume a dipole-like magnetic field, which in the lowest-order approximation is specified by a parabolic equation, (e.g. see Trottet, Pick, and Heyvaerts 1979).

$$B(s) = B_0 \left[1 + (R - 1) \frac{s^2}{s_M^2} \right] , \tag{91}$$

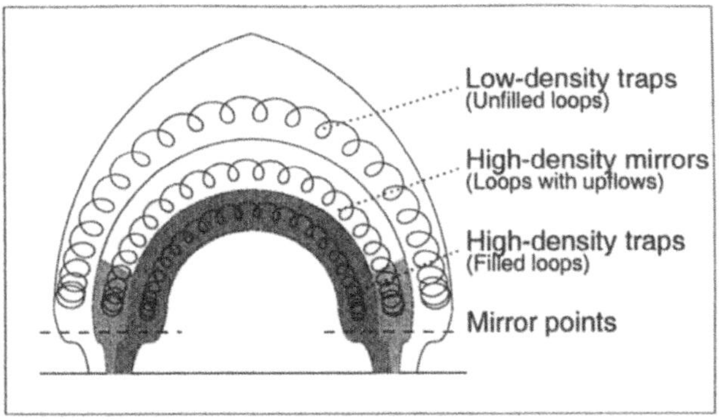

Figure 76. Three different trap configurations: (1) low-density trap (outermost loop), (2) high-density trap (innermost loop), and (3) a low-density trap with high-density mirror points (intermediate loop). The shaded areas indicate dense upflowing plasma driven by chromospheric evaporation [from Aschwanden et al. 1997a].

where $B_0 = B(s = 0)$ represents the (minimum) magnetic field strength at the loop top at $s = 0$, and s_M is the length from the loop top to the mirror point. The location of the mirror point $s = s_M$ is given by the interface from the collisionless to the collisional regime, it could be as low as in the transition region, or somewhere higher up, e.g. at the instantaneous location of the chromospheric evaporation front (in the latter case varying with time). The *magnetic mirror ratio R* is defined by the ratio of the magnetic field strengths at the mirror point and looptop,

$$R = \frac{B(s = s_M)}{B_0} = \frac{1}{\sin^2(\alpha_0)} . \tag{92}$$

The losscone angle α_0 is defined by the critical pitch angle α_0 of a particle at the looptop that decides whether a particle is mirrored and trapped (if it has a larger pitch angle, $\alpha > \alpha_0$), or whether it escapes through the losscone and becomes untrapped (if it has a smaller pitch angle, $\alpha < \alpha_0$). This losscone angle simply follows from comparing the magnetic moment (Eq. 90) at the loop top ($\alpha(s = 0) = \alpha_0$) with that at the mirror point ($\alpha(s = s_M) = \pi/2$),

$$\alpha_0 = \arcsin\sqrt{\frac{1}{R}} . \tag{93}$$

We will see later on that the mirror ratio R in flare loops can be determined from measurements of the losscone angle α_0, using the ratios of directly-precipitating to trapped electrons from the corresponding hard X-ray fluxes.

An important quantity is also the *bounce time* of a mirroring particle forth and back a mirror trap, which is, for our parabolic field (Eq. 91) and loop half length L,

$$t_B = 4\int_0^L \frac{ds}{v_\parallel(s)} = 4\int_0^L \frac{ds}{\sqrt{v^2 - v_\perp^2}} = \frac{2\pi L}{v\sqrt{1 + \frac{1}{R-1}}} . \tag{94}$$

The ratio of the travel time $t_0 = 4L/v$ along the magnetic field line to the bounce time t_B is

$$q_\alpha(R) = \frac{t_0}{t_B} = \frac{2}{\pi}(1 - \frac{1}{R})^{1/2} \tag{95}$$

which approaches the value $q_\alpha = 0.64$ for large mirror ratios $R \gg 1$. This ratio q_α yields also the relevant *pitch angle correction factor* to convert an electron-time-of flight distance into the length of a magnetic field line, i.e. $q_\alpha = l''/l = t_0/t_B$ (Eq. 81). Averaging the pitch angles over the entire range $(0 \le \alpha \le \alpha_0)$ of particles, yields a more representative correction factor for the entire trapped distribution,

$$q_\alpha = \frac{1}{2}[1 + q_\alpha(R)] = \frac{1}{2}[1 + \frac{2}{\pi}(1 - \frac{1}{R})^{1/2}] \, . \tag{96}$$

The criterion for the bifurcation of particles trajectories, whether they propagate free-streaming directly to the footpoints or become trapped (Fig. 63), is controlled by the initial pitch angle α_0 at the injection site. Those particles that have a pitch angle larger than the critical losscone angle $(\alpha > \alpha_0)$ become trapped, until they are pitch-angle scattered into the losscone $(\alpha < \alpha_0)$ after an energy-dependent trapping time $t^{trap}(E)$. They subsequently escape from the trap and precipitate (in the so-called *Trap-plus-precipitation* model). In the simplest model without energy loss in the trap, for a δ-like injection, the number of electrons $N_\delta^{trap}(E, t)$ (with kinetic energy E) in the trap decreases exponentially, with an e-folding time constant that corresponds to the trapping time $t^{trap}(E)$ (Melrose and Brown 1976), i.e.

$$N_\delta^{trap}(E, t) = N_\delta^{trap}(E, 0) \, \exp[-t/t^{trap}(E)] \, . \tag{97}$$

The precipitation rate, $n_\delta^{trap}(E, t)$, is defined by the time derivative of $N_\delta^{trap}(E, t)$, i.e.

$$n_\delta^{trap}(E, t) = -\frac{dN_\delta^{trap}(E, t)}{dt} = n_\delta^{trap}(E, 0) \, \exp[-t/t^{trap}(E)] \, , \tag{98}$$

with $n_\delta^{trap}(E, 0) = N_\delta^{trap}(E, 0)/t^{trap}(E)$. It is consequently also exponentially decreasing. For a general injection function $f(E, t)$, the precipitation rate from the trap, $n^{trap}(E, t)$, can be described by a convolution with the trapping time $t^{trap}(E)$,

$$n^{trap}(E, t) = \frac{1}{t^{trap}(E)} \int_0^t f(E, t') \, \exp[-\frac{(t - t')}{t^{trap}(E)}] \, dt' \, . \tag{99}$$

We compose now the total electron precipitation rate $n(E, t)$ with a combination of two components (Fig. 77): (1) a fraction q^{prec} of electrons that precipitates directly, and (2) the complementary fraction $(1 - q^{prec})$ that precipitates after some temporary trapping,

$$n(E, t) = q^{prec} f(E, t) + (1 - q^{prec}) n^{trap}(E, t) \tag{100}$$

$$= q^{prec} f(E, t) + \frac{(1 - q^{prec})}{t^{trap}(E)} \int_0^t f(E, t') \, \exp[-\frac{(t - t')}{t^{trap}(E)}] \, dt' \, . \tag{101}$$

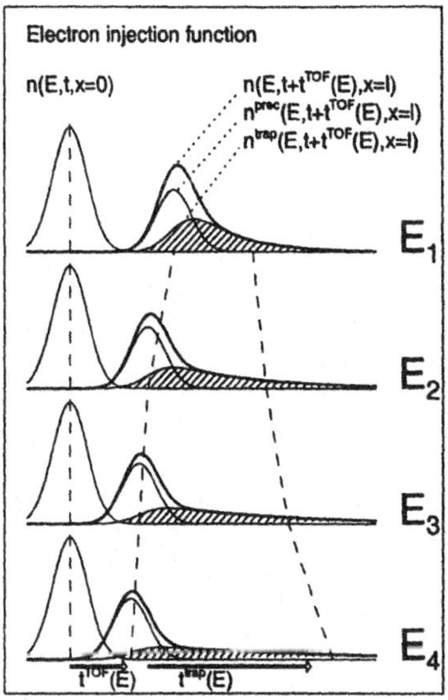

Figure 77. Temporal relation of the electron injection function $n(E, t, x = 0)$ at the injection site and at the hard X-ray emission site $n(E, t + t^{TOF}(E), x = l)$, delayed by an energy-dependent electron time-of-flight interval $t^{TOF}(E)$, schematically shown for 4 different energies $E_1 < E_2 < E_3 < E_4$. The injection function at the hard X-ray emission site (thick curve) is broken down into a directly-precipitating component (thin curve) and the trap-precipitating component (hatched curve). Note that the (e-folding) trapping time $t^{trap}(E)$ increases with energy E, whereas the time-of-flight delays $t^{TOF}(E)$ decreases.

Using the Bethe-Heitler bremsstrahlung cross section, the resulting hard X-ray thick-target bremsstrahlung photon spectrum $I(\varepsilon, t)$ at the chromospheric precipitation site can be expressed according to Brown (1971)

$$I(\varepsilon, t) = I_0 \frac{1}{\varepsilon} \int_{\varepsilon}^{E_0} n(E, t) \int_{\varepsilon}^{E} \ln \frac{1 + \sqrt{1 - \varepsilon/E'}}{1 - \sqrt{1 - \varepsilon/E'}} dE' \, dE, \tag{102}$$

with E_0 the high-energy cutoff of the electron injection spectrum and I_0 being a scaling constant for the mean photon count rate at the Earth's distance.

If a spatial separation between the electron injection site (at $x = 0$) and the hard X-ray emission site (at $x = l$) is included (indicated with a propagation delay $t^{TOF}(E)$ in Fig. 77), the hard X-ray photon spectrum can be expressed by the following approximation (Aschwanden and Schwartz 1996a),

$$I(\varepsilon, t + t^{TOF}[E(\varepsilon)], x = l) \approx I(\varepsilon, t, x = 0) \tag{103}$$

if spectral changes of the electron injection spectrum $n(E, t)$ are negligible during the electron time-of-flight propagation time $t^{TOF}(E)$. The function $E(\varepsilon)$ denotes the inversion of electron energies E from the observed photon energies ε for a time-dependent electron injection spectrum (Aschwanden and Schwartz 1996a). The propagation time $t^{TOF}(E)$ is defined by the electron travel time (with relativistic

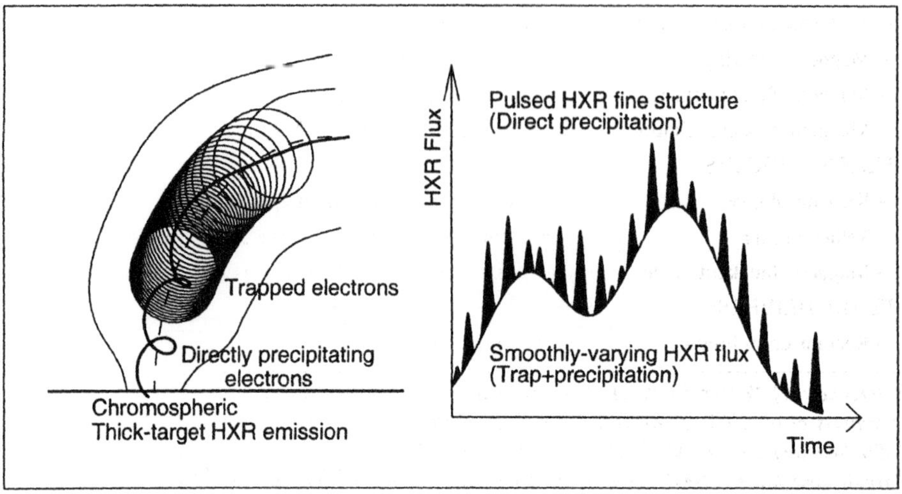

Figure 78. Relation between electron propagation and observed hard X-ray time structures: Electrons with small pitch angles (injected near the flare loop top) precipitate directly and produce rapidly-varying hard X-ray pulses with time-of-flight delays, while electrons with large pitch angles become trapped and produce a smoothly-varying hard X-ray flux when they eventually precipitate, with a timing that corresponds to trapping time scales.

speed v) between the coronal acceleration site and the (chromospheric) hard X-ray emission site, separated by the electron trajectory length l, defined by Eqs. 65 and 80.

The trapping time (in the weak-diffusion limit) is given by the electron collisional deflection time $t^{defl}(E)$ (Trubnikov 1965; Spitzer 1967; Schmidt 1979; Benz 1993)

$$t^{trap}(E) \approx t^{Defl}(E) = 0.95 \cdot 10^8 \big(\frac{E_{keV}{}^{3/2}}{n_e}\big) \big(\frac{20}{\ln \Lambda}\big) , \tag{104}$$

where $\ln \Lambda$ is the Coulomb logarithm,

$$\ln \Lambda = \ln[8.0 \cdot 10^6 \; (T_e \, n_e^{-1/2})] \, , \qquad T_e > 4.2 \cdot 10^5 \; K \; , \tag{105}$$

with typical temperatures of $T_e = 15 \cdot 10^6$ K for the ambient plasma in flare loops.

Data analysis of some 100 flares simultaneously observed with *CGRO* and *Yohkoh* are highly consistent with this model, where the observed hard X-ray flux represents a convolution of directly-precipitating and trapped-plus-precipitating components. The two components can be separated either by subtracting a lower envelope to the fast (sub-second) time structures (Fig. 78), or by a deconvolution as expressed with Eqs. 99–103 (Fig. 77). In Fig. 79 we show a selection of hard X-ray energy-dependent time delay measurements of the slowly-varying hard X-ray flux in 20 flares (Aschwanden et al. 1997a). The delays have been fitted with the expression for the collisional deflection time (Eq. 104), which has an energy

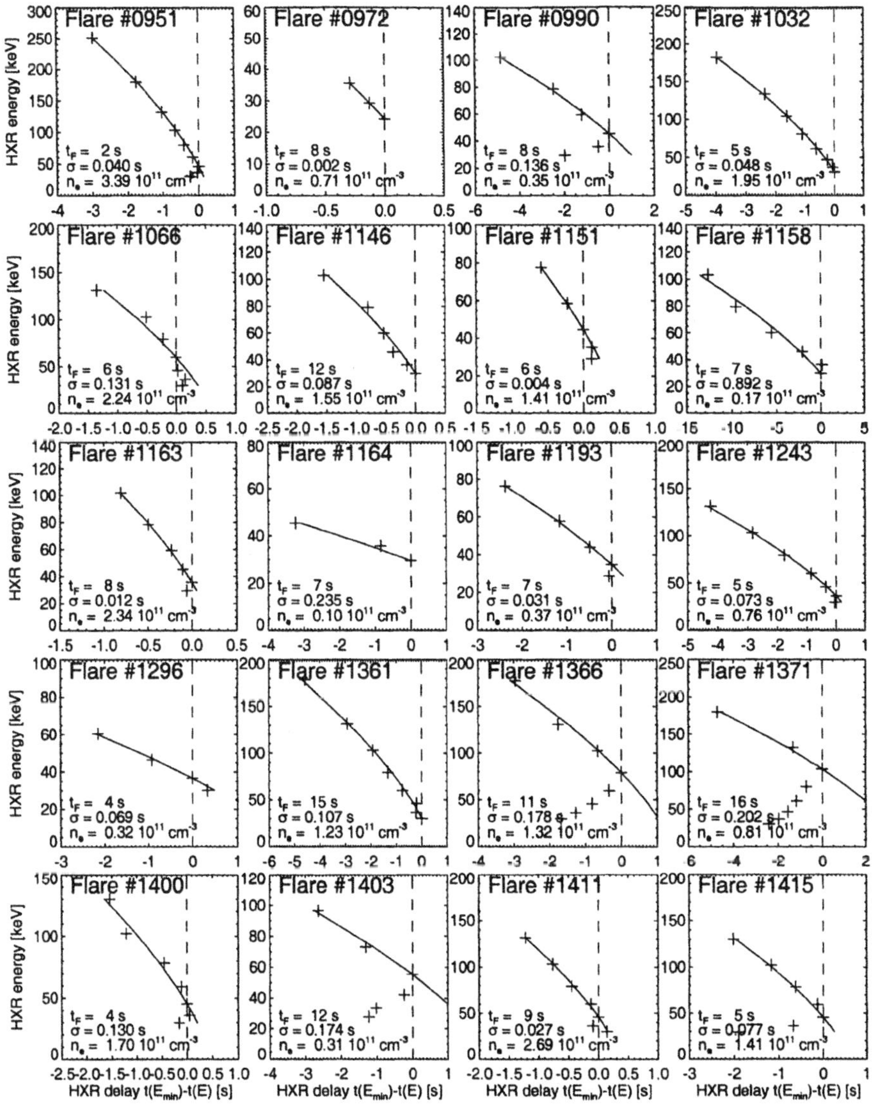

Figure 79. Energy-dependent time delay measurements of the *smooth hard X-ray component*, after filtering out the *pulsed hard X-ray component* with a Fourier filter time scale t_F (indicated in 3rd line from bottom in each panel) from all 44 flare events. The flares are identified (in chronological order) with the *BATSE/CGRO* burst trigger number (indicated at top of each panel). Each cross indicates a time delay measurement of one of the 16 *MER* energy channels with respect to a low-energy reference channel (indicated with a vertical dashed line at zero delay). The best model fit is drawn with a solid line, quantified by the trap electron density n_e (lowest line in each panel) and the mean standard deviation σ (second lowest line in each panel).

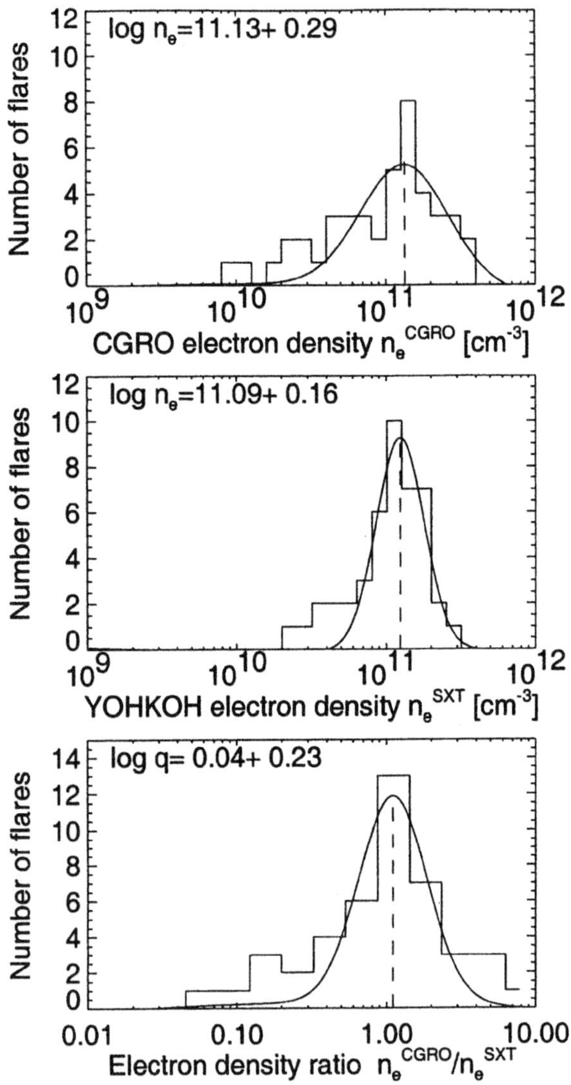

Figure 80. Distributions of the *CGRO*-inferred trap densities n_e^{CGRO} (top), the *Yohkoh*-inferred electron densities n_e^{SXT} (middle), and their ratios $q_{n_e} = n_e^{CGRO}/n_e^{SXT}$, fitted with gaussians (neglecting the asymmetric tails). [from Aschwanden et al. 1997a]

dependence of $\tau(E) \propto E^{-2/3}$ and seems to be consistent for all analyzed data at energies $E \gtrsim 25 - 200$ keV.

The fits of the collisional deflection times to the trapping delays yield also an electron density measurement n_e^{trap} in the trap region, typically in the range of $n_e^{trap} \approx 3 \times 10^{10}$ cm^{-3} $- 3 \times 10^{11}$ cm^{-3} (Fig. 80 top panel). A comparison of simultaneous electron density measurements in the soft X-ray bright flare loops with *Yohkoh/SXT* yields similar densities (Fig. 80, middle). This result indicates that the trap density is in the average comparable with the flare loops that have been filled with chromospheric upflows. It does not necessarily mean that the trap regions have to be co-spatial with the filled soft X-ray bright flare loops. The trapping loops could be collisionless and only the mirror points could have a density comparable with the chromospheric upflows, producing efficient collisional deflection near the mirror points, consistent with our fits. The fact that the trapping times are consistent with collisional deflection times is a strong argument that trapping is controlled in the weak-diffusion limit. For the strong-diffusion limit, where pitch-angle scattering is controlled by wave turbulence, the trapping times would be expected to be significantly shorter.

6.7. Kinematics of Particle Precipitation

6.7.1. *Symmetric Traps*

The fraction of directly-precipitating electrons q^{prec} can self-consistently be related to the critical losscone angle α_0. For sake of simplicity, we may consider first the case of an isotropic pitch-angle distribution at the injection site and for symmetric loop geometries,

$$q^{prec}(\alpha_0) = \frac{\int_0^{\alpha_0} \sin\alpha \, d\alpha}{\int_0^{\pi/2} \sin\alpha \, d\alpha} = (1 - \cos\alpha_0) \ . \tag{106}$$

This case corresponds to a double-sided losscone distribution, as it occurs in symmetric flare loops. With a deconvolution method we can measure q^{prec} directly. The corresponding losscone angle α_0 is then (for isotropic pitch angle distributions and symmetric loops),

$$\alpha_0(q^{prec}) = \arccos(1 - q^{prec}) \ , \tag{107}$$

leading to the mirror ratio $R(\alpha_0)$ defined in Eq. 92. The distributions of the precipitation fractions q^{prec} (Eq. 106), loss cone angles α_0 (Eq. 107), and magnetic mirror ratios R (Eq. 108) measured in 46 flares (Aschwanden et al., 1998b) are shown in Fig. 81.

The inference of the magnetic mirror ratio $R = B^{loss}/B^{inj}$ together with the (projected) time-of-flight distance l^{TOF} between the injection site and chromospheric energy loss site (which is presumably close to the losscone site), yields a measure of the magnetic scale height λ_B. Defining the scale height by an exponential model

$$B^{inj} = B^{loss} \exp[-\frac{(h^{inj} - h^{loss})}{\lambda_B}] \approx B^{loss} \exp[-\frac{l^{TOF}}{\lambda_B}] \ , \tag{108}$$

and using the definition of the mirror ratio $R = B^{loss}/B^{inj}$, we obtain

$$\lambda_B = \frac{l^{TOF}}{\ln(R)} \ . \tag{109}$$

The distributions of the magnetic scale heights λ_B inferred from TOF distances l^{TOF} measured in 46 flares (Aschwanden et al., 1998b) are also shown in Fig. 81.

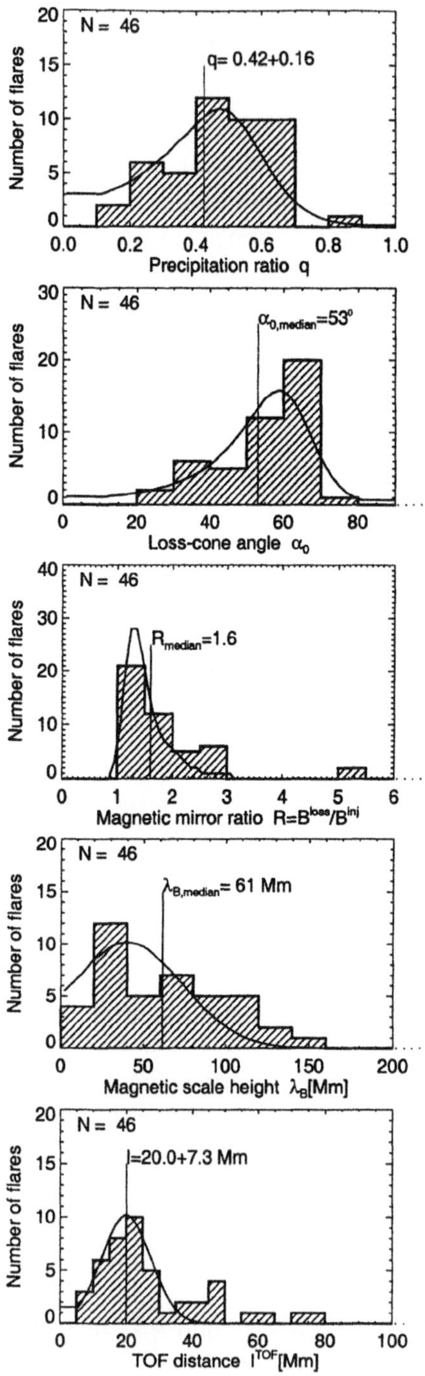

Figure 81. The distributions of measured precipitation fractions q^{prec} (top panel), losscone angles α_0 (second panel), magnetic mirror ratios R (third panel), magnetic scale heights λ_B (forth panel), and TOF distances (fifth panel). The smooth distributions represent a theoretical model based on normal distributions of (observed) TOF values ($l^{TOF} \pm \sigma_l = 20.0 \pm 7.3$ Mm) and dipole depths ($h_D \pm \sigma_{h_D} = 100 \pm 100$ Mm) in a model with dipole fields.

The magnetic field can be approximated with a dipole field, parametrized with the dipole depth h_D and the photospheric field B^{ph},

$$B(h) = B^{ph}(1 + \frac{h}{h_d})^{-3} , \tag{110}$$

to which the magnetic mirror ratio can be related by $R = B^{inj}/B^{loss} \approx B(h)/B^{ph}$. The resulting magnetic scale height λ_B is then

$$\lambda_B(h) = -\frac{B}{\nabla B} = \frac{(h_D + h)}{3} . \tag{111}$$

We find that a distribution of dipole depths with $h_D \pm \sigma_{h_D} \approx 100 \pm 100$ Mm can approximately reproduce (smooth distributions in Fig. 81) the measured parameters in Fig. 81 (histograms).

6.7.2. *Asymmetric Traps*

The timing analysis of *CGRO* data, based on the total hard X-ray flux without spatial information, provides a global electron trapping time scale. The spatial structure, however, can often be described by two magnetically conjugate footpoint sources, which often have asymmetric hard X-ray fluxes according to the *Yohkoh/HXT* images. These unequal double footpoint sources indicate electron precipitation sites in a flare loop with asymmetric magnetic field geometry. We need therefore to develop an asymmetric trap model to relate the trapping time information from *CGRO* data to the asymmetric double footpoint sources seen in *Yohkoh/HXT* data.

In order to mimic an asymmetric trap model we rotate the reference system of a symmetric dipole-like magnetic field by an angle ψ, as shown with three examples in Fig. 82 (top): the symmetric case with $\psi = 0$ (left), a weakly asymmetric case where the dipole coil is rotated by $\psi = 30^0$ (middle), and strongly asymmetric case where the coil is rotated by $\psi = 60^0$. (A spherically symmetric sunspot with a "unipolar" vertical field would correspond to the extreme case of $\psi = 90^0$.) The acceleration or injection site into the trap is assumed to be midway (with a magnetic field B_A) between the two footpoints (with magnetic fields B_1 and B_2). These three magnetic field values B_A, B_1, B_2 are decisive what fraction of electrons are trapped or precipitate to the two footpoints. Because of conservation of the magnetic moment, $\mu = \frac{1}{2} m_e v_\perp^2 / B \propto \sin\alpha(s)^2/B(s) = const$, the critical pitch angles that separate precipitating from trapped particles at the two footpoints are defined by the *magnetic mirror ratios*

$$R_1 = \frac{B_1}{B_A} = \frac{1}{\sin(\alpha_1)^2} \tag{112}$$

$$R_2 = \frac{B_2}{B_A} = \frac{1}{\sin(\alpha_2)^2} \tag{113}$$

For positive rotation angles $\psi > 0$ the magnetic field at footpoint B_1 is stronger ($B_1 > B_2$, $R_1 > R_2$), and the critical angle, also called *losscone angle*, is smaller

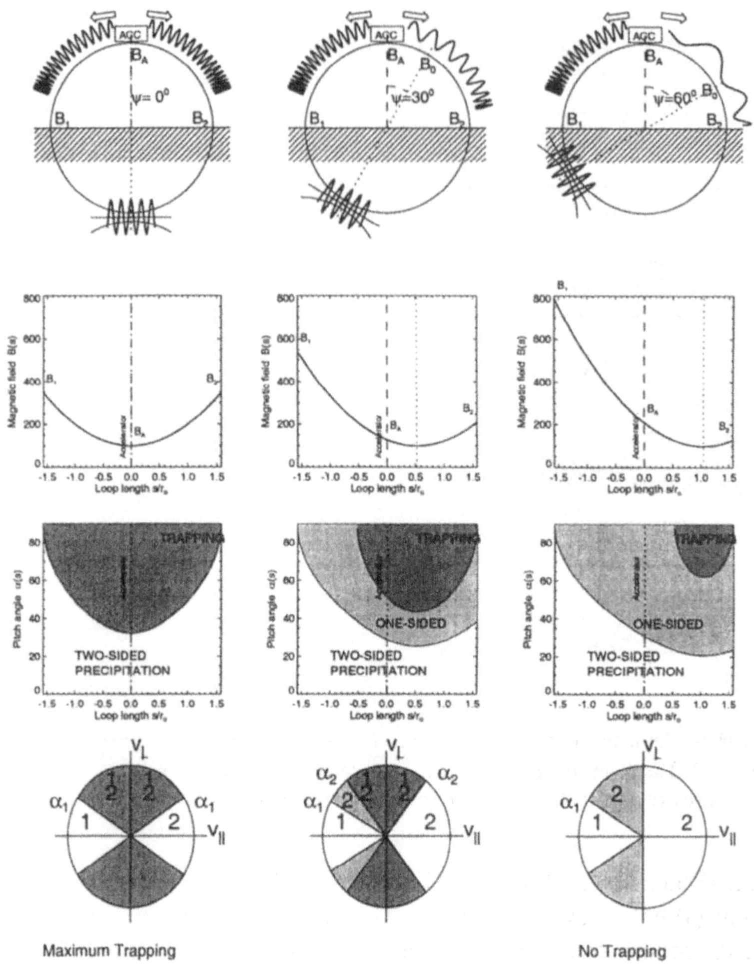

Figure 82. Three model scenarios for a symmetric (left column), a slightly asymmetric (middle column), and strongly asymmetric magnetic trap (right column). The spatial configuration of a buried dipole and the resulting pitch-angle motion of trapped and/or precipitating electrons is sketched (in top row), the magnetic field $B(s)$ is parametrized as function of the loop coordinate s (second row), the pitch angle variation $\alpha(s)$ as function of the loop coordinate and three pitch angle regimes are shown (third row), and the corresponding pitch-angle regimes in velocity space ($v_{\parallel}$, $v_{\perp}$) (bottom row). The numbers 1 and 2 correspond to the left and right losscone site, with the stronger magnetic field located at footpoint 1.

($\alpha_1 < \alpha_2$) than at the opposite footpoint 2. The asymmetric magnetic field $B(s)$ along a flare loop is visualized in Fig. 82 (second row) with a quadratic model. Note that in the case with strong asymmetry ($\psi = 60^0$), trapping is not possible because $B_2 < B_A$, corresponding to a mirror ratio $R_2 < 1$.

The pitch angle variation $\alpha(s)$ along the loop according to conservation of the magnetic moment is shown in Fig. 82 (3rd row). Generally, three regimes can be distinguished in the α-s plane: (1) a direct precipitation regime for initial pitch angles $0 < \alpha(s=0) < \alpha_1$, (2) a secondary precipitation regime after one mirror bouncing for initial pitch angles $\alpha_1 < \alpha(s=0) < \alpha_2$, and (3) a trapping regime for initial pitch angles $\alpha_2 < \alpha(s=0) < \pi/2$. These three regimes are clearly discernible in the case with weak asymmetry ($\psi = 30^0$, Fig. 82 middle column). For the symmetric case, the secondary precipitation regime collapses to zero because $\alpha_1 = \alpha_2$ (Fig. 82 left column). For the strongly asymmetric case ($\psi = 60^0$) no trapping is possible because there is no solution for $\alpha_2 < \pi/2$ with $R_2 < 1$.

After we have a quantitative description of the pitch angle ranges that contribute to trapping and precipitation at both footpoints, we can now calculate the relative fractions of precipitating electrons at the two footpoints and obtain quantitative expressions for the hard X-ray flux asymmetry A. We visualize the pitch angle regimes in velocity space ($v_\parallel, v_\perp$) (in Fig. 82 bottom) and label the different regimes with the footpoint numbers 1 and 2, to which the electrons precipitate, either directly, after one mirror bounce, or after intermediate trapping. We determine now the relative fractions of precipitating electrons by integration over the corresponding pitch angle ranges in velocity space. Here and in the following we assume an isotropic pitch-angle distribution at the acceleration/injection site $[f(\alpha) = const]$. The fraction q_{DP1} of directly precipitating electrons at footpoint 1, which has the smaller losscone ($\alpha_1 \le \alpha_2$), is

$$q_{P1} = q_{DP1} = \frac{\int_{\pi-\alpha_1}^{\pi} f(\alpha)\sin(\alpha)\,d\alpha}{\int_0^{\pi} f(\alpha)\sin(\alpha)\,d\alpha} = \frac{(1-\cos\alpha_1)}{2} \ . \tag{114}$$

The fraction of directly precipitating electrons at footpoint 2, which has the larger losscone angle (α_2), includes not only those electrons which precipitate without bouncing (q_{DP2}) but also those that bounce once at the mirror site 1 (q_{MP2}) and precipitate then at footpoint 2, i.e. with initial pitch angles of $(\pi - \alpha_2) < \alpha(s = 0) < (\pi - \alpha_1)$,

$$q_{P2} = q_{DP2} + q_{MP2} \tag{115}$$

$$= \frac{\int_0^{\alpha_2} f(\alpha)\sin(\alpha)\,d\alpha + \int_{\pi-\alpha_2}^{\pi-\alpha_1} f(\alpha)\sin(\alpha)\,d\alpha}{\int_0^{\pi} f(\alpha)\sin(\alpha)\,d\alpha} \tag{116}$$

$$= \frac{(1+\cos\alpha_1 - 2\cos\alpha_2)}{2} \ . \tag{117}$$

From the spatially unresolved *CGRO* data we measured the combined fraction q_{prec} of directly-precipitating electrons at both footpoints, which is equivalent to the sum of both footpoint components q_{P1} and q_{P2},

$$q_{prec} = q_{P1} + q_{P2} = 1 - \cos\alpha_2 \tag{118}$$

The total fraction of trapped electrons is determined by the pitch angle range of the larger losscone, i.e. $\alpha_2 < \alpha(s=0) < (\pi - \alpha_2)$,

$$q_T = \frac{\int_{\alpha_2}^{\pi-\alpha_2} f(\alpha)\sin(\alpha)\,d\alpha}{\int_0^{\pi} f(\alpha)\sin(\alpha)\,d\alpha} = \cos\alpha_2 = (1 - q_{prec}) \ . \tag{119}$$

Trapped electrons are randomly scattered, but their pitch angle increases statistically with time, until they diffuse into a losscone. Because collisional deflection is an accumulative process of many small-angle scattering deflections that can add up to a small or large net value after each loop transit, there is some probability that trapped electrons can escape on either losscone side. The escape probability on the side with the larger losscone is higher than at the opposite side with the smaller losscone. In first order we estimate that the relative escape probabilities from the trap towards the two losscone sides be proportional to the probabilities for direct precipitation at the two losscone sides (for particles with pitch angles $\alpha \leq \alpha_2$),

$$q_{T1} = q_T \, (\frac{q_{P1}}{q_{prec}}) \tag{120}$$

$$q_{T2} = q_T \, (\frac{q_{P2}}{q_{prec}}) \tag{121}$$

The proportionality implies also symmetric escape probabilities for symmetric losscones. We emphasize that this proportionality *ansatz* $q_{T1}/q_{T2} \approx q_{P1}/q_{P2}$ represents an approximation we employ for mathematical convenience. A more rigorous treatment of the asymmetric escape probabilities can be derived with statistical theory or with numerical simulations. Numerical simulations using the Fokker-Planck equation are currently investigated (Lyndsay Fletcher, private communication). Preliminary results from these numerical simulations indicate that the proportionality assumption (Eqs. 120–121) represents a reasonable approximation.

To estimate the relative hard X-ray fluxes at both footpoints, we have to sum the precipitating and trapped contributions at both sides. We denote the combined fractions at both footpoints by q_1 and q_2,

$$q_1 = q_{P1} + q_{T1} = q_{P1}(1 + \frac{q_T}{q_{prec}}) = q_{P1}(\frac{1}{q_{prec}}) \tag{122}$$

$$q_2 = q_{P2} + q_{T2} = q_{P2}(1 + \frac{q_T}{q_{prec}}) = q_{P2}(\frac{1}{q_{prec}}) \tag{123}$$

Neglecting differences in the spectral slope (e.g. arising from asymmetric accelerators or asymmetric coronal energy loss), the hard X-ray flux at a given energy ε is proportional to the number of (non-thermal) electrons with energies $E \gtrsim \varepsilon$. Assuming a similar spectral slope of the electron injection spectrum towards the two opposite directions 1 and 2, the hard X-ray fluences $F_1 = \int f_1(t)dt$ and $F_2 = \int f_2(t)dt$ observed at the two footpoints are then expected to be proportional to the precipitating electron fluxes q_1 and q_2. This constitutes a relation between the observed hard X-ray flux asymmetry A and the loss cone angle α_2,

$$A = \frac{F_2}{(F_1 + F_2)} = \frac{q_2}{(q_1 + q_2)} = q_2 = (1 - q_1) \tag{124}$$

With these relations we have a simple method to determine the two losscone angles α_1 and α_2 from the observables q_{prec} and A,

$$\alpha_1 = \arccos\left[1 - 2(1 - A)q_{prec}\right], \tag{125}$$

$$\alpha_2 = \arccos\left[1 - q_{prec}\right] \tag{126}$$

and the corresponding magnetic mirror ratios R_1 and R_2 with Eqs. 112 and 113, or the ratios of the magnetic fields B_1/B_A and B_2/B_A, respectively.

Most of the observed double footpoint flares show indeed asymmetric hard X-ray fluxes. A selection of 54 double footpoint flares observed with *Yohkoh/HXT* is shown in Fig. 83. The asymmetry of the fluxes is plotted along a profile that intersects the two footpoints (Fig. 84). The inferred fractions of precipitating fluxes at the two footpoints are $q_{P1} = 0.14 \pm 0.06$ and $q_{P2} = 0.26 \pm 0.10$, and the complementary fraction of trapped electrons is $q_T = q_{T1} + q_{T2} = 0.60 \pm 0.13$ (see distributions in Fig. 85 top). The inferred losscone angles at the two conjugate footpoints have means of $\alpha_1 = 42^0 \pm 11^0$ and $\alpha_2 = 52^0 \pm 10^0$ (see distributions in Fig. 85 middle). The associated mirror ratios at conjugate footpoints are $R_1 = 1.6...4.0$ and $R_2 = 1.3...2.5$ (see distributions in Fig. 85 bottom). The asymmetry of magnetic field configurations in flare loops produces a higher precipitating electron flux on the side with higher hard X-ray fluxes, but increases the trapping efficiency and related emission on the other side, such as gyrosynchrotron emission from trapped high-relativistic electrons. Such asymmetric radio emission has therefore been modeled with a so-called *cornupia model* (Li et al. 1997), and the complementarity of asymmetric hard X-ray and radio emission has been verified in several observations (e.g. Wang et al. 1995).

6.8. Kinematics of Particle Energy Loss

Regarding the collisional energy loss of nonthermal particles in the solar corona and chromosphere, there are two models that represent limiting cases: the *thick-target model*, where energy-loss occurs in a localized region that is small compared with the particle propagation path, and the *thin-target model*, where particles continuously lose energy along their propagation path. In the ideal thick-target model, loss times are by definition much smaller than propagation times, i.e. $t^{loss}(E) \ll t^{prop}(E)$, and thus can be neglected in the energy-dependent timing equation (Eq. 60). This essentially reflects the fact that the geometric length of the (collisionless) coronal propagation path is anyway much larger than the (collisional) chromospheric stopping depth.

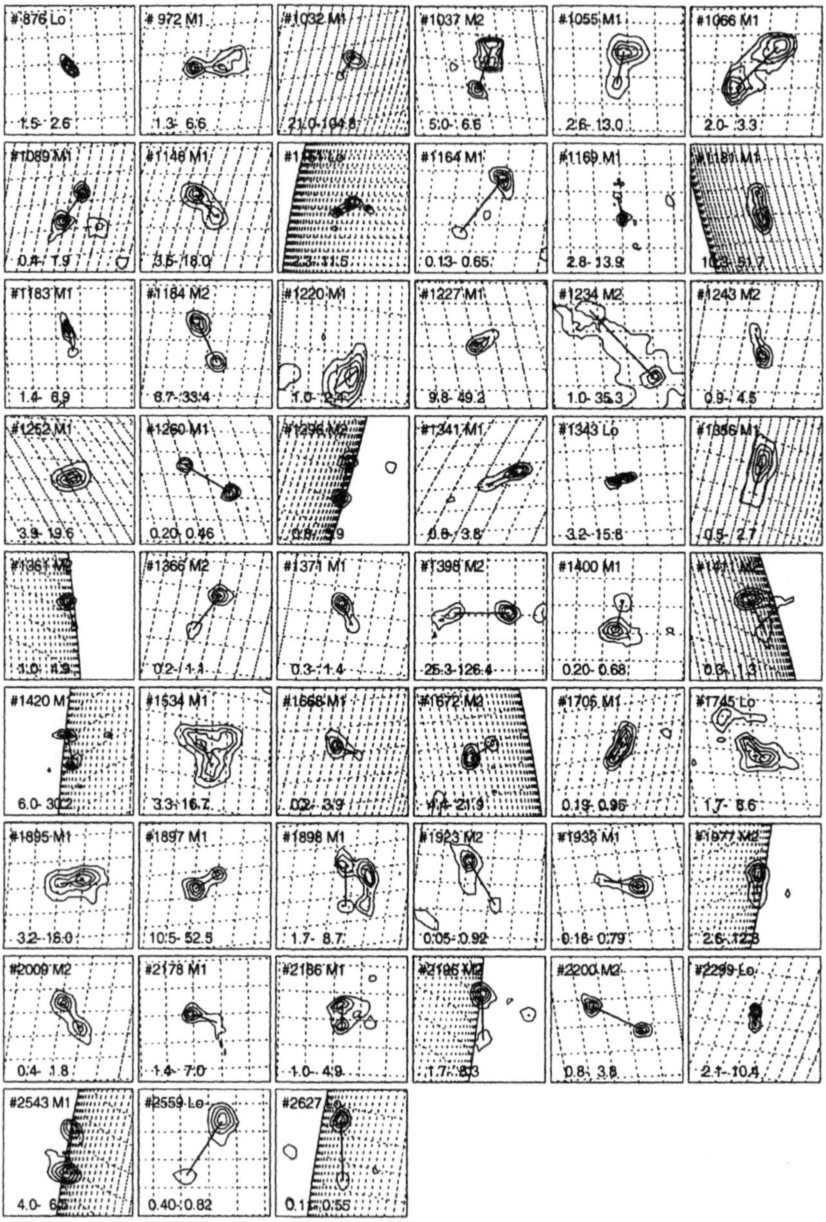

Figure 83. Yohkoh/HXT maps of 54 analyzed solar flares, labeled by the *BATSE* burst number (#876,...,#3124) for which the observing dates, times and measured parameters are given in Aschwanden et al. (1999a). From the 3 lower *HXT* energy channels (Lo=14-23 keV, M1=23-33 keV, M2=33-53 keV) we show for each flare that *HXT* image with the highest energy in which the double footpoint structure was analyzable. The footpoint separation (defined by the distance between the two fitted gaussian functions) is marked with a thick line. The contours of each *HXT* image are linearly spaced between a lower threshold F_{min} and the flux maximum F_{max} (typically 20%, 40%, 60%, 80%, 100%), with the range $[F_{min}, F_{max}]$ in cts/s per subcollimator indicated at the bottom of each frame. All frames have the same spatial scale of 32 *HXT* pixels in x- and y-axis (i.e. 78.5" or 55,000 km). A heliographic coordinate system with a spacing of 1^0 heliographic degree (=12,150 km) is overlaid [from Aschwanden et al. 1999a].

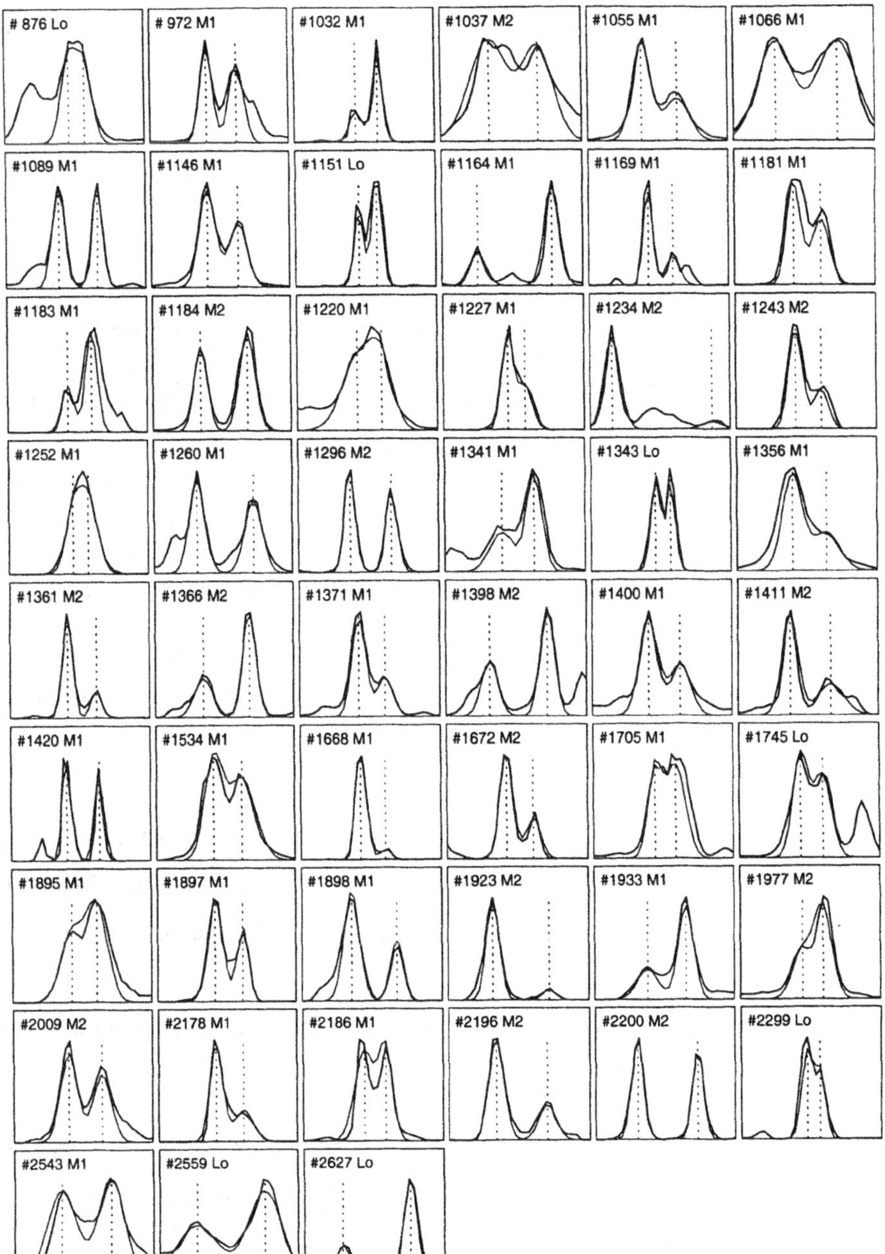

Figure 84. HXT flux profiles (bilinearly) interpolated along the footpoint baseline computed from the gaussian fits as shown in Fig. 83. The observed flux is shown with a thick solid line (normalized), while the fit of the two-component gaussian model is shown with a thin solid line (and the center positions of the gaussians are marked with dashed lines). Note that the fits are optimized in the 2-dimensional image plan, and thus do not necessarily coincide with the best conceivable fit along a 1-dimensional scan line, especially for source components which are not spherically symmetric. [from Aschwanden et al. 1999a].

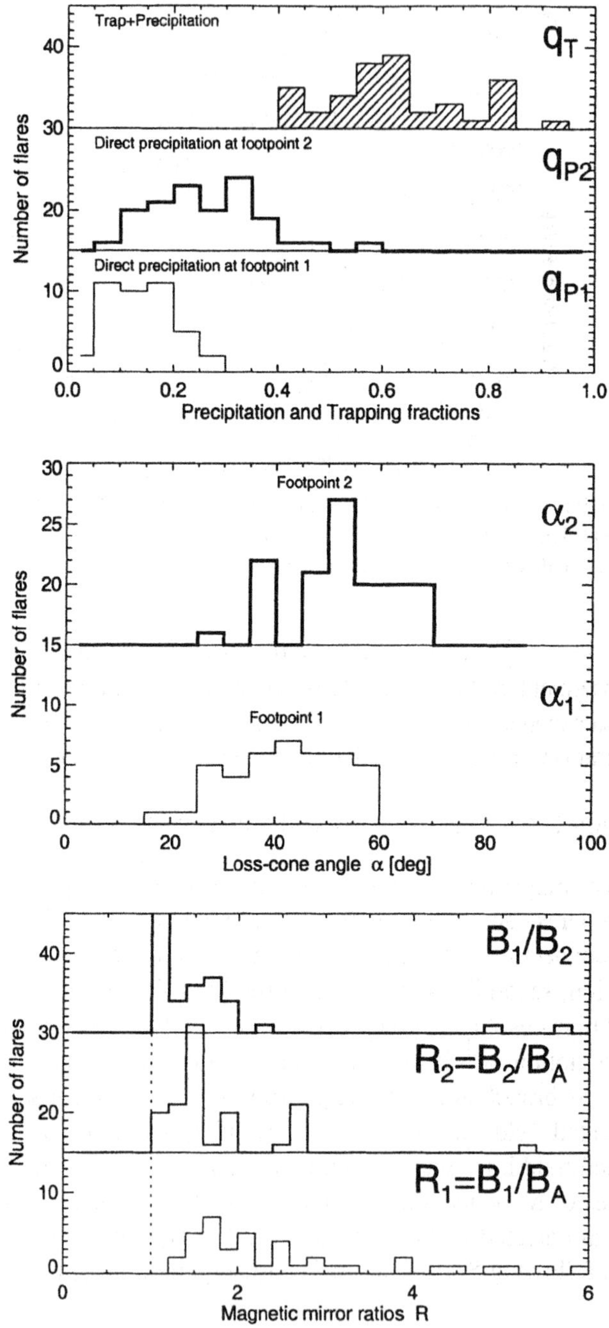

Figure 85. Distributions of precipitation and trapping fractions (top), conjugate loss cone angles (middle), and magnetic mirror ratios (bottom), calculated for an isotropic pitch-angle distribution at the acceleration/injection site [from Aschwanden et al. 1999a].

Let us calculate the stopping depth and thick-target density for a specific flare where observations were made with constraining parameters. In the 1993-Feb-17, 1035:44 UT, flare, which was observed above the western solar limb, altitudes of $h_{Lo} = 5900$ km, $h_{M1} = 4000$ km, and $h_{M2} = 3500$ km were measured for the centroids of one hard X-ray footpoint observed with the *Yohkoh/HXT (Lo, M1, M2)* channels (Aschwanden et al. 1999a). Similar altitude differences were measured for other limb flares. To determine the density and scale height of the thick target we use the definition of the stopping depth (including scattering) given by Brown (1972) or Emslie (1983),

$$N_s = \mu_0 \frac{E_0^2}{3K} \approx 0.92 \cdot 10^{17} E_{keV}^2 \ \ [\mathrm{cm}^{-2}] \, , \tag{127}$$

with E_0 the initial electron energy, μ_0 the initial pitch angle, and the constant $K = 2\pi e^4 \Lambda \approx 9.3 \cdot 10^{-36}$ (cgs-units). For the characteristic electron energies $E_i \approx 1.1\varepsilon_i$ we find for the lowest three *HXT* channels ($\varepsilon_1 = 14$ keV, $\varepsilon_2 = 23$ keV, and $\varepsilon_3 = 33$ keV) the following stopping depths: $N_s = 2.2, 5.9, 12.1 \times 10^{19}$ cm^{-2}. Using an exponential density model for the stopping depth,

$$N_s = \int_{-\infty}^{h_s} n_p(h')\, dh' = \int_{-\infty}^{h_s} n_{p,0} \exp(-\frac{h}{\lambda})\, dz' = n_{p,0} \lambda \exp(-\frac{h_s}{\lambda}) \tag{128}$$

and the observed heights of $h_s = 5900$, 4000, and 3500 km, we find that a base density of $n_{p0} = 2.6 \cdot 10^{12}$ cm^{-3} and a scale height of $\lambda = 1900$ km reproduces the stopping depths (defined by Eq. 127). The inferred thick-target densities at the three stopping heights are then $n_p(h_s)$= 1.1, 3.1, 4.1 $\times 10^{11}$ cm^{-3}. Note that Matsushita et al. (1992) found somehwat larger mean altitudes (h_2=9700, 8700, 7700 km) for the same three *HXT* energy channels. In both observations, nonthermal electrons with increasing energy are stopped in progressively lower altitudes.

Hard X-ray emission $\gtrsim 25$ keV is observed at loop footpoints in most of the flares, which indicates that the propagation path is essentially collisionless during the propagation time. For nonthermal electrons with energies $\lesssim 25$ keV, energy loss in the corona is generally not negligible. Setting the propagation time equal to the collisional deflection time,

$$t^{prop}(E) = \frac{L}{v_{\parallel}(E)} \le t^{defl}(E) = 2 \times 10^8 \left(\frac{E_{keV}^{3/2}}{n_e} \right) \left(\frac{20}{\ln \Lambda} \right) \ [s] \tag{129}$$

yields a lower limit for the particle energy required to cross a propagation path with length L across a density n_e,

$$E \ge 20 \sqrt{ \left(\frac{L}{10^9 \ \mathrm{cm}} \right) \left(\frac{n_e}{10^{11} \ \mathrm{cm}^{-3}} \right) \left(\frac{0.7}{\cos\alpha} \right) } \ [\mathrm{keV}] \tag{130}$$

Therefore, hard X-ray images from *Yohkoh/HXT* show usually in the lowest energy channel (Lo=14-23 keV) hard X-ray emission along the entire length of the flare loops, because electrons at these low energies (below the threshold given in Eq. 130) lose their entire energy before they reach the footpoints.

Energy loss is also significant for trapped particles. So far we treated particle kinematics mainly in the collisionless regime, where adiabatic particle motion is applied. This is justified in low-density regions in the corona, where particle propagation times are much shorter than energy loss times. For particle trapping, however, the collisional time scale for pitch-angle scattering into the losscone, τ^{defl},

$$\tau^{defl} = \frac{v^2}{< \Delta v_{\perp}^2 >} \tag{131}$$

is defined as half of the energy loss time τ^{loss} (Trubnikov 1965; Spitzer 1967; Schmidt 1979),

$$\tau^{loss} = \frac{E^2}{< \Delta E^2 >} \approx 2\tau^{defl} \; . \tag{132}$$

Thus, trapped particles lose a significant fraction of their energy in the trap before they precipitate, a second-order effect that has been ignored in the previous deconvolution of trapping times (Eqs. 97–103) and trap densities (Fig. 80).

7. Gamma Ray Emission

In this Section we review briefly theory and observations at gamma-ray wavelengths that are relevant for the diagnostics of particle acceleration and propagation in solar flares. Recent reviews on this subject can be found in Ramaty (1986), Ramaty and Mandzhavidze (1994, 2001), Hudson and Ryan (1995), Chupp (1996), Share, Murphy, and Ryan (1997), Trottet and Vilmer (1997), Vestrand and Miller (1999), and Share and Murphy (2000).

7.1. Gamma Ray Emission Processes

The probably most concisest overview of gamma-ray emission processes occurring in solar flares is given in Ramaty and Mandzhavidze (1994), listing all the relevant high energy photon and neutron production mechanisms that have been identified in flares. We reproduce it here in Table IX.

There are at least six different types of emission seen in gamma-ray wavelengths: continuum emission, nuclear deexcitation lines, neutron capture line, positron annihilation line, pion decay radiation, and neutrons. An example of a gamma-ray spectrum is shown in Fig. 86, where the different line and continuum components have been fitted together to the observed gamma ray spectrum.

TABLE IX

High Energy Photon and Neutron Production Mechanisms (from Ramaty and Mandzhavidze, 1994)

Emissions	Processes	Observed photons or neutrons	Primary ion or electron energy range
Continuum	Primary electron bremsstrahlung	20 keV - 1 MeV >10 MeV	20 keV - 1 GeV
Nuclear deexcitation lines	Accelerated ion interactions, e.g. ^{4}He(α,n)^{7}Be* ^{4}He(α,p)^{7}Li* ^{20}Ne(p,p')^{20}Ne* ^{12}C(p,p')^{12}C* ^{16}O(p,p')^{12}O*	Lines at e.g. 0.429 MeV 0.478 MeV 1.634 MeV 4.438 MeV 6.129 MeV	1-100 MeV/nucl
Neutron capture line	Neutron production by accelerated ions followed by ^{1}H(n,γ)^{2}H	Line at 2.223 MeV	1-100 MeV/nucl
Positron annihilation radiation	β^+ Emitter or π^+ production by accelerated ions, e.g. ^{12}C(p,pn)$^{11} \mapsto ^1 1B + e^+ + \nu$ p+p$\mapsto \pi^+ ..., \pi^+ + \mapsto \mu^+ \mapsto e^+$ followed by $e^+ + e^- \mapsto 2\gamma$ $e^+ + e^- \mapsto Ps + h\nu$ or $e^+ +^1 H \mapsto Ps + p$ Ps$\mapsto 2\gamma, 3\gamma$	Line at 0.511 MeV Orthopositronium Continuum $<$ 511 keV	1-100 MeV/nucl
Pion decay radiation	π^0 and π^+ production by accelerated particles, e.g. p+p$\mapsto \pi^0, \pi^{pm}, ...$ followed by $\pi^0 \mapsto 2\gamma, \pi^\pm \mapsto \mu^\pm \mapsto e^\pm$ $e^+ \mapsto \gamma_{brems}, \gamma_{ann}$ in flight $e^- \mapsto \gamma_{brems}$	10 MeV-3 GeV	0.2-5 GeV
Neutrons	Accelerated particle interactions, e.g. ^{4}He(p,pn)^{3}He $p + p \mapsto \pi + n + ...$ $^{22}Ne(\alpha, n)^{25}Mg$	neutrons in space (10-500 MeV)	10 MeV - 1 GeV
		neutron induced atmospheric cascades (0.1-10 GeV)	0.1-10 GeV
		Neutron decay protons in space (20-200 MeV)	20-400 MeV

1. *Bremsstrahlung continuum:* Continuum emission, which can extend up to 1 GeV, is produced primarily by bremsstrahlung from 20 keV-1 GeV electrons, which precipitate from the corona and lose their energy by collisions with chromospheric protons, ions, and electrons. The similarity of the time profiles in hard X-rays and gamma-rays suggests that they come from the same coronal acceleration source.
2. *Nuclear deexcitation lines:* Most of the line emission ($\epsilon \approx 0.5 - 8$ MeV) is produced by protons and ions accelerated in the corona which precipitate to the chromosphere, where they collide with other ions and produce nuclear deexcitation lines (e.g. ^{56}Fe at 0.847 MeV, ^{24}Mg at 1.369 MeV, ^{20}Ne at 1.634 MeV, ^{28}Si at 1.779 MeV, ^{12}C at 4.439 MeV, or ^{16}O at 6.129 MeV), or capture neutrons. Narrow lines result from the bombardment of chromospheric nuclei by accelerated protons and α particles, while broad lines result from the inverse reaction in which accelerated C and heavier nuclei collide with ambient H and He. The broadening of the deexitation lines is a consequence of the Doppler shift in the rest frame of the excited nuclei.
3. *Neutron capture line:* The 2.2 MeV line emission is not a prompt process, it is emitted only after neutrons become thermalized in the chromosphere and become captured by protons to produce deuterium, which has a binding energy of 2.223 MeV, so that a delayed photon with energy of $h\nu = 2.223$ MeV is emitted.
4. *Positron annihilation:* Positrons in solar flares are produced by the decay of radioactive nuclei and charged pions. Annihilation of positrons with electrons produces the 511 keV line emission.
5. *Pion decay radiation:* Above 10 MeV there is, besides the electron bremsstrahlung, also significant pion decay radiation detected. Charged and neutral pions (π^+, π^- mesons) are produced by collisions among protons and ions (with energies $\gtrsim$ 300 MeV/nucleon) in the chromosphere, which then decay subsequently into muons (μ^+, μ^-). The secondary electrons and positrons produce bremsstrahlung, while the neutral pions decay electromagnetically into two γ rays, each with a (Doppler-broadened) energy of 67 MeV.
6. *Neutrons:* Neutrons are produced by interactions between accelerated ions and protons, mostly in chromospheric precipitation sites. Neutrons that escape the chromospheric flare site can propagate directly to the Earth, unimpeded by the heliospheric magnetic field, and produce atmospheric showers (cascades) that are detectable on ground.

7.2. Acceleration and Propagation of Protons

If gamma-ray emission is detected during flares, both electrons and ions are involved, so the question arises what are the differences in the acceleration properties for both species, regarding the acceleration times, efficiency, and maximum obtained energies. In well-observed flares, the data clearly show that both species

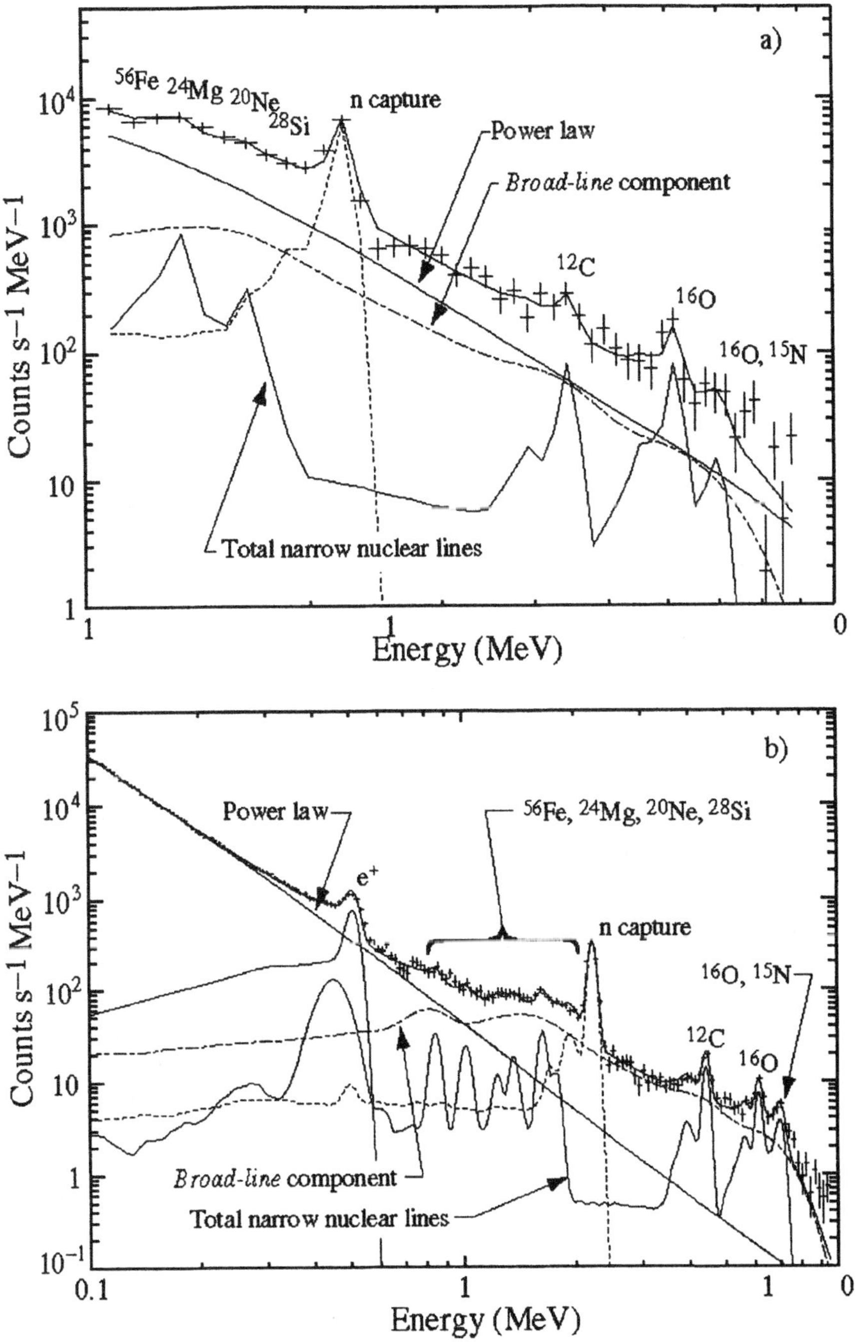

Figure 86. Count spectra and best fits obtained using the photon models described in Murphy et al. (1997). (a) An 8 s *CGRO/OSSE* spectrum accumulated soon after the peak of emission in the first orbit by detector 2 while pointing 45^0 off the Sun. The spectrum has been rebinned into larger energy intervals to improve the statistics for fitting. (b) A 2-minute spectrum accumulated late in the first orbit by detector 1 while pointing at the Sun [from Murphy et al. 1997].

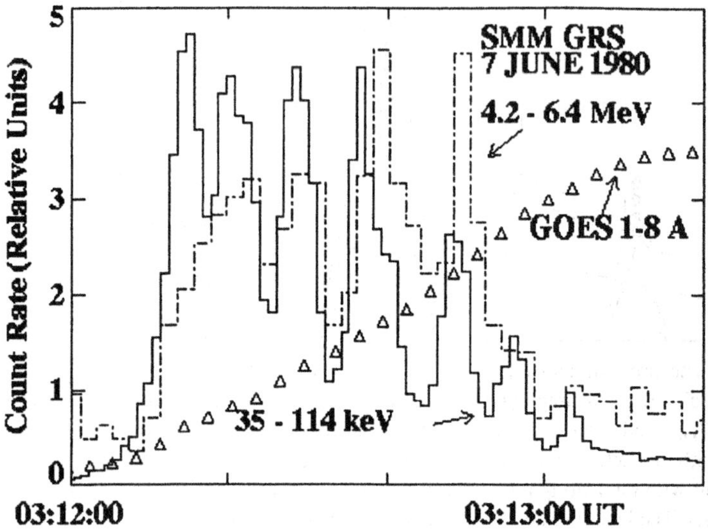

Figure 87. Time profiles for a flare showing near- coincidence impulsive peaks in 35-114 keV hard X-rays (from energetic electrons) and 4.2-6.4 MeV gamma-rays (from energetic ions). The time binning is 1 s. Note that the gamma-ray pulses are delayed by only $\approx 1 - 2$ s each [from Forrest and Chupp, 1983].

of particles, electrons and ions, are accelerated to relativistic velocities relatively promptly (within a few seconds), and nearly simultaneously (Forrest and Chupp 1983). This is illustrated in Fig. 87 for the 1980-Jun-7 flare, which consists of a sequence of 7 hard X-ray pulses, each one (except for the first one) followed by a gamma-ray pulse detected in the energy range of 4.2-6.4 MeV, with only a slight delay of about $\approx 1 - 2$ seconds. Forrest and Chupp (1983) found that the peaks in hard X-rays and gamma rays during the two intense flares of 1980-Jun7 and 1980-Jun-21 coincided within ± 2.2 and ± 0.8 s. Also in the 1992-Feb-8 flare, individual peaks between 40 keV and 40 MeV coincided within ± 1 s (Kane et al. 1986). It is generally argued that this observed delay of $\lesssim 2$ s represents an upper limit for the acceleration time scale of protons and ions (that are responsible for the 4.2-6.4 MeV gamma-ray line emission). In Section 6.4 we found that only acceleration times that are significantly smaller than the electron propagation times (from the coronal acceleration site to the chromospheric hard X-ray emission site), $t^{acc}(E) \ll t^{prop}(E)$, can satisfy the observed energy-dependent hard X-ray delays. This necessarily also implies that the spatial extent x^{acc} of the acceleration region is significantly smaller than the propagation path length L, i.e. $x^{acc} \ll L$. Based on this argument we can assume that both electrons and ions are accelerated in the same coronal acceleration region, regardless of the type of acceleration mechanism, such as small-scale electric fields, stochastic acceleration, or shocks (Section 5).

For co-spatial acceleration, say in the cusp near a coronal reconnection point, accelerated electrons and protons have to propagate about the same distance L to the flare loop footpoints, where electrons produce thick-target bremsstrahlung in hard X-rays, while the protons and α-particles produce nuclear deexcitation lines with ambient chromospheric ions in gamma-rays. What is the relative timing between hard X-rays and gamma-rays for this simplest scenario? For hard X-rays, e.g. for the $\epsilon = 35 - 114$ keV channel as shown in Fig. 87, we know that the hard X-ray photons with energy ϵ are produced by relativistic electrons with kinetic energies of $E = \epsilon \times q_E \approx (35 - 140 \text{ keV}) \times 1.124 \approx 40 - 160$ keV (from the photon-to-electron energy conversion factor for bremsstrahlung, defined in Eq. 69). For gamma-ray emission, however, which are dominated by nuclear deexcitation lines of ^{12}C and ^{16}O ions in the $\epsilon = 4.2 - 6.4$ MeV energy range (e.g. Hudson et al. 1980), hit by precipitating protons (see Table IX), the kinetic energy of the accelerated protons can have a large possible range. Because the standard composition of coronal plasma is made of 90% protons and 10% helium (but can be up to 50%, see Share and Murphy, 1997), we consider only protons here. As a first approximation, let us assume equipartition for the kinetic energy of the accelerated electrons and protons, i.e. $E_p = E_e$,

$$E_p = m_p c^2 \gamma_p = E_e = m_e c^2 \gamma_e \,, \tag{133}$$

which yields the following velocity ratio β_p/β_e for protons and electrons,

$$\frac{\beta_p}{\beta_e} = \frac{\sqrt{1 - [\frac{E_p}{m_p c^2} + 1]^{-2}}}{\sqrt{1 - [\frac{E_e}{m_e c^2} + 1]^{-2}}} \approx \sqrt{\frac{E_p}{E_e} \cdot \frac{m_e}{m_p}} = \frac{1}{43}\sqrt{\frac{E_p}{E_e}} \,, \tag{134}$$

where the right-hand approximation applies to the non-relativistic (or mildly relativistic) case. For equipartition, the velocity ratio is $\beta_p/\beta_e \approx \sqrt{m_e/m_p} = \sqrt{1/1836}$ $\approx 1/43$. Therefore, the time-of-flight difference τ_{ep} between electrons and protons over a distance l' is

$$\tau_{ep} = \frac{l'}{c}\left(\frac{1}{\beta_e} - \frac{1}{\beta_p}\right) = \frac{l'}{\beta_e c}\left(1 - \sqrt{\frac{E_e}{E_p} \cdot \frac{m_p}{m_e}}\right) \,, \tag{135}$$

which, in the case of equipartition, reduces to

$$\tau_{ep}^{equip} \approx -42 \frac{l'}{\beta_e c} \; [s] \,. \tag{136}$$

Given the example of the flare shown in Fig. 87, the relevant electron velocities for $\epsilon = 35 - 140$ keV are $\beta_e = 0.37 - 0.65$. We do not know the flare loop size for the 1980-Jun-7 flare (Fig. 87), but if we substitute typical flare loop radii observed with *Yohkoh* ($r = 3000 - 25,000$ km, Fig. 12) and the canonical scaling law for electron time-of-flight distances, $l' \approx 1.5r \times (\pi/2)$ (Eqs. 5, 6), we obtain propagation distances of $l' \approx 7000 - 60,000$ km, with a geometric mean of $l' = 20,000$ km. In the case of equipartition we therefore expect an electron-proton delay of $\tau_{ep}^{equip} \approx -42(l'/\beta_e c) \approx -42 \times (20,000 \text{ km}/[0.37\text{-}0.65]300{,}000 \text{ km}) \approx -[4.3 - 7.5]$ s. For

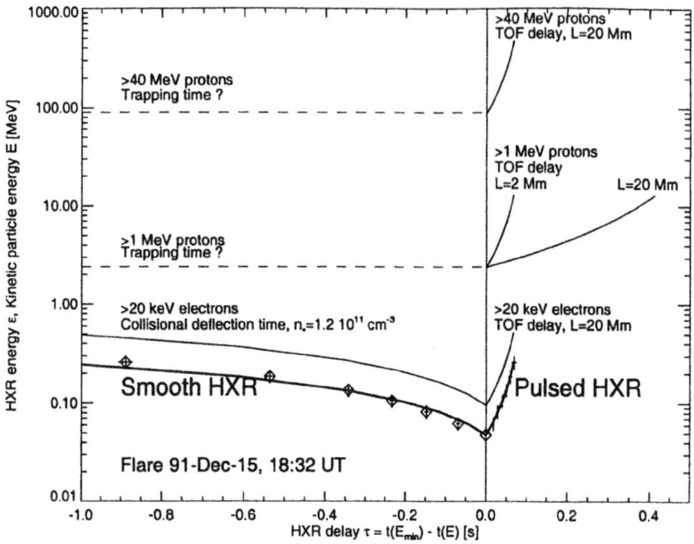

Figure 88. The energy-dependent time delays $\tau(E) = t(40 \text{ keV}) - t(E)$ are shown for a flare observed on 1991 Dec 15, separately measured for the the *pulsed* hard X-ray flux (crosses on right-hand side) and the *smooth* hard X-ray flux (diamonds on left-hand side), in the energy range of 40-300 keV. The delays can be fitted with two models: in terms of electron time-of-flight (TOF) differences (thick line on right-hand side bottom), and in terms of electron collisional deflection time differences (thick line on left-hand side bottom). The electron energies ($E \approx 2\varepsilon$) with the same timing delays as the hard X-ray pulses (observed at energy ε) are indicated with thin lines (bottom part). For comparison, we show also the required timing of > 1 MeV protons (left-hand side middle) and > 40 MeV protons (left-hand side top), if they would be responsible for the > 20 keV hard X-ray emission. The trapping time of > 1 MeV protons (dashed line on left-hand side middle) or > 40 MeV protons (dashed line on left-hand side top) are so large that no energy-dependence can be seen on the displayed time scale of 1 s [from Aschwanden 1996e].

a smaller flare loop with a radius of $r = 3000$ km, the relative delay would amount to $\tau_{ep}^{equip} \approx -[1.5 - 2.5]$ s, which is consistent with the observed delay between hard X-rays and gamma rays. Therefore, we can predict that we expect for a typical range of flare sizes generally gamma-ray delays of $|\tau_{ep}| \approx 1-10$ s, just based on the proton time-of-flight delay from the coronal acceleration site to the chromosphere. So, the gamma-ray delay is a proton propagation delay rather than an acceleration delay. If shorter gamma-ray delays are observed, this could be explained by faster protons than the equipartition principle predicts, and the measured delay can then be used to put an upper limit on the maximum energies of the accelerated protons (with Eq. 135). More stringent time delays could be predicted by estimating the primary proton or ion energies involved in a given nuclear deexcitation line from the nuclear cross-sections, rather than using the equipartition model.

7.3. Inverse Bremsstrahlung of Protons

Both electrons and ions are accelerated in solar flares and carry nonthermal energy from the acceleration site to the chromospheric energy loss site, but the relative amount of energy carried by electrons versus ions is subject of debate. Moreover, proponents of proton-dominated models argue that even the ≥ 20 keV hard X-ray bremsstrahlung is produced either by protons directly (Emslie and Brown 1985; Heristchi 1986; 1987) or by proton-energized electrons (Simnett and Haines 1990). The process where protons produce bremsstrahlung by Coulomb collisions with electrons, is also called *inverse bremsstrahlung*. Models of chromospheric heating by electron beams versus proton beams have also been addressed in Brown et al. (1990).

Early measurements of inelastic-scattering γ-rays (4.4 MeV from ^{12}C and 6.1 MeV from ^{16}O) during the 1972-Aug-4 flare were used to estimate the 20-200 MeV proton fluxes, and it was found these protons would produce a six orders of magnitude too small hard X-ray flux in the 10-100 keV energy than observed, which clearly rules out protons as producers of hard X-rays (Hudson 1973).

A kinematic test whether the observed energy-dependent timing delays of 20-200 keV hard X-ray emission can be explained in terms of propagating electrons versus protons is discussed in Aschwanden (1996e). For typical flares, the time delays of fast ($\lesssim 1$ s) *hard X-ray pulses* are consistent with time-of-flight differences of directly precipitating electrons, while the timing delays of the *smooth hard X-ray* flux is consistent with collisional deflection times of trapped electrons. In Aschwanden (1996e) it is shown (see also Fig. 88) that these hard X-ray timing delays cannot be explained either by ≤ 1 MeV protons (as proposed in a model by Simnett and Haines 1990), because of their longer propagation and trapping times, or by ≈ 40 MeV protons (which have the same velocity as ≈ 20 keV electrons), because of their longer trapping times and the excessive fluxes required to generate the hard X-ray. Thus, the hard X-ray timing results clearly rule out protons as the primary generators of ≥ 20 keV hard X-ray emission.

7.4. Relative Proton-to-Electron Acceleration Ratios

The relative ratio of accelerated electrons to protons is not very well known. There are so-called *electron-rich events* that are so intense in electron bremsstrahlung that they obscur gamma-ray line emission (Rieger and Marschhäuser 1990). On the other hand, there are arguments that protons dominate the energy budget (Simnett 1986). So, what is the relative proton-to-electron acceleration ratio? The bulk of the γ-ray line emission is believed to be produced by ions with energies of 10-100 MeV/nucleon that contain only a small fraction of the energy in the > 20 keV electrons (e.g. Lin et al. 2000). However, systematic studies of γ-ray lines observed with *SMM* (Share and Murphy 1995) shows that the 1.634 MeV ^{20}Ne line is unexpectedly enhanced. Because the cross-section for ^{20}Ne has an unusually low energy threshold (≈ 2.5 MeV), this effect may be due to large fluxes of low-

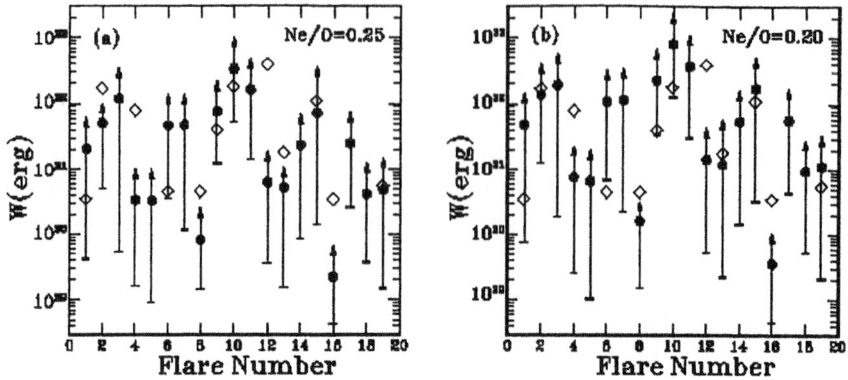

Figure 89. Calculated energy contents in ions and electrons. The solid circles and the lower limits indicated by horizontal bars are, respectively, the ion energy contents assuming low energy cutoffs at 1 MeV/nucleon and at the maximum E_C allowed by the observed Ne/O line fluence ratios. The open diamonds are electron energy contents assuming a cutoff at 20 keV [from Ramaty and Mandzhavidze, 2000].

energy ions with a total energy content perhaps comparable to that in accelerated electrons (Ramaty et al. 1995; Emslie, Brown, and MacKinnon 1997). To estimate the relative proton-to-electron ratio, or the relative energy content of their spectra, an extrapolation of the spectra has to be made in the $E < 1.6$ MeV range. Earlier spectral modeling with Bessel functions yielded a small proton-to-electron energy content, $W_i \ll W_e$ (Ramaty 1986). However, when more gamma-ray data that constrain the relative Ne/O abundance became available, the spectral fits in the gamma-ray range favored an unbroken powerlaw, which led to a much larger ion energy content, comparable with equipartition, $W_i \approx W_e$ (Ramaty et al. 1995). The total energy contained in ions and electrons were calculated for 19 flares, based on a single-powerlaw spectrum with the slope constrained by the Ne/O ratio (Fig. 89). For a range of Ne/O=0.2-0.25, the energy in ions seems to be larger than the energy in ≥ 20 keV electrons in about 5 out of 19 flares.

7.5. Long-Term Trapping of High-Energy Particles

In Section 6.6 we described the trapping of hard X-ray-producing electrons. The main result from observations is that electron trapping seems to be governed by collisional pitch-angle scattering in coronal parts of flare loops, with typical densities of $n_e \approx 10^{11}$ cm^{-3} near the mirror points. The trapping times scale with the collisional deflection time, which can be expressed in dimensionless parameters by (Eq. 104),

$$\tau^{trap}(E) \approx \tau^{defl}(E) \approx 0.95 \left(\frac{E}{100 \text{ keV}} \right)^{3/2} \left(\frac{10^{11} \text{ cm}^{-3}}{n_e} \right) \left(\frac{20}{\ln \Lambda} \right) \text{ [s]} . \quad (137)$$

Although the collisional deflection time represents an upper limit on trapping times, which could be significantly shorter if strong wave turbulence would be present, the

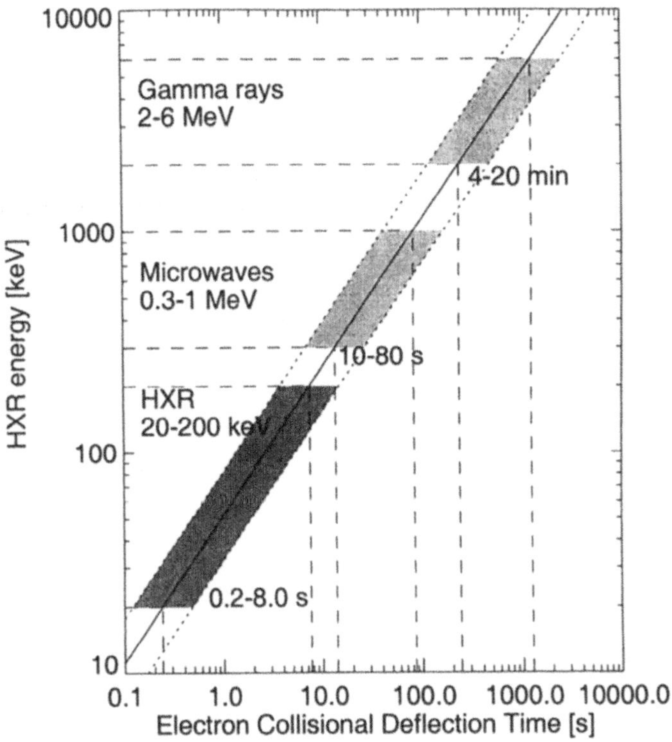

Figure 90. Upper limits of trapping times as function of energy, based on the collisional deflection time, extrapolated from our hard X-ray delay measurements in the 20-200 keV range. The shaded areas apply to trap densities of $n_e \approx 10^{11}$ cm^{-3} (within a factor of 2). The electron energies responsible for the indicated hard X-rays, microwave gyrosynchrotron, and γ-ray emissions are approximately a factor of 2 higher than the indicated photon energies (y-axis) [from Aschwanden et al. 1997a].

observed delays do not support the model of pitch-angle scattering in the *strong-diffusion regime* (e.g. as modeled for turbulent trapping by Ryan and Lee 1991). Thus, trapping times for $E = 20-200$ keV hard X-ray-producing electrons amount only to $\tau^{trap} \approx 0.2-2.0$ s in typical flare loops (see also measurements in Fig. 79). If we extrapolate this trapping time to particles with higher energies, e.g. to gyro-synchrotron-producing high-relativistic electrons ($E \approx 0.3 - 1.0$ MeV) detected at microwaves, we expect trapping times of $\tau^{trap} \approx 3 - 10$ s, and for gamma-ray-producing electrons ($E \approx 1 - 100$ MeV) we expect $\tau^{trap} = 10 - 1000$ s, i.e. up to a maximum of 20 minutes (Fig. 90). In large-scale traps in the upper corona, where the density drops down to $n_e \approx 10^9$ cm^{-3}, gamma-ray-producing electrons with $E = 1 - 100$ MeV could be trapped up to $\tau^{trap} \approx 0.3 - 30$ hours, if there is no other pitch-angle scattering mechanism present.

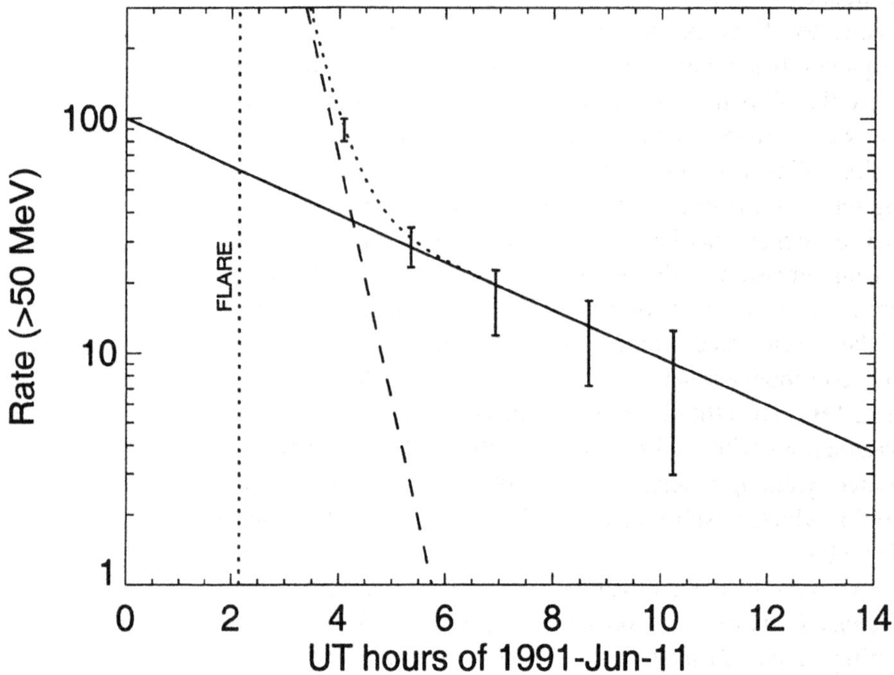

Figure 91. Time profile of the solar emission > 50 MeV on 1991-Jun-11. The two-component fit to the time profile yield e-folding times of 25 and 255 minutes [from Kanbach et al. 1993].

However, for such long trapping times, there may be some containment problems due to the magnetic gradient and curvature drifts (e.g. Sturrock 1994, p.53),

$$\mathbf{v}_D = \frac{\gamma mc}{qB} \left(\frac{v_\parallel^2}{B} (\mathbf{B} \times \mathbf{k}) + \frac{1}{2} \frac{v_\perp^2}{B^2} (\mathbf{B} \times \nabla \mathbf{B}) \right) , \tag{138}$$

where **k** is the curvature vector. This confinement problem can be cured, like in *tokamaks*, by a compensating electric field that produces a counter-acting $\mathbf{E} \times \mathbf{B}$-drift, or by a twisted field, that satisfies the force-free equilibrium equation $\mathbf{B} \times (\nabla \times \mathbf{B}) = 0$, or

$$\nabla \times \mathbf{B} = \lambda \mathbf{B} . \tag{139}$$

Particle orbits in such force-free fields have been simulated by Lau, Northrop, and Finn (1993), and it was found that long-term containment of energetic protons in a coronal loop is possible if the magnetic field lines have enough twist, i.e. $\approx 2\pi$ between the mirror points of a bounce orbit. It was also calculated that the amount of matter encoutered by a 1 GeV proton (the grammage) is not that high that the protons loses too much kinetic energy during several hours trapping time to become incapable for pion production.

Observations of long-duration gamma-ray flares with >25 MeV emission extended over several hours, much longer than the impulsive phase seen in hard X-

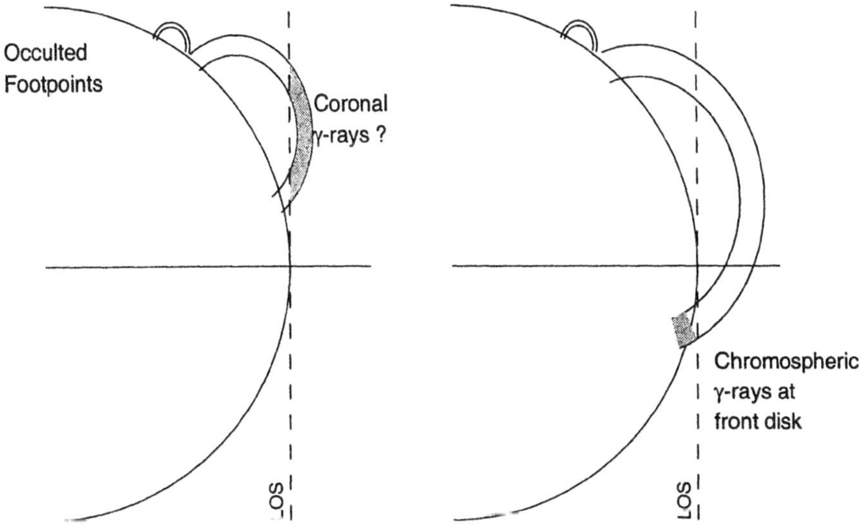

Figure 92. Two scenarios to explain γ-ray emission from behind-the-limb flares. *Left:* Because all footpoints of the flare loops are occulted behind the limb, γ-ray emission was postulated to originate from a non-occulted coronal part of a large-scale trapping flare loop. *Right:* Although the primary flare site is behind the limb, some large-scale trapping flare loop could connect to the front side, where γ-ray emission is produced in the chromospheric footpoint.

rays (typically a few minutes), was interpreted as evidence for prolonged acceleration of high-energy particles, requiring a distinct second-phase acceleration process (Bai 1982) in addition to the impulsive phase. However, an interpretation in terms of trapping in large-scale (low-density) loops is actually more plausible. In the 1991-Jun-11 flare, 50 MeV-2 GeV γ-ray emission was observed by *CGRO/EGRET* (Fig. 91) for more than 8 hours (Kanbach et al. 1993). Since gamma-ray emission in this energy range is dominated by pion decay, either a continuous source of proton acceleration or a long-term trapping of accelerated protons is needed. Ramaty and Mandzhavidze (1994) summarize that the following conditions are required for an interpretation in terms of long-term trapping: (1) low level of plasma turbulence and relatively high mirror ratio ($W_A < 2 \times 10^{-8}$ ergs cm^{-3} for $B_p/B_c = 50$) to prevent the fast precipitation of the particles through the loss cones; (2) matter density in the coronal part of the loop $n_c < 5 \times 10^{10}$ cm^{-3} to prevent Coulomb and nuclear losses. Continuous or second-step acceleration over 8 hours can be fairly excluded, because neither the lack of hard X-ray emission from lower-energy particles nor the smooth exponential decay of the 8-hour gamma-ray emission could be explained in a natural way.

7.6. Gamma-rays from Behind-the-Limb Flares

Because gamma-rays in the 2-7 MeV range is generally dominated by nuclear deexcitation lines which require chromospheric ion densities, some observations of occulted (behind-the-limb) flare sites raised the problem how the gamma-ray emission could be explained for such flares. The issue is whether gamma-rays are produced in an extended coronal trap region connected to the flare site (Fig. 92 left), or whether a large-scale loop was connecting the behind-the-limb flare site with a remote footpoint on the front side of the solar disk, where it could produce the usual chromospheric gamma-ray emission (Fig. 92, right). A special class of gradual hard X-ray emission originating above occulted flares was already reported earlier (Hudson 1978) before gamma-ray observations were available, interpreted as large-scale trapping structures that are capable to trap electrons and protons for extended periods of time.

Gamma-ray emission from an occulted flare was observed on 1989-Sep-29, 10:47 UT, which was associated with NOAA region 5698 at position W105^0, so about 15^0 behind the limb (Vestrand and Forrest, 1993). The gamma-ray emission showed electron bremsstrahlung continuum, a positron annihilation line, prompt nuclear emission in the 1-8 MeV range, and a neutron capture line at 2.22 MeV that was surprisingly strong. Since the bulk of the prompt gamma-ray emission requires densities of $n_H > 10^{12}$ cm^{-3} to efficiently capture neutrons within their 900 s lifetime, it was concluded that the observations require a spatially extended flare loop that extends from behind-the-limb to the front side, covering a distance of $\approx 30^0$ heliographic degrees (i.e. $\approx 370,000$ km).

Intense gamma-ray emission of prompt lines in the 1.2 and 4-7 MeV energy range was also reported from PHEBUS on GRANAT during the 1991-Jun-1, 14:46 UT, GOES class X12.0 event, associated with NOAA region 6659, located 7^0 behind the East limb (Barat et al. 1994). This corresponds to occultation heights ranging from 3000 to 7000 km above the photosphere. Although prompt gamma-ray line emission require densities of $n_H > 10^{12}$ cm^{-3}, which places the gamma-ray line emission region at a height of $<$ 1500 km, it was concluded that the gamma-ray emission comes from coronal heights $>$ 3000 km, because no 2.2 MeV neutron capture line was observed, that could reveal a possible front-side footpoint connected with the behind-the-limb flare site. Modeling of gamma-ray line emission with a thin-target model with densities of $n_e \approx 1 - 5 \times 10^{11}$ cm^{-3} could reproduce the observed gamma-ray line fluxes (Trottet et al. 1996), although a very hard spectrum for the accelerated particles is required (Murphy et al. 1999).

An electron-dominated event was observed during the 1991-Jun-30 behind-the-limb flare, with significant emission in the 10-100 MeV range from an occultation height of $\gtrsim 10^4$ km, but no gamma-ray line emission in the 2-7 MeV range was detected (Vilmer et al. 1999).

7.7. Pitch-Angle Distribution of High-Energy Particles

The angular (pitch-angle) distribution of accelerated particles contains information on the directivity of the acceleration process or on particle propagation in magnetic mirror fields. For instance, the different spectral line profiles of the 0.429 and 0.478 MeV gamma-ray lines from ^{7}Be and ^{7}Li (produced by accelerated α-

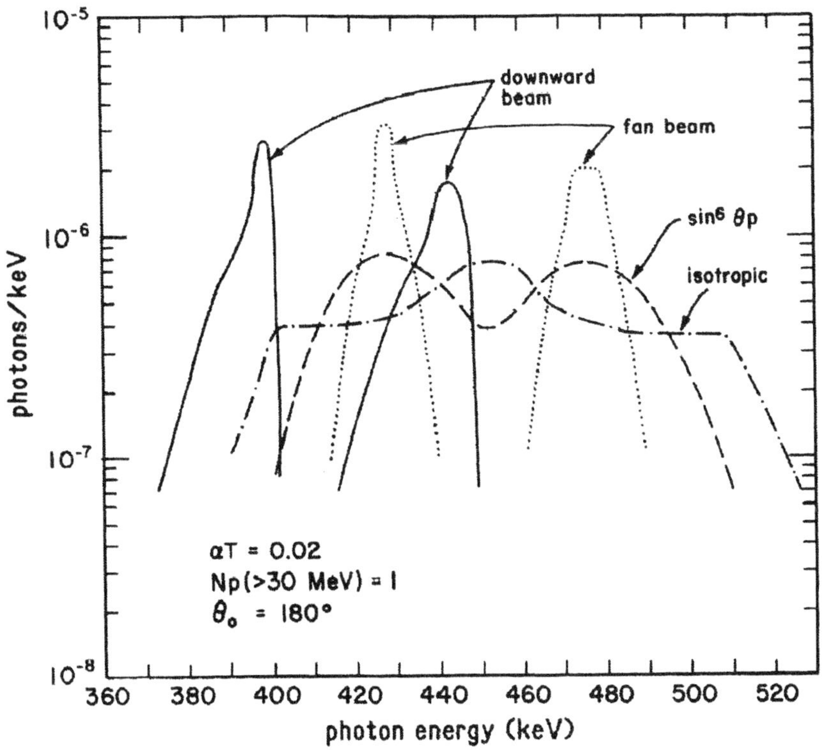

Figure 93. Gamma-ray lineprofiles from $\alpha - \alpha$ fusion reactions in flares occurring at the center of the solar disk for four different angular distributions of accelerated α-particles [from Murphy, Kozlovsky, and Ramaty, 1988].

particles colliding with ambient α particles in the chromosphere) have been calculated for different pitch-angle distributions by Murphy, Kozlovsky, and Ramaty (1988). Fig. 93 shows the different line shapes for four different pitch-angle distributions: for isotropic, downward beam, losscones (also called *fan-beam*), and for a $\sin^6$-distribution. Clearly, the line widths and centroids are quite different for these four distributions and thus provide a sensitive diagnostic for the anisotropy of accelerated α-particles. Recent fits to the $\alpha - \alpha$ lines in the 0.3-0.7 MeV range have been performed with this method in 19 flares by Share and Murphy (1997) and it was found that a downward beam distribution could be ruled out with high confidence in 4 flares, while losscone distributions and isotropic distributions provide acceptable fits to the data. The interpretation of this result depends on magnetic field models at the loop footpoints. For a high magnetic mirror ratio one would expect that the precipitating α-particles have large perpendicular velocities, i.e. losscone distributions, which is consistent with the observed anisotropy. Since beam distributions can be excluded, one might conclude that α-particles with large pitch angles do not lose their energy at higher mirror points, so that a beam distribution would result for lower altitudes in the chromosphere.

Another measure of the anisotropy of accelerated particle distributions is the center-to-limb detection rate of gamma-ray flares. If there is an anisotropic pitch-angle distribution, the resulting gamma-ray emission will also have some directivity. Although we have little simultaneous measurements of the gamma-ray emission from different observing angles, the anisotropy should also be reflected in the center-to-limb distribution of detected events on a statistical basis. Vestrand et al. (1987) performed statistics on 150 gamma-ray flares detected at > 300 keV with *SMM Gamma-Ray Spectrometer (GRS)* and found an excess of gamma-ray events detected near the limb, as well as slightly harder spectrum for limb flares. This observational result is consistent with losscone distributions, and inconsistent with beam distributions, just as it was concluded from $\alpha - \alpha$ line diagnostics (Share and Murphy, 1997). Similar results were found by Li et al. (1994) and Li (1995), who also established that the directivity increases with increasing energy.

A third diagnostic method sensitive to the anisotropy of accelerated particles is the polarization of bremsstrahlung radiation (e.g. Bai and Ramaty, 1978), but measurements are very difficult, so that not much progress has been made with this method (McConnell et al. 1996).

7.8. Interplanetary High-Energy Particles

A small fraction of flare-accelerated high-energy particles can escape into interplanetary space, depending on the magnetic connectivity of the flare site to *open field lines*. The small fraction of escaping particles is due to the fact that majority of the magnetic field lines rooted in active regions close back to the Sun near or further away from active regions, even crossing the equator and connecting with an active region on the opposite hemisphere (so-called *trans-equatorial loops*). However, not all high-energy particles observed in interplanetary space were originally accelerated in the primary energy release site (magnetic reconnection region) of the coronal flare site. Larger flares often produce a coronal mass ejection, which harbors a moving shock front that itself is an efficient particle accelerator. Therefore, high-energy particles detected in-situ in interplanetary space, even when the timing is perfect, are not necessarily witnesses of the primary flare acceleration site. This dichotomy led to two classes of solar energetic particle (SEP) events: *impulsive* and *gradual particle events*. Impulsive events, for which the associated soft X-ray emission is of relatively short duration, have large electron-to-proton ratios (e/p), large ^{3}He/^{4}He ratios, large heavy ion (particularly Fe) to C or O ratios, and charged states corresponding to hotter temperatures than the corona (1-2 MK). High-energy particles from such impulsive SEP events are therefore believed to be accelerated in the primary coronal flare site. Gradual events, on the other hand, for which the soft X-rays last longer, are clearly associated with CMEs, and thus may be accelerated in a CME-associated shock front. The analysis of the 1981-Apr-27 flare, for instance, showed that the composition of accelerated ions resembles more closely to that of impulsive interplanetary particle events with large ^{3}He abundances than that of long-duration particle events (Murphy et al. 1991). This supports the idea that interplanetary events with impulsive time profiles and unusual compositions originate in the flare region itself, whereas gradual interplanetary particle events arise from a shock well separated from the flare. Maximum energies of interplanetary ions have been measured in-situ up to 100 MeV/nucl (Reames, Richardson, and Wenzel 1992; Mazur et al. 1992), but it is often not clear whether these high-energy protons were accelerated in the coronal flare region itself or in interplanetary shocks. For a characterization of flare-related (impulsive) and CME-related (gradual) particle events we refer to Reames (1996).

8. Hard X-Ray Emission

Hard X-ray emission, which is produced by bremsstrahlung from nonthermal electrons with energies $E \gtrsim 20$ keV, undoubtly provides us a key diagnostic of accelerated particles in solar flares. The impulsive flare phase tells us the times of acceleration episodes, the hard X-ray spectrum reveals us the energy spectrum of accelerated particles, energy-dependent time delays reveal us information of time-of-flight and magnetic trapping, and time structures yield us insights into the spatial fragmentation and intermittency of magnetic reconnection processes. Reviews on flare-related hard X-ray emission can be found in Emslie and Rust (1980), Dennis (1985, 1988), De Jager (1986), Dennis, Orwig, and Kiplinger (1987), Vilmer (1987), Dennis and Schwartz (1989), Bai and Sturrock (1989), Brown (1991), Culhane and Jordan (1991), Hudson and Ryan (1995), Aschwanden (1999, 2000b), and Lin (2000).

8.1. Hard X-Ray Bremsstrahlung Emission

The energy range of hard X-ray emission includes $\epsilon \approx 10$ keV-1 MeV, while the lower energy range of $\epsilon \approx 1 - 10$ keV is referred to as soft X-rays, and the higher energy range of $\epsilon \approx 1 - 100$ MeV as gamma rays (Fig. 94).

Bremsstrahlung (also called *braking radiation* or *free-free radiation*) is produced when an electron is deflected in the Coulomb electric field of another charged particle, whereby electromagnetic radiation is emitted as consequence of the acceleration of the electron during the swing-by around the ion (Fig. 95). For mildly relativistic electrons, the loss of energy by bremsstrahlung is small ($\approx 10^{-5}$) compared with the collisional energy loss. Based on the Rutherford (1911) formula for elastic scattering and the Larmor formula for emitted electromagnetic power by an accelerated charged particle, the radiation cross-section for hydrogen, derived by Bethe-Heitler (see e.g. Jackson 1962, p.701ff), is

$$\chi(\epsilon, E) = \frac{8}{3} \frac{r_0^2}{137} \frac{m_e c^2}{\varepsilon E} \log \frac{1 + \sqrt{1 - \epsilon/E}}{1 - \sqrt{1 + \epsilon/E}} , \tag{140}$$

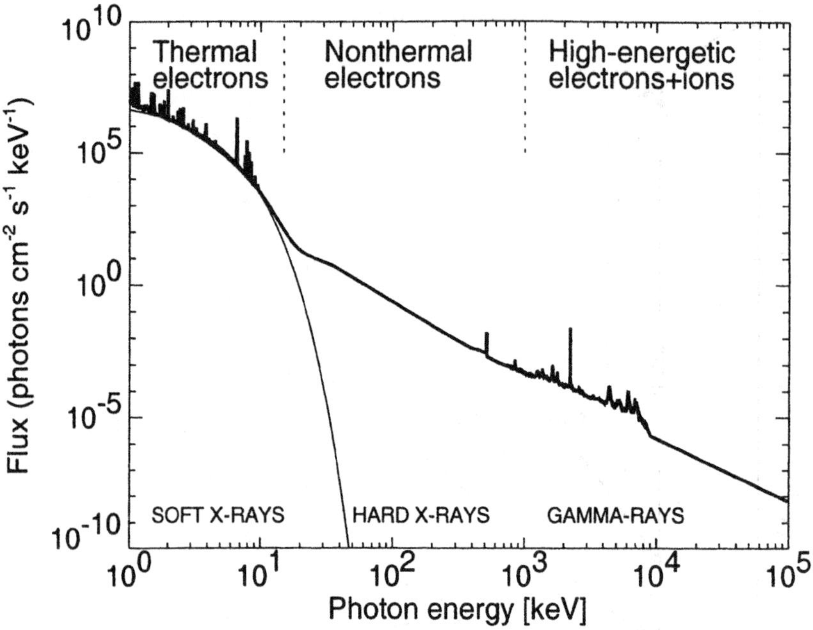

Figure 94. Composite photon spectrum of a large flare, extending from soft X-rays (1-10 keV), hard X-rays (10 keV-1 MeV), to gamma rays (1 MeV-10 GeV), mostly produced by thermal, nonthermal (energetic), or high-energetic electrons. Gamma-ray line emission and parts of the gamma-ray continuum is produced by interactions of protons, neutrons, ions, and pion decay.

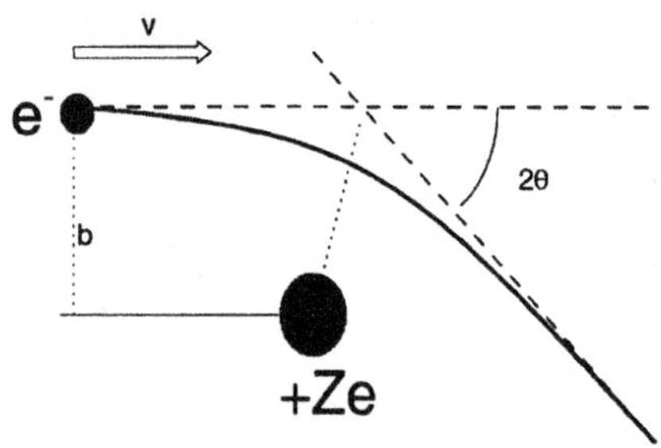

Figure 95. Elastic scattering of an electron (e^-) off a positively charged ion ($+Ze$). The electron moves with velocity v on a path with impact parameter b and is deflected by an angle of 2ϑ, with $\tan \vartheta/2 = Ze^2/(m\mathrm{v}^2 b)$, according to Rutherford.

where $\chi(\epsilon, E)$ is the cross-section differential in photon energy ϵ, while m_e, r_0, and E are the rest mass, radius, and kinetic energy of the electron, and c the speed of light (in cgs units). The total X-ray emission from a volume V is then,

$$\frac{dN_{photons}}{dt\ \Delta\varepsilon} = \int_{\epsilon}^{\infty} \chi(\epsilon, E)\mathrm{v}(E) \left(\int n_p n(E) dV \right) dE \ , \tag{141}$$

where $n(E)$ is the number density of electrons, n_p the number density of ambient protons, and v(E) the velocity of an electron with kinetic energy E. The mean photon count rate (per unit ϵ) at Earth's distance ($R = 1.5 \times 10^{13}$ cm) is then

$$I(\epsilon) = \frac{dN_{photons}}{dt\ \Delta\ \varepsilon\ 4\pi R^2} = \frac{1}{4\pi R^2} n_p \int \chi(\epsilon, E) \mathrm{v}(E) N(E) dE \ , \tag{142}$$

in units of [photons cm^{-2} s^{-1} erg^{-1}].

In the *thick-target model*, electrons with an injection spectrum of $N(E)$ lose their entire kinetic energy in an observed target region (with density n_p) and produce a photon (bremsstrahlung) spectrum $I(\epsilon)$. Often the observed photon spectra in the energy range of $\epsilon \approx 20 - 100$ keV can be approximated with a single powerlaw function,

$$I(\varepsilon) = I_{\epsilon_1} \frac{(\gamma - 1)}{\epsilon_1} \left(\frac{\epsilon}{\epsilon_1} \right)^{-\gamma} \quad . \tag{143}$$

Brown (1971) calculated the inversion of the bremsstrahlung photon spectrum $I(\varepsilon)$ and found a solution for the electron injection spectrum $N(E)$, which can be approximated by a powerlaw function too, but with a steeper slope,

$$N(E) = 2.68 \times 10^{33} \gamma^2 (\gamma - 1)^3 B(\gamma - \frac{1}{2}, \frac{3}{2}) \frac{I_{\epsilon_1}}{\epsilon_1^2} \left(\frac{E}{\epsilon_1} \right)^{-(\gamma+1)} \tag{144}$$

where $B(p, q) = \int_0^1 u^{p-1}(1-u)^{q-1} du$ is the Beta-function. In the standard thick-target model, the corona is approximated with a collisionless plasma, while the electrons lose most of their energy in chromospheric thick-target regions.

The model of thin-target bremsstrahlung, where the electrons lose only a small part of their energy (Brown and Melrose, 1977; Kane et al. 1980), while they escape from the corona without depositing all their energy, is not relevant for most of the flares, because most of the magnetic field lines emanating from a flare site are *closed field lines*, so that both directly-precipitating as well as trapped particles ultimately produce chromospheric thick-target bremsstrahlung. Open field lines are only involved in *tripolar reconnection events*, which seem to produce preferentially soft X-ray jets, but not flares. The only exception are behind-the-limb flares, where the thick-target emission site is occulted, so that only thin-target emission from the non-occulted trapping region is observed (e.g. Trottet et al. 1996).

In earlier years, it was not clear whether hard X-rays were produced by bremsstrahlung from thermal or nonthermal electrons (see discussion in Vilmer 1987). If a plasma has temperatures of $T \approx 10^8$ K, bremsstrahlung from thermal electrons produces about the same flux as bremsstrahlung produced by nonthermal electrons incident on a cold target (Brown, 1971, 1972). Today, given the many redundant temperature diagnostics in flares, which all show consistently typical temperatures of $T \approx 10 - 30$ MK, which is 1-2 orders of magnitude below the requirement for thermal bremsstrahlung, we can largely rule out the thermal interpretation of hard

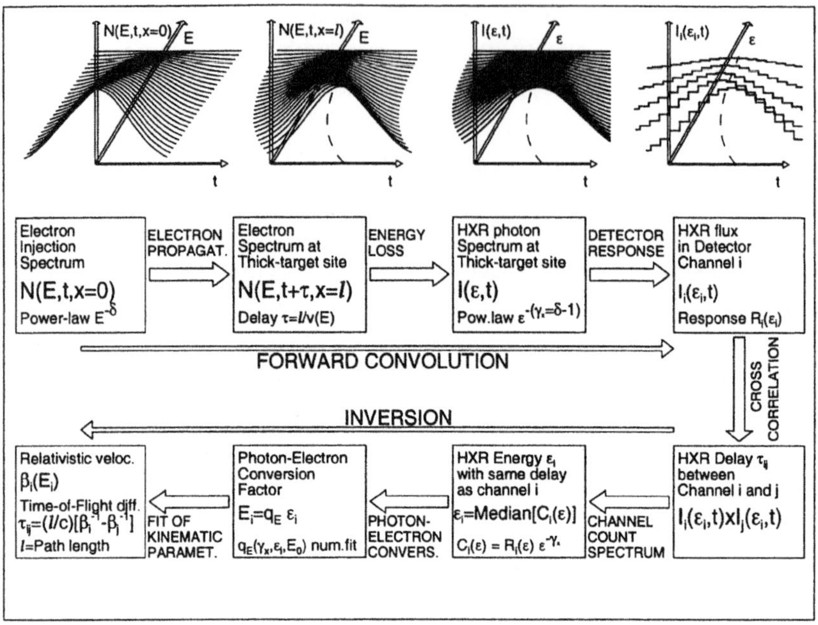

Figure 96. Technique to determine time-of-flight distance l from hard X-ray delay measurements. The upper row of this flow chart illustrates the forward convolution of time-dependent electron spectra N(E,t), while the lower row sketches the steps of the inversion procedure from the hard X-ray delays τ_{ij} to the time-of-flight path length l [from Aschwanden and Schwartz, 1996a].

X-rays for energies $\gtrsim$ 20 keV. A recent study with detailed modeling of the differential emission measure distribution of one of the largest flares (i.e. the Bastille-day flare shown in Figs. 3, 43, 44, 103–106), constrained by comprehensive temperature coverage, exhibited dominant nonthermal emission in the *HXT/M1* channel and higher, i.e. $\gtrsim$ 23 keV (Aschwanden and Alexander 2001).

8.2. Time-dependent Hard X-ray Spectra

8.2.1. *Energy-Dependent Time Delays*

Major progress in the analysis of hard X-ray data has been made over the last five years by deconvolving physical parameters from time-varying hard X-ray spectra $N(E, t)$. The majority of all flares shows fast time structures, e.g. sub-second pulses with a significance of $> 5\sigma$ have been detected in 73% of 640 analyzed flares that were recorded by a burst trigger on *CGRO* (Aschwanden, Schwartz, and Alt 1995a). These fast time structures seen in hard X-rays reflect the intermittent injection of accelerated electrons, probably controlled by bursty and intermittent magnetic reconnection episodes (Section 4.2). In addition, there is a bifurcation in the propagation of electrons from the coronal acceleration site to the chromospheric thick-target site: depending on the initial pitch angle, electrons stream directly to the chromosphere or become temporarily trapped (Fig. 63). The propagation times

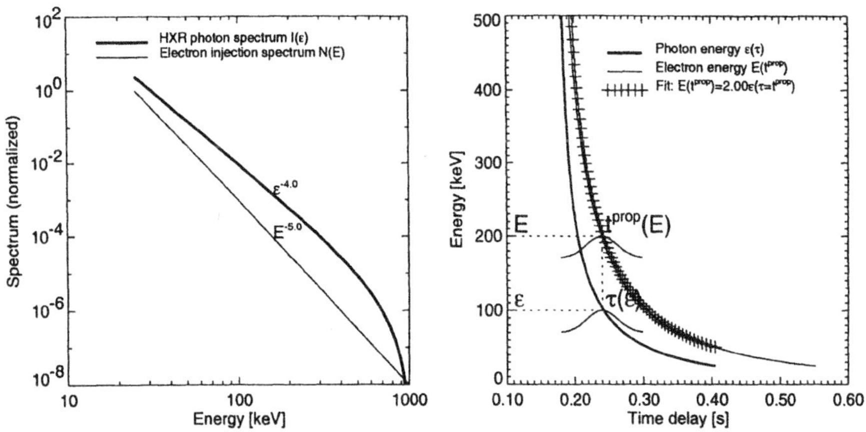

Figure 97. A time-dependent hard X-ray photon spectrum $I(\epsilon, t)$, computed form a time-dependent electron injection spectrum $N(E, t)$, represented in the spectral (left) and temporal domain (right panel). *Left:* The electron injection spectrum $N(E, t) \propto E^{-5}$ (with an upper cutoff energy of $E_0 = 1$ MeV) and the numerically computed hard X-ray photon spectrum $I(\epsilon, t = 0) \propto \epsilon^{-4}$. *Right:* The peak time $\tau(\epsilon)$ of the hard X-ray photon spectrum $I(\epsilon, t)$ is marked with the thick curve. The propagation delay $t^{prop}(E)$ of electrons is shown with the thin curve, which has a similar functional dependence as the hard X-ray delay $\tau(\epsilon)$, and can be fitted by multiplying the photon energies $\epsilon(\tau)$ with a factor of $q_E \approx 2.0$ (crosses). The Gaussians shown at $\epsilon = 100$ keV and $E = 200$ keV symbolize coincident peak times of photons and electrons, but the width of the Gaussian pulses is reduced by a factor of 100 for clarity [from Aschwanden and Schwartz, 1996a].

along these two paths have a different energy-dependence: free-streaming electrons reach the chromosphere after a time-of-flight $t^{TOF} \propto L/v \propto E^{-1/2}$, while trapped electrons encounter an additional delay that is controlled by the collisional deflection time, $t^{trap} \approx t^{defl} \propto E^{3/2}/n_e$. These energy-dependent timing relations introduce temporal variations in hard X-ray spectra that require a forward-fitting or inversion technique to analyze hard X-ray data $N(E, t)$ quantitatively. First, we describe such an inversion technique to analyze the velocity dispersion from electron time-of-flight delays (Aschwanden and Schwartz 1996a).

Let us approximate the nonthermal hard X-ray spectrum with a power law, $I(\epsilon) \propto \epsilon^{-\gamma}$ (Eq. 143). The electron injection spectrum also obeys a powerlaw form $N(E) \propto E^{-\delta}$ with a slope $\delta = \gamma + 1$ (Eq. 144), according to the thick-target model (Brown, 1971). Now we consider a time-dependent electron injection spectrum, which we characterize with a power law in energy and with a Gaussian pulse shape in time, i.e.

$$N(E, t, x = 0) = N_0 E^{-\delta} \exp\left[-\frac{(t - t_0)^2}{2w^2}\right] . \tag{145}$$

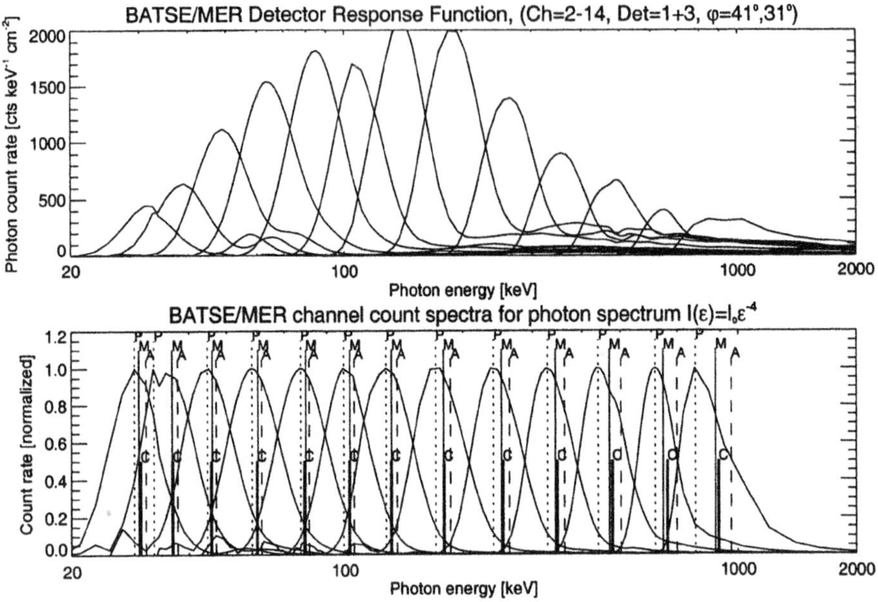

Figure 98. The *BATSE* detector response function $R_i(\epsilon)$ is shown for the 16-channel spectra (MER data type). Only the channels 2-14 useful for time-of-flight analysis are shown. The corresponding channel count spectra $C_i = R_i\epsilon^{-4}$ are shown for an incident photon spectrum of $I(\epsilon) \propto \epsilon^{-4}$ (bottom panel). The peak (P, dotted lines), and average (A, dashed lines) energies of the channel count spectra $C_i(\epsilon)$ are indicated with vertical lines [from Aschwanden and Schwartz, 1996a].

After the electrons propagate over a distance l to the thick-target site, they have a velocity dispersion corresponding to the time-of-flight $t^{TOF} = l/\text{v(E)}$, and the instantaneous electron injection spectrum at $x = l$ is,

$$N(E, t, x = l) = N_0 E^{-\delta} \exp\left[-\frac{(t - t_0 - l/\text{v}[E])^2}{2w^2}\right] . \tag{146}$$

From this instantaneous electron injection spectrum $N(E, t, l)$ at the thick-target site, the hard X-ray photon spectrum $I(\epsilon, t, l)$ can be calculated by convolution with the Bethe-Heitler bremsstrahlung cross-section according to Brown (1971),

$$I(\epsilon, t, l) = I_0 \frac{1}{\epsilon} \int_\epsilon^{E_0} N(E, t, l) \left(\int_\epsilon^E \ln \frac{1 + \sqrt{1 - \epsilon/E'}}{1 - \sqrt{1 - \epsilon/E'}} dE' \right) dE \, , \tag{147}$$

with E_0 being the high-energy cutoff of the electron injection spectrum.

Next, we convolve the hard X-ray photon spectrum at each time t with the instrumental response functions $R_i(\epsilon)$ of the energy channels i to obtain the count rate profiles $C_i(\epsilon, t)$,

$$C_i(\epsilon, t) = I(\epsilon, t, x = l) \otimes R_i(\epsilon) . \tag{148}$$

An example of an instrumental response function is shown in Fig. 98, for the 16-channel spectra of *CGRO/MER*. These four steps to simulate count rate data with

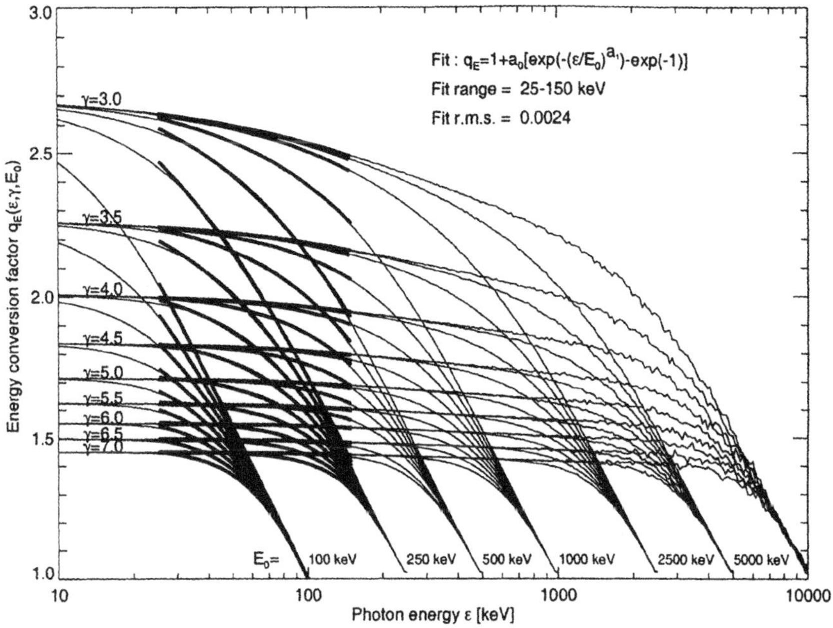

Figure 99. Numerically determined values of the *electron-to-photon conversion factor* $q_E(\gamma, \epsilon, E_0)$ (thin lines) as function of the photon energy ϵ for different spectral indices $\gamma = 3.0, 3.5, ..., 7.0$ and electron cutoff energies $E_0 = 0.1, 0.25, 0.5, 1.0, 2.5, 5.0, 10.0$ MeV. Analytical approximations $q_E(\gamma, \epsilon, E_0)$ described in Aschwanden and Schwartz (1996a) are optimized for the energy range of $\epsilon = 25 - 150$ keV (thick lines).

built-in velocity dispersion are illustrated in the top sequence of Fig. 96, obtained with a forward-convolution method. For analysis of real count rate data, we need to develop an inversion technique that allows us to recover the physical input parameters, in particular the time-of-flight distance l. These steps are sketched in the lower row of the diagram in Fig. 96.

The first step of the analysis is the measurement of relative time-delays between different energy channels i and j by cross-correlation of the count rate time profiles, seeking the maximum of the cross-correlation coefficient $CCC(\tau)$,

$$CCC(\tau) = C_i(\epsilon_i, t) \otimes C_i(\epsilon_i, t + \tau) , \tag{149}$$

which yields the energy-dependent time delays $\tau(\epsilon_i)$. As the response functions in Fig. 98 show, each channel i has a finite energy width. Depending on the photon spectrum, the contribution function (i.e. the product of the response function $R_i(\epsilon)$ with the photon spectrum $C_i(\epsilon)$), peaks at slightly different photon energies ϵ_i. In order to assign to each energy channel a characteristic mono-energetic photon energy, ϵ_i, we use the median value of the contribution function, implicitly defined by

$$\frac{\int_{\epsilon_{median}}^{\infty} C_i(\epsilon) d\epsilon}{\int_0^{\infty} C_i(\epsilon) d\epsilon} = \frac{1}{2} . \tag{150}$$

In Fig. 97 we show the peak time of the gaussian time profile $N(E, t, x = l)$ as function of the electron energy, $t^{prop}(E)$, as well as the peak times of the photon spectrum $I(\epsilon, x)$ as function of the photon energy, $\tau(\epsilon)$, which essentially contains the energy-dependent time delay that is measured with the cross-correlation function (Eq. 149). A convenient concept for our analysis is to specify a quantitative relation between the electron energy E and photon energy ϵ. Of course such a relation depends on the energy ϵ as well as on spectral parameters, such as the powerlaw slope γ and high-energy cutoff E_0 of the photon spectrum. Thus, we define such a *electron-to-photon energy conversion factor* q_E,

$$E(t = t_{peak}) = \epsilon(t = t_{peak}) \times q_E(E, \gamma, E_0) \ . \tag{151}$$

which we already used in Eq. 69. The example shown in Fig. 97, for which an electron injection spectrum with a slope of $\gamma = 4$ and a cutoff of $E_0 = 1$ MeV was used, shows at an electron energy of $E = 200$ keV a ratio of $q_E \approx 2.0$. To quantify this ratio q_E for general applications, this ratio has been numerically computed in a large parameter space, i.e. $E = 10 - 10,000$ keV, $E_0 = 100 - 10,000$ keV, and $\delta = 3.0 - 7.0$. The results are shown in Fig. 99, and quantified with analytical approximation formulae in Aschwanden and Schwartz (1996a).

Finally, we can relate the measured energy-dependent time delays τ_{ij} between energy channels ϵ_i and ϵ_j to the time-of-flight distance l,

$$l = c\tau_{ij} \left(\frac{1}{\beta_i} - \frac{1}{\beta_j} \right)^{-1} \tag{152}$$

using the median energies of the convolution function, ϵ_i (with Eqs. 148 and 150), the *electron-to-photon energy conversion* $E_i = \epsilon_i \times q_E$ (Eq. 151), and the relativistic relation $\beta_i = \mathrm{v}(E_i)/c$ (Eq. 65). These three analysis steps are illustrated in the bottom row of the flow chart (Fig. 96), which represent the inversion steps that are equivalent to the forward convolution shown in the top row of the flow chart (Fig. 96). The self-consistency of this inversion procedure to infer the time-of-flight distance l from hard X-ray time profiles obtained in different energy channels has been tested in Aschwanden and Schwartz (1996a) and was found to be accurate within a few percents.

8.2.2. *Spectral Variation due to Velocity Dispersion*

Equivalently, one can express the velocity dispersion in terms of spectral variations (Brown et al. 1998), rather than as energy-dependent time delays. In the following, we calculate such time-dependent changes in the spectral slope caused by electron time-of-flight delays in order to test whether these effects are observable or related to reported spectral changes.

We consider a Gaussian pulse with width w (as defined in Eq. 145) and a modulation depth q of the total hard X-ray flux. The time-dependent hard X-ray

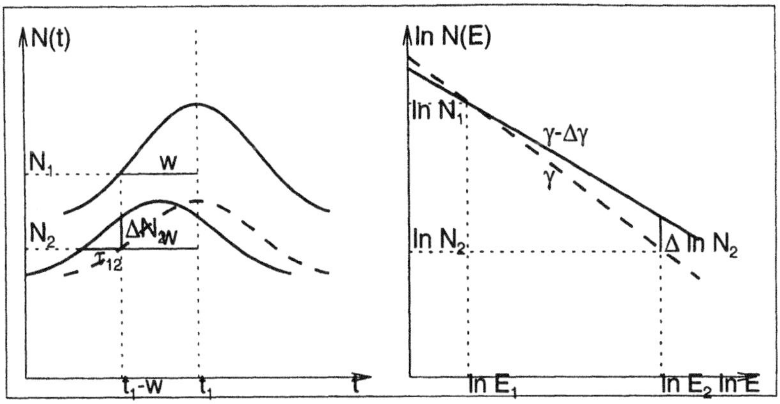

Figure 100. Temporal-spectral relation between time delay (τ_{12} between energies E_1 and E_2) and spectral variation $\Delta\gamma$ for a Gaussian pulse with Gaussian width w. The time profile (left) and spectrum (right) of the undelayed pulse are shown with dashed lines for comparison [from Aschwanden and Schwartz 1996a].

photon spectrum of such a pulse, produced by electrons with a propagation delay of $t^{prop}(E)$, is then

$$N(\epsilon, t) = N_0 \epsilon^{-\delta} \left((1-q) + q \, \exp\left[-\frac{[t - t_0 - t^{prop}(E)]^2}{2w^2} \right] \right) . \tag{153}$$

In Fig. 100, we illustrate the relation of the observed hard X-ray delay $\tau_{12} = t^{prop}(E_1) - t^{prop}(E_2)$ between two energies E_1 and E_2 and the associated variation of the spectral slope γ. For pulses with no time delay $t^{prop}(E) = 0$ (indicated with a dashed time profile at the higher energy E_2 on the left-hand-side of Fig. 100), the hard X-ray spectrum has an identical slope γ during all times. We quantify now the spectrum during the rise time ($t = t_1 - w$) of the Gaussian pulse, shifted by a time interval w with respect to the peak time $t_1 = t_0 + t^{prop}(E_1)$ at the lower energy E_1. Denoting the count rates by $N_1 = N(t_1 - w, E_1)$ and $N_2(t_1 - w, E_2)$ at the two energies E_1 and E_2, the powerlaw slope can be expressed (Fig. 100, right) by

$$\gamma = -\frac{\ln(N_2) - \ln(N_1)}{\ln(E_2) - \ln(E_1)} . \tag{154}$$

Because time-of-flight effects advance the pulse at higher energies, the spectral slope is expected to be slightly flatter (harder) during the rise time, and slightly steeper (softer) during the decay time than during the peak time t_1. We evaluate this spectral change $\Delta\gamma$ during the rise time at time $t = t_1 - w$, as indicated in Fig. 100 (left), where the count rate is increased by an amount ΔN_2 at energy E_2, compared with an undelayed pulse. The corresponding change in the spectral slope is thus (Fig. 100, right)

$$\gamma - \Delta\gamma = -\frac{\ln(N_2 + \Delta N_2) - \ln(N_1)}{\ln(E_2) - \ln(E_1)} \tag{155}$$

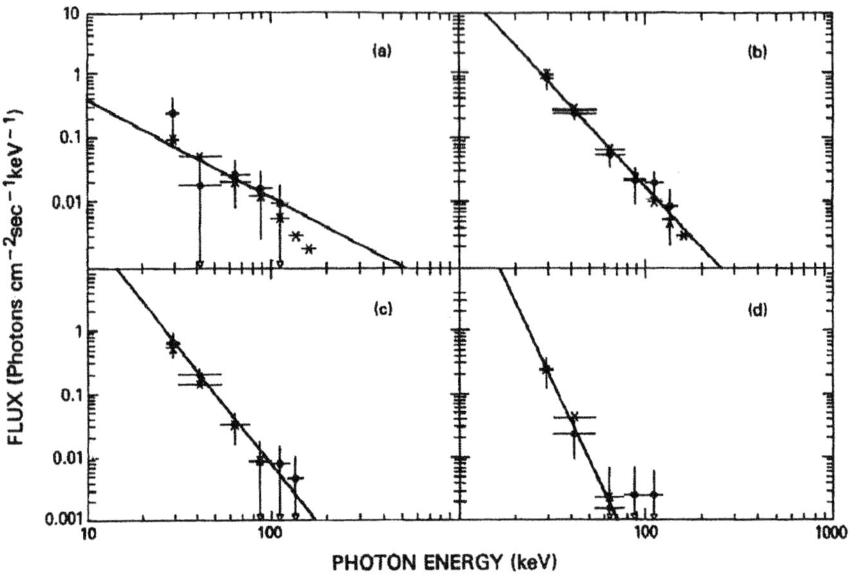

Figure 101. Observed spectra (open circles) and theoretical spectra (crosses) on the rise (a), peak (b), and on subsequent points of the decline (c and d) of a fast hard X-ray spike with a duration of ≈ 0.3 s duration, observed during the 1980-May-10, 1757:46 UT, in the 27-496 keV energy range. Horizontal bars represent the *hard X-ray Burst Spectrometer (HXRBS)* channel widths, while the vertical bars on the observed points reflect 1σ statistical uncertainties [from Kiplinger et al. 1984].

$$= \gamma - \frac{\ln(1 + \Delta N_2/N_2)}{\ln(E_2) - \ln(E_1)} \ . \tag{156}$$

The flux gradient dN/dt during the rise time can be approximated by the ratio of the pulse height qN_2 to the half-pulse width w,

$$\frac{dN}{dt} = \frac{\Delta N_2}{\tau_{12}} \approx \frac{qN_2}{w} \ , \tag{157}$$

which yields an expression for $\Delta N_2/N_2$ that can be inserted into Eq. 153. In this way, for the relative change in the spectral slope $\Delta\gamma/\gamma$ we find the following simple relation:

$$\frac{\Delta\gamma}{\gamma} = \frac{\ln(1 + \Delta N_2/N_2)}{\gamma[\ln(E_2) - \ln(E_1)]} \approx \frac{\Delta N_2/N_2}{\gamma \ln(E_2/E_1)} \approx \frac{\tau_{12}}{w} \frac{q}{\gamma \ln(E_2/E_1)} \ . \tag{158}$$

This relation tells us that the spectral change is proportional to the modulation depth q of the pulse and to the ratio of the time delay to the pulse width, τ_{12}/w. Because both factors are generally small compared with unity, the expected spectral change is even smaller.

As an example, we insert the observed values for the Masuda flare 1992-Jan-13 (Aschwanden et al. 1996b). The hard X-ray pulses were found to have an average modulation depth of $q = 0.1$, a Gaussian width of w=FWHM/2.35 = 3.0 s /2.35 =

1.3 s, a hard X-ray delay of $\tau_{12} = 0.10$ s between the energies $E_1 = 37$ keV and $E_2 = 61$ keV, and a spectral slope of $\gamma = 3.74$. Inserting these values into Eq. 158, we find an expected spectral change of $\Delta\gamma/\gamma = 0.004$ during the rise time of the pulses. This is a very small change that is not detectable within the uncertainties of the photon noise and the unknown true background of a hard X-ray pulse.

There are very few reports on fast spectral changes on time scales of $\lesssim 1$ s, mainly because of the limitations of instrumental sensitivity. One of the fastest spectral changes was reported by Kiplinger et al. (1984), with four successive spectra in time steps of 128 ms (Fig. 101). The hard X-ray spectra of a prominent pulse in the energy range of 25-100 keV was found to be harder during the rise and softer during the decay with respect to the spectrum at the peak time, as expected for a time-of-flight delay. From Figs. 1 and 2 in Kiplinger et al. (1984), we measure a pulse width of $FWHM = 0.3$ s, a modulation depth of $q = 0.83$, and spectral slopes of $\gamma - \Delta\gamma \approx 2.0$ during the rise time and $\gamma \approx 3.5$ during the peak time. If we insert these values (w=0.3 s /2.35 = 0.13 s and $\Delta\gamma = 1.5$) into our relation (Eq. 158), we find that the observed spectral change is consistent with a time-of-flight delay of $\tau_{12} = 120$ ms (between the energies $E_1 = 37$ keV and $E_2 = 61$ keV), similar to that found during the Masuda flare 1992-Jan-13. Kiplinger et al. (1984) fitted two thick-target beam models of Emslie (1983) that include time-of-flight effects and obtained a separation of $l = 14,000 - 30,000$ km between the acceleration site and the thick-target site, which is comparable to the time-of-flight distance found in the Masuda flare within a factor of 2. The main difference between Kiplinger's spike and the hard X-ray spikes in Masuda's flare is that the modulation depth is about 8 times larger and the pulse duration is about 10 times shorter, and this leads to an 80 times stronger spectral variation of the spectral slope, which is clearly detectable. However, we consider Kiplinger's spike as an exception in this respect, since most of the hard X-ray fine structure shows less modulation and longer timescales, leading to a less pronounced (and often not detectable) spectral variation.

In summary, we have shown how energy-dependent (e.g. time-of-flight) delays τ_{ij} translate into spectral variations $\Delta\gamma$ (Eq. 158), and that both descriptions are equivalent. However, energy-dependent time delays are easier to measure (e.g. with simple cross-correlation techniques) than systematic spectral variations before and after pulses (which would require first a break down into pulses and the corresponding rise and decay times). The analysis method described in this Section is therefore a tool of choice to extract energy-dependent time scales $\tau(E)$ from hard X-ray data $N(E, t)$. Although we described the analysis technique in this Section only for time-of-flight delays, it can equally be applied to trapping delays (Eq. 104), or to combined time-of-flight plus trapping delays (Eqs. 97–102). Such an analysis technique with a combined deconvolution of electron time-of-flight and trapping delays is described in Aschwanden (1998a) and performed in Aschwanden, Schwartz, and Dennis (1998b). Alternative interpretations of time-of-flight delays in terms of spectral variations due to the acceleration mechanism are discussed in Brown, Conway, and Aschwanden (1998).

8.3. Hard X-Ray Spectra

A typical hard X-ray spectrum is shown in Fig. 94, which essentially can be characterized by a featureless powerlaw function in the hard X-ray range of $\epsilon \approx 25-400$ keV, bordered by a thermal spectrum at the lower energy, and by gamma-ray lines at higher energies $\epsilon \gtrsim 430$ keV. In small flares, hard X-ray spectra are as soft as with powerlaw slopes of $\gamma \gtrsim 7$, while larger flares show harder spectra with $\gamma \approx 2-3$ (Dennis 1988). The hardest possible spectral slope is given by the high-relativistic limit of the Bethe-Heitler bremsstrahlung cross-section, which is around $\gamma = 1.5$. There is also the typical evolution of the soft-hard-soft pattern during flares (Dennis 1988). The highest temperature of spectrally resolved thermal emission was measured at $T \approx 34$ MK, using germanium-cooled detectors on a balloon-borne experiment (Lin et al. 1981). The thermal component dominates the hard X-ray spectrum only at low energies, up to $\epsilon \lesssim 25$ keV.

So, the hard X-ray spectrum between 25 and 400 keV is generally dominated by thick-target bremsstrahlung from nonthermal electrons. In larger flares, when photon statistics is sufficiently high, deviations from a simple powerlaw spectrum were noticed. Dulk, Kiplinger, and Winglee (1992) characterize hard X-ray spectra as broken-down powerlaws at the time of the peak flux, i.e. the spectra at higher energies $\epsilon \gtrsim 100$ keV are steeper than those at lower energies. They interpret the spectral break point as a possible manifestation of an electric DC field that has a maximum potential drop of 150-200 keV. However, we caution that the exact spectral shape of the electron injection spectrum cannot be sufficiently constrained by the limited energy resolution of the used sodium-iodid detectors from *HXRBS* on *SMM*. The spectral fits in Dulk et al. (1992) were done with *a priori* double powerlaws, without comparisons of other spectral fits, e.g. with Bessel functions, as measured in interplanetary particle spectra. Therefore, spectral shapes are not much constrained at this point, but will be explored with high spectral resolution with the germanium-cooled detectors of *HESSI* (Lin et al. 1998).

Nevertheless, for future modeling of hard X-ray spectra, we might ask why powerlaw (or powerlaw-like) spectra are observed so frequently, there must be some fundamental physics behind it. For thermal spectra, we know that the exponential or Poissonian function of the spectrum is explained with the Maxwellian distribution of random collisions. In contrast, nonthermal particles that produce 25-400 keV emission propagate in a collisionless plasma, and thus obviously are not subject to Maxwellian thermalization. So, what is the fundamental physics that underlies powerlaws? It is a fundamental property of nonlinear processes that their energy distribution is a powerlaw if: (1) the energization process is exponentially growing with time (an "ideal nonlinear" process), and (2) the energization process (or acceleration process here) is interrupted at random time intervals. We can prove that simply with two mathematical steps. An exponentially growing process has a time evolution of

$$E(t) = E_1 \exp\left(\frac{t}{t_G}\right) , \tag{159}$$

where t_G is the growth time constant, E_1 is the initial energy, and t is a variable time interval (here the acceleration time of an individual particle). The second condition of a random (or Poissonian) process for energization times (here acceleration times) has the normalized distribution ($\int_0^{t_{max}} N(t)dt = n$),

$$N(t)\, dt = (\frac{n}{\tau_A}) \frac{1}{[1 - \exp(-t_{max}/\tau_A)]} \exp\Big[- \frac{t}{\tau_A}\Big]\, dt \text{ for } 0 \le t \le t_{max} \tag{160}$$

where τ_A is an e-folding time constant for energization (here acceleration), t_{max} an upper limit, and n the total number of events. From these two equations, the distribution of energies, $N(E)$,

$$N(E)dE = N[t(E)]\, |\frac{dt}{dE}| dE , \tag{161}$$

can directly be derived, by substituting the time dependence from Eq. 159, $t(E) = t_G \ln(E/E_1)$, and its derivative, dt/dE. This leads to a powerlaw distribution for energies,

$$N(E) = \frac{2n}{E_1} \frac{1}{[1 - \exp(-t_{max}/\tau_A)]} (\frac{\tau_G}{\tau_A})(\frac{2E}{E_1} - 1)^{-\delta} , \tag{162}$$

$$\delta = (1 + \frac{\tau_G}{\tau_A}) . \tag{163}$$

This purely mathematical derivation of a powerlaw distribution has been used for many other applications, e.g. for statistics of cosmic rays or flare energies (Rosner and Vaiana 1978; Aschwanden, Dennis, and Benz 1998), but it can equally be applied to the statistics of particle energies, which make up an electron injection spectrum. This general model contains two pieces of information: (1) it explains the observed powerlaws in a natural way in terms of nonlinear statistics, and (2) the value of the powerlaw slope has a relation to the ratio of two physical time scales. For hard X-ray spectra, for which photon spectra with slopes of $\gamma \approx 2 - 7$ have been measured, which correspond to powerlaws of $\delta = \gamma + 1 \approx 3 - 8$ for electron energies, the ratio of time scales is $\tau_G/\tau_A = \delta - 1 = \gamma \approx 2 - 7$. In other words, even in the hardest spectra ($\gamma \approx 2$), the e-folding acceleration time of the acceleration process is only a half growth time, which means that particles very rarely become accelerated exponentially over several growth times. The relatively steep spectra rather suggest highly incoherent acceleration steps, such as stochastic acceleration processes.

8.3.1. *Spectra from First-Order Fermi Acceleration*

Fermi (1949, 1954) derived a powerlaw spectrum for cosmic-ray particles, based on statistical arguments that a head-on collision of a particle with a randomly moving magnetic field is more likely than an overtaking one, so that the average particle will be accelerated. He derived the spectrum from the law of conservation of momentum. If **u** is the velocity of a region of high magnetic field (a magnetic mirror), then the change in particle energy ΔE for one collision of a relativistic particle is (see e.g. Lang 1980, p.474),

$$\Delta E = -2E \frac{\mathbf{u} \cdot \mathbf{v}_{\parallel}}{c^2} , \tag{164}$$

where E is the particle energy and $\mathbf{v}_{\parallel}$ is the parallel velocity of the particle. Since the probability of a head-on collision is proportional to v+u, and that for an overtaking one is proportional to v-u, the average energy gain, $< \Delta E >$, per collision is

$$< \Delta E > \approx \frac{\mathrm{v} + u}{2\mathrm{v}} \Delta E - \frac{\mathrm{v} - 2}{2\mathrm{v}} \Delta E \approx 2 \frac{u^2}{c^2} E \tag{165}$$

This leads to an exponential growth in mean energy,

$$\frac{dE}{dt} \approx \frac{2u^2}{t_F c^2} E = \frac{E}{\tau_G} , \tag{166}$$

where t_F is the mean time between collisions, and the growth time τ_G is

$$\tau_G = t_F \frac{c^2}{2u^2} \tag{167}$$

Given this exponential growth in energy (Eq. 159), together with the assumption of random acceleration time intervals (Eq. 160), both criteria are fulfilled to yield a powerlaw spectrum (Eqs. 161–163). For cosmic ray spectra, powerlaw slopes of $\gamma \approx 2$ are observed (e.g. see Lang 1980, p.471-472), which is similar to the hardest spectra observed in solar flares.

First-order Fermi acceleration has been proposed for solar flares by Somov and Kosugi (1997) and by Tsuneta and Naito (1998), see discussion in Section 5.3, applied to particles in a mirror trap between the reconnection outflow of a coronal current sheet and the underlying standing fast shock interface above the soft X-ray flare loop. In order to explain the observed slopes of $\gamma = \delta - 1 \approx 2 - 4$ we obtain from the powerlaw slope definition (Eq. 163) the condition

$$\frac{\tau_G}{\tau_A} = (\delta - 1) = \gamma \approx 2 - 7 \tag{168}$$

On the other hand, we obtain from Fermi's relation (Eq. 167) an estimate of the energization growth time τ_G, using the collisional deflection time $t_F \approx t^{defl}(E)$ (Eq. 104) and a magnetic mirror speed of $u = \mathrm{v}_A / \cos\vartheta$ with $\mathrm{v}_A \approx 1000$ km s^{-1}

the Alfvénic speed, and ϑ the angle of the magnetic mirror motion to the magnetic field,

$$\tau_G \approx \left(\frac{E}{100\ \mathrm{keV}}\right)^{3/2} \left(\frac{n_E}{10^{11}\ cm^{-3}}\right)^{-1} \left(\frac{c^2 \cos^2 \vartheta}{2v_A^2}\right) [s] . \tag{169}$$

We see that sufficiently short acceleration times can only be achieved in fast shocks, where $\cos \vartheta \ll 1$, as pointed out in Section 5.3. To accelerate an electron from a supra-thermal energy of $E_1 \approx 3$ keV to a typical hard X-ray energy of $E = 100$ keV, an acceleration time of $t = t_G \ln(E/E_1) \approx t_G \ln(100/3) = 3.5 t_G$ is needed (according to Eq. 159). The observed durations of hard X-ray pulses set an upper limit of $t \lesssim 1$ s on the acceleration time, or $t_G \lesssim 0.3$ s on the growth time, requiring a fast shock angle of $\vartheta \lesssim 87^0$. To obtain a spectral slope of $\gamma \approx 2-4$ as observed in larger flares, an e-folding acceleration time of $\tau_A = t_G/\gamma = (0.25-0.5) \times t_G \approx 0.1$ s is needed.

8.3.2. *Spectra from Electric DC-Field Acceleration*

Electric DC-field acceleration has been proposed for reconnection outflows in solar flares by Tsuneta (1995). The idea is that the fast reconnection outflow produces small-scale, time-varying shear flows (vortices) at the loop top and drives an oppositely-directed, magnetic field-aligned current. The field-aligned current then generates a voltage drop ($\approx$ 1 MeV) along the magnetic fields and accelerates runaway electrons. What is the expected electron energy spectrum? For mildly relativistic electrons, accelerated in a constant electric field $\mathcal{E}_\parallel$, the kinetic energy grows quadratically with time,

$$E(t) = \frac{1}{2} m_e v(t) = \frac{e^2 \mathcal{E}_\parallel^G}{G m_e} t^2 \tag{170}$$

so this deviates from the condition of exponential growth (Eq. 159), a necessary condition for pure powerlaw spectra if the acceleration times obey a Poisson distribution (Eq. 160). We can invert $t(E)$ from Eq. 170 and calculate the derivative dt/dE and insert these terms into Eqs. 160 and 161 and obtain the electron energy distribution,

$$N(E) = N_0 \exp\left(-\sqrt{\frac{E}{E_0}}\right) E^{-1/2} \tag{171}$$

with the reference energy value

$$E_0 = \frac{e^2 \mathcal{E}_\parallel^2}{2 m_e} t_A^2 \tag{172}$$

This result is exactly the same as derived in Tsuneta and Naito (1998), although derived somewhat differently therein, with t_A called an escape time. In Fig. 102 we

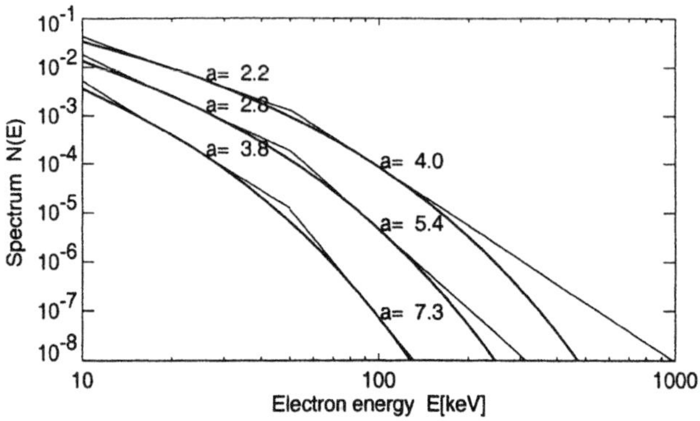

Figure 102. Three electron injection spectra (thick curves) predicted from a DC electric field model (Eq. 171, see also Tsuneta 1995), for values of $E_0 = 0.5, 1, 2$ keV. The powerlaw slopes are measured at energies of $E = 25$ and 100 keV.

show the solutions of three spectra, with $E_0 = 0.5$, 1, and 2 keV. For comparison we show also powerlaw slopes at 25 keV and 100 keV. The powerlaw slopes have a range of $a = 2.2 - 3.8$ at 25 keV and $a = 4.0 - 7.3$ at 100 keV, similarly to the broken-dawn powerlaws measured by Dulk et al. (1992). If this electron injection spectrum is convolved with the bremsstrahlung cross-section and fitted to observed hard X-ray spectra, the best fit would yield a constraint for the reference energy E_0, and together with an estimate of the acceleration time t_A, yields the electric field $\mathcal{E}_\parallel$. The examples in Fig. 102 show that the electron injection spectra have a powerlaw-like shape over a limited energy range (say 25-100 keV), but fall off near exponentially at higher energies.

This model is based on a Poissonian distribution of acceleration times (Eq. 160). The physical mechanism for this could be that electrons are scattered out of the electric field at random times, e.g. when they reach the lateral current sheet boundary in the model of Litvinenko (Fig. 49), or when they are scattered outside the separatrix of the magnetic islands in the bursty reconnection model of Kliem (1994, see Fig. 51–52), or when they leave the fast shock region in the model of Somov and Kosugi (Fig. 60) or Tsuneta and Naito (1998, Fig. 58). While such DC electric field models provide a framework for forward-modeling of hard X-ray spectra, the inversion of hard X-ray spectra has also been attempted to constrain DC electric field models (Johns and Lin 1992; Lin and Johns 1993). Clearly we need measurements of additional physical parameters to discriminate different acceleration mechanisms from an observed hard X-ray spectrum. Because many models with a nonlinear energization characteristics $E(t)$ and random acceleration times statistics $N(t)$ yield powerlaw like spectra, $N(E) \propto E^{-a}$, a determination of the spectral shape (even with high spectral resolution) is not sufficient to discriminate between different acceleration mechanisms.

8.4. Spatial Structure of Hard X-Ray Sources

The spatial morphology of hard X-ray sources is now pretty well understood. Since there is overwhelming evidence that energetic particles are primairily accelerated in magnetic reconnection regions, we can just follow the path that has a magnetic connectivity from the acceleration region to the chromosphere and calculate the collisional energy loss along these field lines to evaluate where most of the hard X-ray photons are emitted by collisional bremsstrahlung. Typical altitudes of coronal reconnection points are $h \approx 5000 - 35,000$ km (Table II) based on electron time-of-flight measurements and imaging of soft X-ray-bright flare loops. Setting the propagation time equal to the energy loss time yields a relation (Eq. 130) between the stopping length L of an electron with initial energy E in a medium with density n_e. According to this rule of thumb we expect that electrons with energies of $E \gtrsim 20$ keV (in flare loops with $L \approx 10^9$ cm and $n_e = 10^{11}$ cm^{-3}) deposit their energy mostly at the footpoints, while lower-energy electrons or trapped electrons deposit their energy in the upper parts of the loops. This explains the numerous observations of single (28.5%), double (43%), and multiple (28.5%) footpoint sources (Sakao, 1994). Because a flare usually consists of multiple acceleration episodes, indicated by multiple peaks in the hard X-ray time profiles, or by multiple loops seen in high-resolution *TRACE* images, the spatial morphology changes in accordance with this episodes, showing dominant footpoint sources during the acceleration episodes (produced by directly-precipitating electrons), but more looptop emission between the peaks (produced by trapped electrons). Such an oscillatory behavior between footpoint and looptop emission is clearly seen in the 22-33 keV energy channel recorded during the 1991-Nov-15, 22:37 UT, flare, analyzed by Sakao et al. (1992).

What is less understood is the temporal evolution of the footpoint hard X-ray locations. Tracking the footpoint separation during a flare, Sakao (1999) found that the footpoint separation increases sometimes, as expected in the Kopp-Pneuman reconnection scenario due to the rise of the reconnection point, but sometimes they decrease, opposite to the Kopp-Pneuman scenario. Interestingly, Sakao (1999) finds that the sign of the footpoint separation correlates with the spectral evolution. Based on this correlation he interprets that flares with separating footpoints have hard X-ray spectra with a super-hot thermal component as expected in the Kopp-Pneuman model, while flares with approaching footpoints have no super-hot thermal component, es expected in the emerging-flux model of Heyvaerts et al. (1977).

Other hard X-ray source components are the so-called *above-the-loop-top sources*, discovered by Masuda et al. (1994a), which are generally an order of magnitude weaker than footpoint sources (Figs. 8–9), and have been explained in terms of nonthermal bremsstrahlung from temporarily trapped electrons during the acceleration phase in magnetic cusp regions (Alexander and Metcalf 1997; Figs. 73–74).

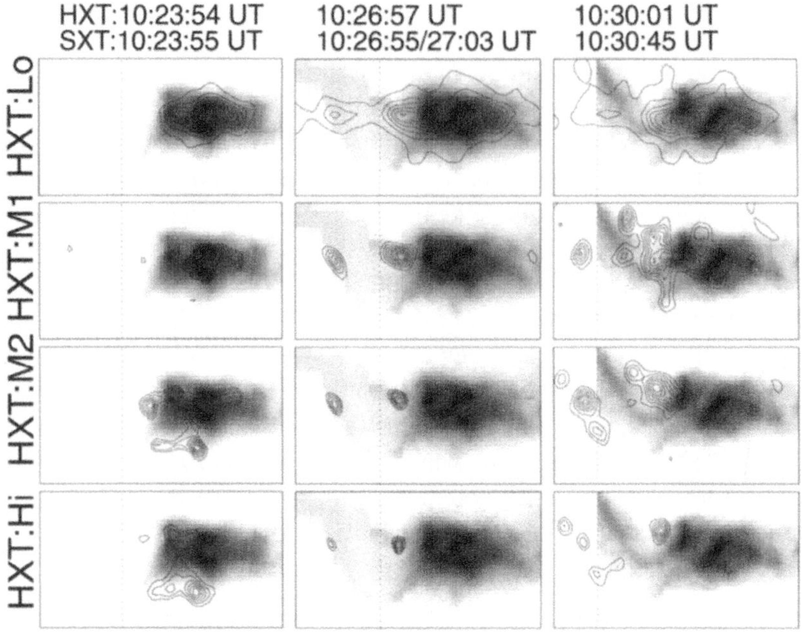

Figure 103. Yohkoh/HXT maps (contours) of energy channels (Lo, M1, M2, Hi) at the three times 2000-Jul-14, 10:23:54 UT, 10:26:57 UT, 10:30:01 UT, marking the beginning, peak, and end of a nonthermal episode. The contours are shown in increments of 10%, starting at 10% for Lo, at 20% for M1, at 30% for M2, and at 40% for Hi. The *HXT* maps are overlaid on *SXT*/Al12 maps (on logarithmic scale). The second image is a composite of a full-resolution *SXT*/Al12 and half-resolution *SXT*/AlMg image. The third *SXT* image was selected somewhat later to show the location of a developing "spine" in the eastern part above the neutral line. The dotted lines demarcate the field-of-view of the *SXT* full-resolution frames. The field-of-view is 74×33 pixels (with pixel size 2.455"), i.e. 182"×81" or 132×59 Mm [from Aschwanden and Alexander 2001].

Occasionally, impulsive brightenings at hard X-ray footpoint sources can also be seen in soft X-rays (McTiernan et al. 1993, Hudson et al. 1994). Impulsive soft X-ray footpoint sources were found to last less than a minute, and cospatial with the hard X-ray footpoints, probably caused by chromospheric plasma heating from precipitating electron beams (McTiernan et al. 1993; Hudson et al. 1994). Such impulsive soft X-ray footpoint brightenings can only be observed at the very beginning of the impulsive flare phase, because the flare area becomes fully swamped by chromospheric evaporation later on.

A new phenomenon, which we like to discuss in more detail here, is the discovery of hard X-ray ribbons (Masuda, Kosugi, and Hudson 2001; Fletcher and Hudson 2001). The classical Kopp-Pneuman model is a 2D cartoon that makes no prediction about the spatial extent along the neutral axis. However, Hα ribbons on both sides of the neutral line have been observed long ago, sometimes with considerable length, but never the matching counterparts of hard X-ray double ribbons have been reported. So, it was not clear whether nonthermal hard X-ray emission

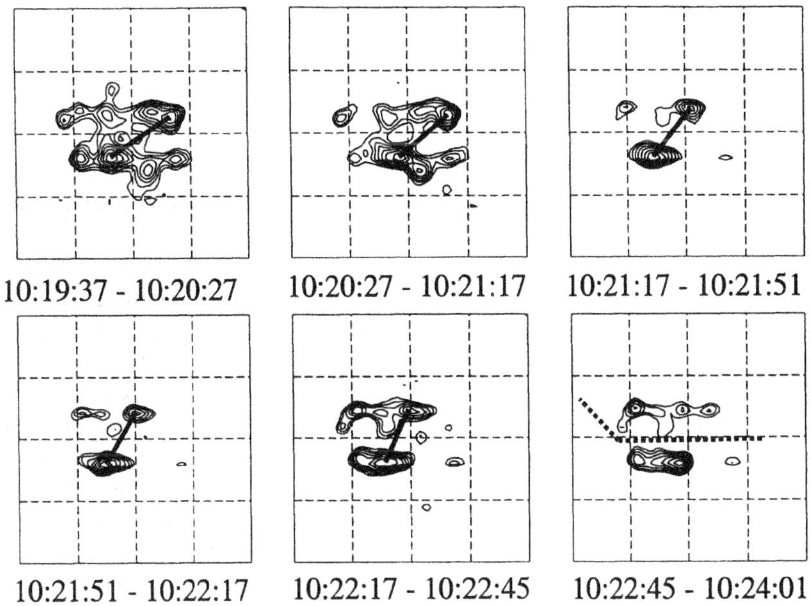

Figure 104. Sequence of *Yohkoh/HXT/Hi* (53-93 keV) images observed during the beginning phase of the 2000-Jul-14 flare, during 10:19-10:24 UT. The field of view of each panel is 157" × 157". The contours levels are 9%, 13%, 18%, 25%, 35%, 50%, 71% of the peak intensity for each panel. A thick line connects the most intense kernels in the two-ribbon structure. In the final panel, the magnetic neutral line is indicated with a thick dotted line according to SOHO/MDI observations [from Masuda et al., 2001].

occurs more localized or whether this is an instrumental sensitivity problem. A set of *Yohkoh/HXT* hard X-ray maps superimposed on soft X-ray maps, which outline the extent of the flare arcade, is shown in Fig. 103, for the Bastille-Day flare on 2000-Jul-14, 10:24 UT. The maps in Fig. 103, clearly show that the low-energy hard X-ray emission (*HXT:Lo*=14-23 keV) is concentrated at the looptop, while all high-energy emission (*HXT*:M2,Hi) is concentrated near the flare ribbon footpoints, but only at partial segments of the arcade, progressing from west to East during the course of this flare (see also Fig. 3, 43, and 44). Nevertheless, these are larger hard X-ray ribbons than ever seen before in flares.

A detailed study on these hard X-ray ribbons has been performed by Masuda et al. (2001). Fig. 104 shows the evolution of the hard X-ray sources during a first spike of the flare, which occurs in the western half of the flare arcade. Connecting the strongest kernel in the northern ribbon with the strongest kernel in the southern ribbon, deemed to be conjugate footpoints, one clearly sees an evolution from a highly sheared hard X-ray footpoint pair (10:19:37 UT) to a less and less sheared footpoint pair (10:22:17). This is the same evolution as outlined in Figs. 43, 44, and 46, where the general pattern evolves from initially highly-sheared, low-lying arcade loops to less-sheared, higher loops. This pattern, which is not predicted in the 2D standard model of Kopp-Pneuman, perhaps indicates some important

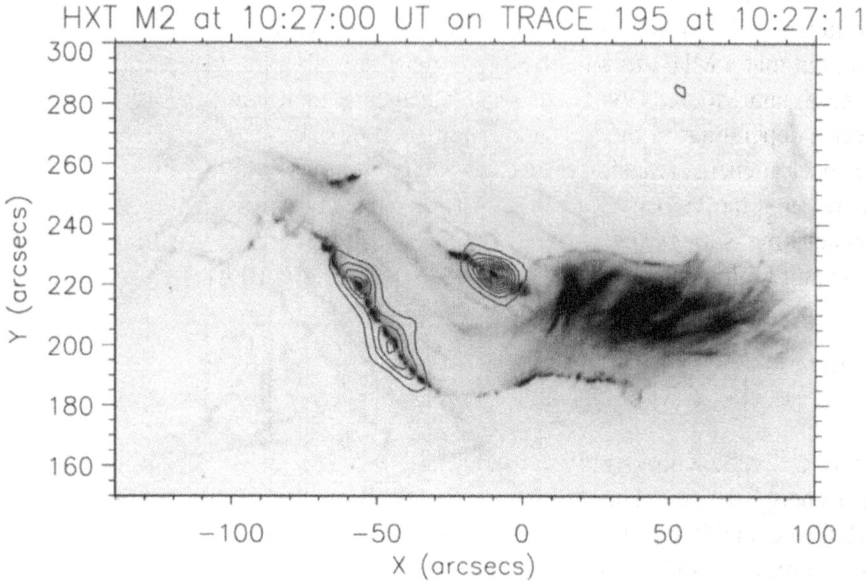

Figure 105. The position of the strongest hard X-ray sources (*Yohkoh/HXT*) at the peak time of the 2000-Jul-14 flare relative to the EUV sources (*TRACE* 171 Å) [from Fletcher and Hudson, 2001].

physics on the trigger mechanism. The highly-sheared arcade part is probably most unstable to tearing mode and triggers the magnetic island formation and subsequent coalescence, as simulated by Karpen et al. (1998) and Kliem et al. (2000), see Section 4.2. This spatio-temporal pattern seems to be preserved from the initial hard X-ray loops (Fig. 104) until to the later appearance of EUV loops (Fig. 43), after the loops have been filled with heated chromospheric upflows and have cooled down again to 1-2 MK.

Another study on this flare concentrated on the relation between the hard X-ray ribbons and the EUV ribbons (Fletcher and Hudson, 2001). While the hard X-ray ribbons mark the chromospheric sites of direct bombardment of nonthermal particles, the EUV ribbons can also be produced by conductive heating, besides beam heating. So, they dont need necessarily to be identical in the two wavelengths, and thus differences may tell us something about the different exciter roles. For instance, Czaykowska et al. (1999, 2001) demonstrated that EUV ribbons were heated dominantly by heat conduction (or possibly by protons) in one flare, without any detectable > 20 keV hard X-ray emission, a puzzling exception to the majority of flares. However, evidence for preflare EUV footpoint brightening preceding hard X-ray emission was also reported by Warren and Warshall (2001). In the 2000-Jul-14 Bastille-Day flare, nevertheless, Fletcher and Hudson (2001) find an extremely good spatial (Fig. 105) and temporal correlation (Fig. 106) between the hard X-ray and EUV fluxes, suggesting that both signatures are excited by the same precipitating electrons. If the EUV ribbons were produced by thermal conduction,

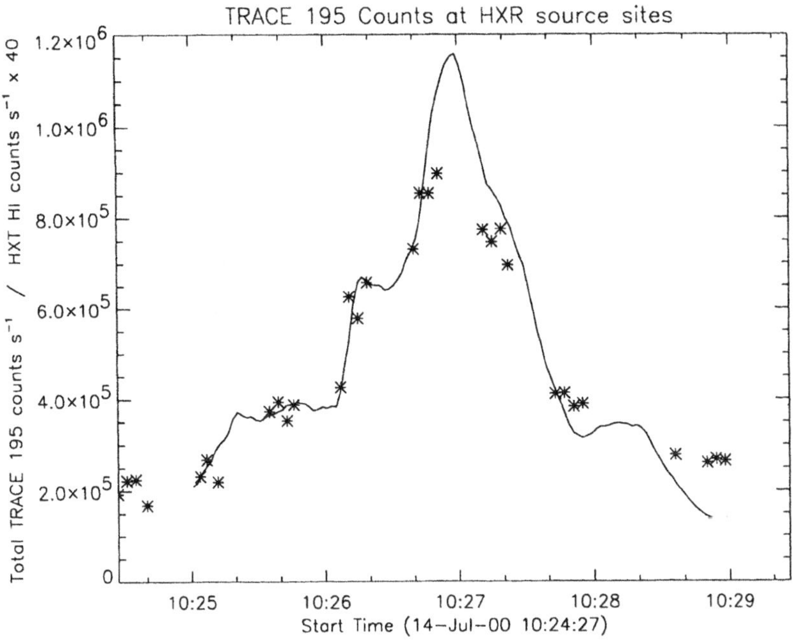

Figure 106. Yohkoh/HXT scaled light curve (solid lines) on *TRACE* 195 Å source counts/s, measured at the locations of the *HXT* sources shown in Fig. 105. The HXT energy range of the HI channel is $\gtrsim$ 50 keV. Note that the hard X-ray light curve represent the total flux [from Fletcher and Hudson, 2001].

one would expect that the peak of the EUV flux is delayed to the peak of the hard X-ray flux, according to the Neupert effect, which was not the case here (see Fig. 106). The "bubbly" hard X-ray images overlaid onto the "linear" EUV ribbons leave the impression that the hard X-ray images are compromised by the limited spatial resolution and insufficient uv-coverage of Fourier imagers. Because linear features, such as flare ribbons, cannot properly be imaged with sparse uv-coverage, we might expect some improvement from instruments with better uv-coverage, e.g. with a rotation-modulated collimator detector such as the *High Energy Solar Spectroscopic Imager (HESSI)* (Lin et al. 1998). If EUV and hard X-ray emission are produced by the same precipitating electrons, we expect to see the same sharp flare ribbons in both wavelengths, once perfect hard X-ray imagers will be available. Some parts of EUV ribbons, however, seem to be activated in the preflare phase without hard X-ray signatures (Warren and Warshall, 2001).

Ancillary information on the spatial structure of chromospheric particle precipitation sites, which are mostly illuminated by hard X-ray and EUV emission, can also be gathered from Hα and occasional white-light emission, but its physics is less understood. The low-energy end of $\gtrsim$5 keV thick-target electron spectrum was found to contain enough energy to supply white-light continuum emission occasionally observed in major flares (Hudson 1972, 1973). The penetration of the nonthermal electrons creates long-lived excess ionization which enhances the free-free and free-bound continuum in the heated medium (Hudson 1972). Recent high-resolution observations show that the locations of white-light, hard X-ray, and soft X-ray footpoint emissions are not co-spatial (Sylwester and Sylwester 2000), as expected in a scenario with electron beam precipitation into denser chromospheric regions.

Also Hα observations, although numerous and frequently reported, are difficult to use in a quantitative way for particle kinematics, because its complicated physics invovling radiative transfer is less understood. Theoretical models consider strong flare heating that penetrates deeply into the chromosphere where elevated free electron densities produce increased Stark emission in the wings of hydrogen Lyman and Balmer lines (Gayley and Canfield 1991). Fast fluctuations of the Hα blue-wing emission during solar flares have been modeled in terms of nonthermal collisional excitation and ionization by small-scale injections of high-energy electrons into the chromosphere (Ding et al. 2001). A momentum balance was established between the upflowing heated plasma seen in blueshifted Ca XIX soft X-rays and the downflowing chromospheric material seen in redshifted Hα, which is consistent with the chromospheric evaporation interpretation (Wülser, Zarro, and Canfield 1992). Recent high-cadence (33 ms) Hα observations can, by virtue of correlated time profiles with hard X-rays, pinpoint the footpoints before and after reconnection, and this way provide a capability to disentangle the change of magnetic topologies during flares (Wang et al. 2000).

9. Radio Emission

Many solar radio bursts are produced by propagating electron beams that excite plasma oscillations and Langmuir waves which are detectable in metric, decimetric, and microwave wavelengths. Such beam-driven radio emission is therefore an excellent diagnostic of accelerated and propagating nonthermal particles. Some other radio emission is produced by losscone distributions or by gyro-synchrotron emission from trapped high-relativistic particles, and thus provides a diagnostic on the process of particle trapping. In this Section we focus mainly on those types of radio emission that provide us a direct diagnostic on particle acceleration, propagation, and trapping. More comprehensive and complementary reviews on flare-related radio emission can be found in Wild and Smerd (1972), Krüger (1979), Marsh and Hurford (1982), Kundu and Vlahos (1982), Wu (1985), McLean and Labrum (1985), Kundu (1985), Dulk (1985), Goldman and Smith (1986), Aschwanden (1987), Benz (1993, 2000), Bastian, Benz, and Gary (1998), and Bastian (2000).

The following Section is organized into three groups, first we discuss radio signatures that tell us something about the physical properties of the acceleration region, then we discuss radio bursts that provide information on the trajectories

TABLE X
Radio Emission Mechanisms During Solar Flares

Emission mechanism	Frequency	Exciter
Incoherent Radio Emission Mechanisms:		
Free-free bremsstrahlung	$\nu \gtrsim 1$ GHz	thermal plasma
– microwave postbursts		thermal plasma
Gyroresonance emission	$\omega = s\Omega_e$	low-relativistic electrons
Gyrosynchrotron emission		high-relativistic electrons
– type IV moving		trapped electrons
– microwave type IV		trapped electrons
Coherent Radio Emission Mechanisms:		
Plasma emission	$\nu_{pe} = 9000\sqrt{n_e}$	electron beams
– type II bursts		beams from shocks
– type III bursts		upward propagating beams
– reverse-slope (RS) bursts		downward propagating beams
– type J bursts		beams along closed loops
– type U bursts		beams along closed loops
– type IV continuum		trapped electrons
– type V		electron beams
Electron cyclotron maser	$\omega = s\Omega_e + k_\parallel v_\parallel$	losscones
- decimetric ms spike bursts		losscones

of nonthermal particles and constrain the magnetic topology of the reconnection region, and finally we discuss radio emission from trapped particles, which characterize the topology of mirroring and confining magnetic fields in the flare region.

9.1. Radio Diagnostic of Electron Acceleration

9.1.1. *Spatial Structure of Acceleration Region*

Do we observe any direct radiation from the particle acceleration region in solar flares that could help us to probe its spatial structure? Previously (in Section 3.2) we discussed hard X-ray emission coming from *above-the-loop-top* locations in the cusp of reconnection region, discovered by Masuda et al. (1994a). However, this coronal hard X-ray emission is relatively faint (typically an order of magnitude weaker than the chromospheric footpoint emission), is rather diffuse so that we cannot resolve the spatial structure better than $\approx 10,000$ km, and seems not to

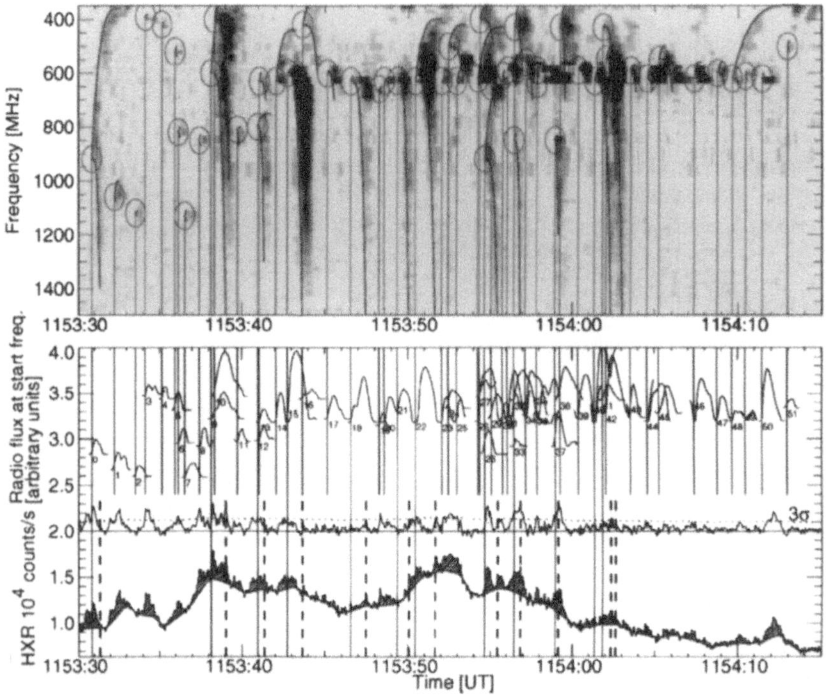

Figure 107. An enlargement of 45 s during the impulsive phase of the 1992-Sep-6 flare. The upper panel shows a radio spectrum recorded by *ETH* Zurich in the 400-1500 MHz frequency range. A total of 52 radio bursts are identified: the start frequencies (measured at the earliest detection time) are marked with circles and connected with a thin line to the hard X-ray data shown in the lower panel (*CGRO/BATSE/DISCSC* data with 64 ms resolution from the two most sunlit detectors). Drifting burst structures are traced out by thick lines; the precipitation times of reverse-drifting bursts are indicated by dashed lines. The time profiles of the start frequency of 52 radio bursts are displayed in the bottom panel (numbered with no. 0-51) to faciliate comparison with hard X-ray peaks. Fast pulses are shaded in the hard X-ray time profile (bottom profile), and a lower envelope is subtracted (second-lowest time profile) [from Aschwanden, Benz, and Schwartz 1993a].

be a comprehensive signature, because it does not reveal the acceleration sites of upward-accelerated electron beams (e.g. the common radio type III bursts).

The next best remote sensing probe of the acceleration region seems to be the start frequency of fast-drifting radio bursts. Recent observations show, at least in some flares, a common frequency band for positively and negatively drifting radio (III and RS) bursts, which suggests that the start frequencies are co-spatial with the acceleration region. Such an example is shown in Fig. 107, where a dynamic radio spectrum in the frequency range of 400-1500 MHz is shown, together with a $>$ 25 keV hard X-ray profile during the 1992-Sep-6, 1153 UT, flare (Aschwanden et al., 1993a). A total of 52 radio bursts is identified during the central 45 s around the flare peak, consisting of 30 narrow-band ($<$ 50 MHz) and 22 broad-band bursts, the latter containing 12 normal-drifting type III bursts and 20 reverse-slope

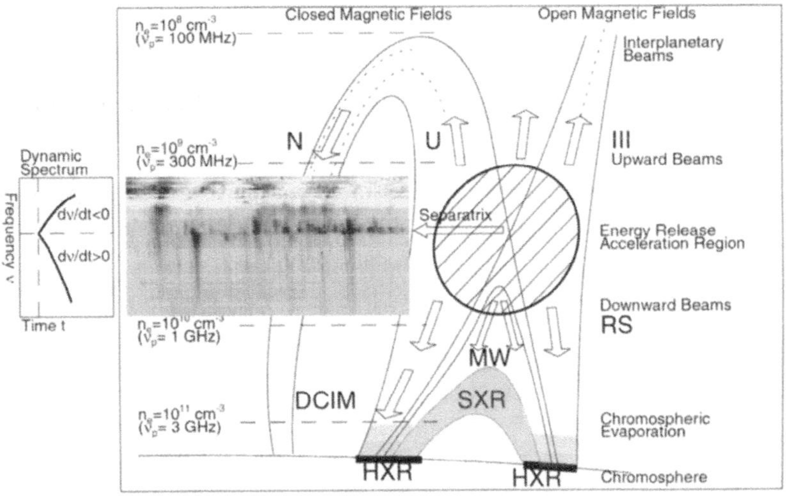

Figure 108. A synthesized standard flare scenario: The acceleration region is located in the reconnection region above the soft X-ray-bright flare loop, accelerating electron beams in the upward direction (type III, U, N bursts) and in the downward direction (type RS, DCIM bursts). Downward moving electron beams precipitate to the chromosphere (producing hard X-ray emission and driving chromospheric evaporation), or remain transiently trapped (producing microwave [MW] emission). Soft X-ray loops become subsequently filled up, with increasing footpoint separation as the X-point rises. The insert shows a dynamic radio spectrum (*ETH* Zurich) of the 92-Sept-06, 1154 UT, flare, showing a separatrix between type III and type RS bursts at $\approx$ 600 MHz, probably associated with the acceleration region [from Aschwanden 1997c].

(RS) bursts. The most interesting finding is that 10 structures consist of pairs of oppositely drifting (type III+RS) bursts, starting practically simultaneous at the same start frequency in upward and downward direction, suggesting a symmetric injection from a common accelerator. Although the start frequencies of all 52 bursts scatter over a relatively large frequency range of 400-1500 MHz, there is a concentration of one half of the bursts to start in a fairly narrow range of $f = 610 \pm 20$ MHz, which is only a variation of $\Delta f \approx 6\%$ in plasma frequency. This corresponds, using a hydrostatic density model ($n_e(h) \propto \exp(-h/\lambda)$ and the plasma frequency ($f \approx f_p \propto \sqrt{n_e}$), to a height difference of

$$\Delta h = \left(\frac{dh}{dn_e}\right)\left(\frac{dn_e}{df}\right)\Delta f = 2\lambda\frac{\Delta f}{f} \tag{173}$$

which is 12% of the density scale height, i.e. $\Delta h \approx 6000$ km for a standard coronal temperature of $T \approx 1$ MK. Thus we conclude that during the main flare phase the majority of electron beams are accelerated in a relatively compact region with a radius of a few 1000 km, while a smaller number of beams originates in a larger volume dispersed over a few 10,000 km. If we associate the acceleration sites with the cusps of reconnection regions, which have typical electron densities of $n_e \approx 10^9 - 10^{10}$ cm^{-3}, based on *Yohkoh* observations, we expect radio start frequencies of $f \approx 0.3 - 1.0$ GHz. The case shown in Fig. 107, with a center frequency of $f = 0.6$ GHz falls middle in this range. Therefore, if coronal conditions are favorable to detect plasma emission in this frequency range, the start frequencies $f_{SF}(n_e)$ of dual type III+RS radio bursts provide a direct density measurement n_e of the reconnection region. Moreover, if a suitable density model $n_e(h)$ is used, the scatter of start frequencies can be mapped to a distribution of altitudes $h(n_e[f_{SF}])$ of elementary acceleration sites. The so-inferred densities of acceleration sites is discussed in Section 3.8 with larger statistics.

The spatio-temporal fragmentation of the acceleration region, for which type III burst rates provide a sensitive measure, and is moreover warranted by the correlation found between type III burst rates and the hard X-ray flux (Aschwanden et al., 1990), has been discussed in Section 4.4. Another implication of dual type III+RS radio burst pairs with simultaneous start time and frequency is the bi-directionality of the acceleration region, which nicely fits into the concept of symmetric current sheets (Fig. 18). These produce newly-reconnected field lines in upward and downward direction, and thus have the potential to accelerate particles in upward and downward direction (Section 3.6). The concept of simultaneous upward and downward acceleration is further corroborated by triple coincidences of type III, RS bursts, and hard X-ray pulses of similar (subsecond) duration with coincident timing (within $\lesssim 0.1$ s), as discussed in Section 3.7 (see synthesized flare scenario in Fig. 108).

9.1.2. *Temporal Patterns of Particle Acceleration*

Radio observations provide the best temporal resolution of elementary acceleration episodes, not only because they have been routinely performed with high time cadence (of 100 ms or better), but also because the confusion of near-simultaneously occurring electron beams can often be resolved spectrally (in dynamic spectra). It is therefore most promising to study temporal patterns of the unknown acceleration mechanisms in flares with radio time series.

Examples of time series of radio flux profiles are shown in Fig. 109, where every peak in a time profile represents an individual type III or J burst, and thus an elementary acceleration episode, supposed that every acceleration episode produces a detectable radio event. Statistics on the time intervals between subsequent radio bursts, also called *waiting times*, should reveal whether this repetitive acceleration process is governed by a random process or by a more regular (periodic or quasi-periodic) process. The analysis of 260 such time series yielded the result that the distribution of periods, normalized by their means, was found to be $\sigma_P/P = 0.37 \pm 0.12$ for type III bursts (Aschwanden et al., 1994a). This result represents an intermediate case between strictly periodic ($\sigma_P/P = 0$) and strictly random behavior ($\sigma_P/P \approx 1$). Another method applied to analyze the periodicity of a time series is auto-correlation (Aschwanden et al. 1994b). The autocorrelation function of four type III time series is shown in Fig. 110, which exhibits quasi-periodic patterns with periods of $P \approx 1.3 - 5.0$ s. For comparison, we show

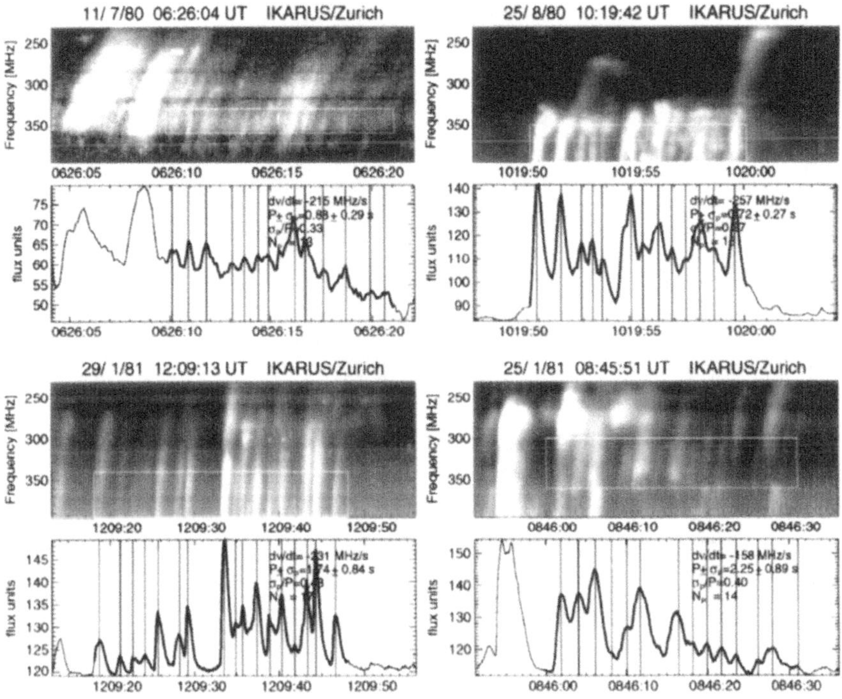

Figure 109. Dynamic spectra (panels with black background) and time series (profiles) of four events with sequences of type III or J bursts, recorded by *ETH* Zurich. The times of the type III peak fluxes are marked with thin vertical lines. Statistics on the intervening time intervals has been performed to evaluate whether acceleration events are governed by periodic or random processes [from Aschwanden, Benz, and Montello (1994a)].

in Fig. 110 also the autocorrelation for a *metric oscillation* event, which is more periodic, and for a *decimetric millisecond spike* event, which turns out to be strictly stochastic. Clearly, the time pattern of type III burst generation is more regular than random, but not close to periodic. Because stochastic time series are equivalent to spatially independent and uncorrelated events, we conclude that subsequent type III bursts are not generated at spatially independent locations, but rather in a nonlinear system with some spatial correlation function. In Section 4.2 we discussed bursty reconnection, where tearing and coalescence instabilities occur along a filamentary current sheet. The spatial structure of interacting magnetic islands along such a filamented current sheet provide such a spatial correlation function for subsequent acceleration episodes, and this way could explain the non-random nature of type III time series.

The waiting time distribution of type III bursts has also been investigated by higher-order moments and it was found that the third moment can be used as a sensible criterion to distinguish between random and periodic processes (Yurovsky and Magun 1998). According to this study, based on the assumption of a low detec-

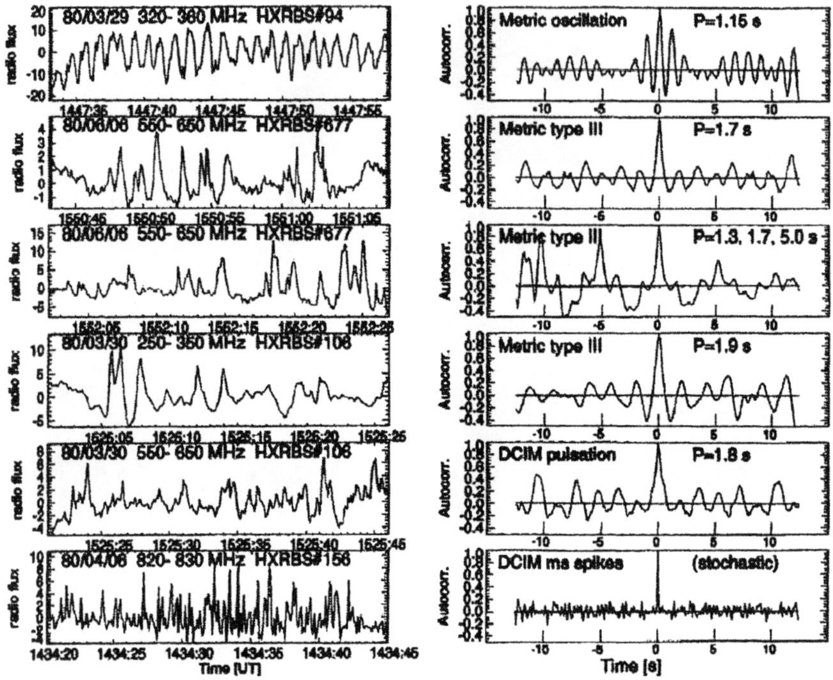

Figure 110. Six time series analysis (left) extracted from radio dynamic spectra and their autocorrelation function (right), after subtraction of a gliding background. Periodicities of coherent patterns in the autocorrelation function and also identified in FFT power spectra are indicated in seconds. Note the periodic pattern for the metric oscillation event, the quasi-periodic petterns for type III events, and the purely stochastic case for the decimetric millisecond spike event (bottom) [from Aschwanden et al. 1994b].

tion efficiency of type III bursts, the observed time series were found to be nearly consistent with random processes. Time series studies on *marked and filtered point processes* have also demonstrated that temporal correlations in stochastic models can lead to "quasi-periodic" time patterns (Isliker 1996). The apparent temporal correlation could be caused by pulse shapes or pulse-detection algorithms. So, at this point it is not clear whether the data analysis technique or physical processes in the solar corona introduce correlation lengths (also called *memory function* in waiting time statistics) in type III burst time series. While current time series analysis techniques seem to be exhausted to solve this problem, future high-resolution mapping in hard X-rays and EUV may reveal the true spatio-temporal dynamics of magnetic reconnection processes, which may reveal self-organized criticaliy in form of "chain reactions" or "avalanches".

9.2. Radio Diagnostic of Electron Propagation

9.2.1. *Detection of Electron Beams*

While radio signatures of electron beams seem to be ubiquitous and omnipresent in solar flares, their detection is not 100% reliable and can be entirely suppressed under certain coronal conditions. A key criterion whether plasma emission is detectable from electron beams is the opacity of free-free absorption. The free-free optical depth for fundamental and harmonic plasma emission is (e.g. Dulk 1985; Benz 1993, p.263)

$$\tau = \begin{cases} 4.6\left(\frac{f}{1\ \mathrm{GHz}}\right)^2 \left(\frac{T_e}{1\ \mathrm{MK}}\right)^{-3/2} \left(\frac{H}{1\ \mathrm{Mm}}\right) & \text{for } f = f_p \\ 0.12\left(\frac{f}{1\ \mathrm{GHz}}\right)^2 \left(\frac{T_e}{1\ \mathrm{MK}}\right)^{-3/2} \left(\frac{H}{1\ \mathrm{Mm}}\right) & \text{for } f = 2f_p \end{cases} \tag{174}$$

where f is the observed frequency, T_e the electron temperature, and H the exponential density scale height in direction to the observer. This shows us that the stratified corona ($T_e \approx 1$ MK, H=46,000 km) is optically thick for fundamental plasma emission at $f \gtrsim 100$ MHz, and for harmonic plasma emission at $f \gtrsim 500$ MHz. Indeed, most of the radio type III bursts are seen at metric frequencies of $f \lesssim 300$ MHz. Thus we expect that fundamental plasma emission is only observed from electron beams propagating at altitudes $h \gtrsim 46,000$ km, while harmonic plasma emission is observable down to the transition region (in regions with up to 4 times higher densities than fundamental plasma emission is detectable).

During flares, however, fast-drifting bursts are observed at decimetric and even microwave frequencies. Some of the highest frequencies where type III bursts have been observed are in the range of $f = 6.2-8.4$ GHz (Benz et al. 1992). How is this possible? From *Yohkoh/SXT* observations we know that flare loops have densities up to $n_e \approx 10^{11}$ cm^{-3} (e.g. Fig. 80) which corresponds to a plasma frequency of $f_p \approx 3$ GHz, so we expect to see harmonic plasma emission up to $f \approx 2f_p \approx 6$ GHz, as it was observed in the case of Benz et al. (1992). In order to overcome free-free absorption at this frequency, either the temperature of the source has to be higher and/or the density scale height of the source has to be shorter (Aschwanden et al. 1985). For harmonic plasma emission at $f \approx 6$ GHz, it is sufficient to have an overdense loop with a density of $n_e \approx 10^{11}$ cm^{-3}, a temperature of $T_e \approx 10$ MK, and a transverse scale height of $H \approx 1000$ km, which are all fairly typical flare loop parameters. So, the generation of harmonic plasma emission is almost a requirement to detect electron beams in flare loops, because fundamental plasma emission cannot escape from flare loops below altitudes of $H \lesssim 46,000$ km (at disk center), and this lower altitude limit is even higher for limb flares due to the longer column depth along the line-of-sight. A favorable condition to detect electron beams are high-arching flare loops filled with hot plasma, along which plasma emission can be ducted (by total wave refraction) with low free-free absorption along its way. This mechanism of ducting, however, leads to a displacement of the

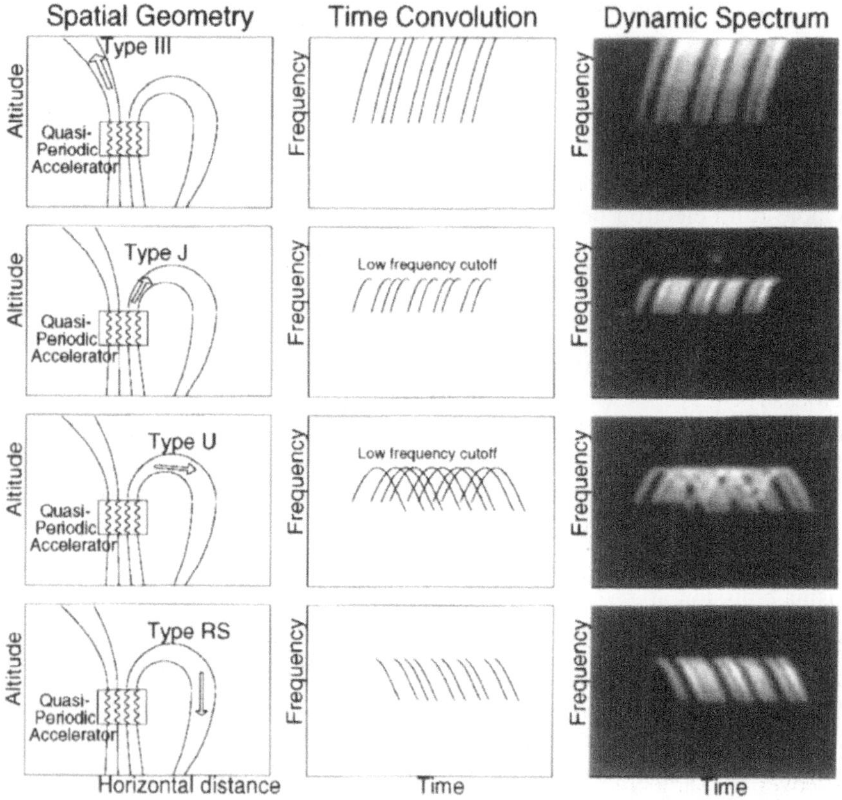

Figure 111. Spatial geometry and schematic dynamic spectrum of quasi-periodically injected electron beams, for the case of injection into open (top row) or closed (second to forth row) magnetic field lines. Note the typical low-frequency cutoff and curvature in the drift rate for electron beams propagating along closed field lines, for the case of J, U, or RS bursts. The flux and quasi-periodicity is randomly modulated to simulate a realistic representation of a dynamic spectrum (right).

observed apparent source locations for fundamental and harmonic emission, which has been noted in earlier Culgoora measurements.

Because the column depth increases as $H = \lambda_T / \cos\vartheta$ with center-limb distance, we expect that plasma emission is more suppressed for limb flares. In a study by Simnett and Benz (1986) it was noticed that about 15% of the flares show no detectable radio emission at metric and decimetric frequencies. Inquiring the 9 cases of gamma-ray flares without type III radio emission listed in this study, we notice that 6 out of these 9 cases were located very close to the limb, at longitudes of $71^0 - 86^0$ off disk center. The lack of beam-driven radio emission was interpreted as lack of electron beams, while proton beams were invoked for the production of hard X-rays (Simnett and Benz, 1986), but a more likely explanation is just the enhanced free-free absorption for plasma emission near the limb. Thus, the flare location is another important factor for the detectability of electron beams.

A systematic study on the detectability of upward, downward, and bi-directional electron beams within a realistic coronal density model has been recently performed by Robinson and Benz (2000). The authors come to the same conclusion that harmonic emission is expected to dominate at plasma frequencies above ≈ 100 MHz, where the efficiency for fundamental emission falls off steeply, while its free-free reabsorption rises. The model explains the observed trends in the likelyhood of occurrence of normal, reverse-slope, and bidirectional coronal type III bursts: (1) at metric frequencies ($f < 300$ MHz) open-field lines, and thus normal type III bursts are dominant; (2) at decimetric frequencies ($f = 2f_p \gtrsim 600$ MHz) many closed loops exist and type J, U, and RS-bursts occur frequently; (3) above 1 GHz one sees mainly RS-bursts, because they occur preferentially below acceleration regions. The relation between electron beam trajectories and the appearance of type III, J, U, and RS radio bursts in (frequency-time) dynamic spectra is illustrated in Fig. 111.

9.2.2. *Propagation Speed of Electron Beams*

The propagation of electron beams produces a frequency-time drift rate df/dt in radio dynamic spectra, which can be translated into a beam velocity if a suitable density model is given. Hydrostatic equilibrium is usually a good assumption in the quiet corona, which applies to electron beams propagating in the upper corona. In the lower corona, electron beams are produced in parts of flare loops or connected active region loops, which are overdense with respect to the quiet corona and moreover are subject to dynamic processes that cause deviations from hydrostatic equilibrium. Statistically, it was found that the frequency-time drift rate of type III bursts can be approximated by a powerlaw in frequency, over the range of 75 – 550 MHz,

$$\frac{df}{dt} = -A\ f_{MHz}^{\alpha} , \tag{175}$$

with the constants $A = 0.01$ and $\alpha = 1.84$ (Alvarez and Haddock, 1973). If we want to adopt a density model that is consistent with the powerlaw approximation of Alvarez and Haddock (1973), we have to assume a powerlaw in density too. Therefore, we construct a composite density model that has a powerlaw function in the lower corona (at $h < h_1$), and a continuous transition to a hydrostatic exponential function in the upper corona (at $h > h_1$) (Aschwanden and Benz 1997), see also Fig. 112,

$$n_e(h) = \begin{cases} n_1 (h/h_1)^{-p} & \text{for } h < h_1 , \\ n_Q \exp(-h/\lambda_T) & \text{for } h > h_1 . \end{cases} \tag{176}$$

If we choose the base density n_Q and thermal scale height λ_T of the hydrostatic corona as independent parameters, the other parameters h_1 and n_1 follow from the requirement of a smooth transition (steady derivative dn_e/dh at $h = h_1$),

$$h_1 = p\lambda , \qquad n_1 = n_Q \exp(-p) . \tag{177}$$

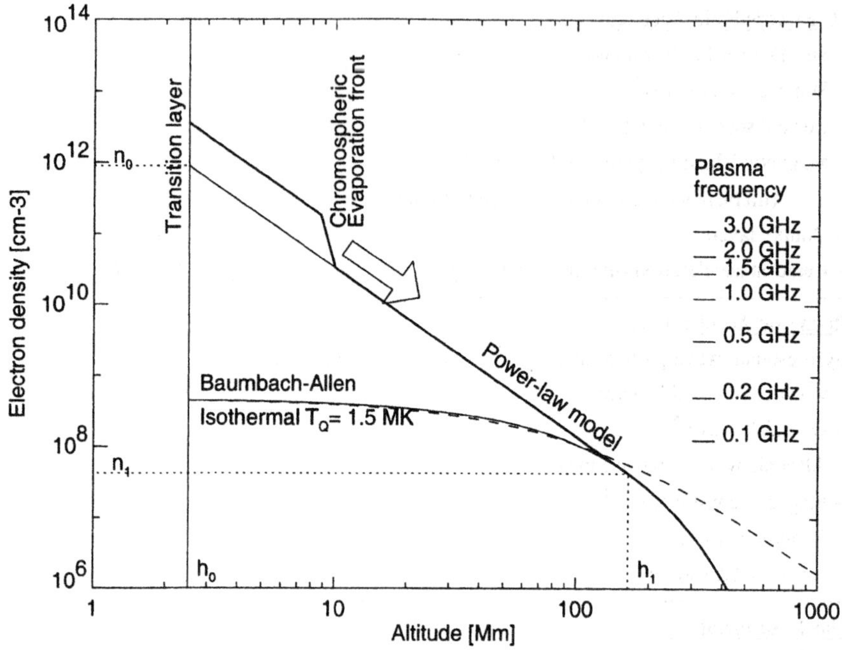

Figure 112. A density model (thick line) for type III-emitting (overdense) flux tubes, consisting of a hydrostatic component for ($h > h_1$), a powerlaw component for $h < h_1$, and a (time-dependent) enhanced density component associated with upflowing chromospheric plasma (indicated here at a height of $\lesssim$ 10 Mm). The Baumbach-Allen model (dashed line) matches the hydrostatic model (thin line) for heights $h < h_1$. The plasma frequency scale is indicated on the right [from Aschwanden and Benz 1997].

For emission at the $s - th$ harmonic of the plasma frequency, $f = s\ f_p$, we have a plasma frequency f_Q at the coronal base,

$$f_Q = \left(\frac{n_Q q_e^2}{\pi m_e} \right)^{1/2} = 8980 \times n_Q^{1/2} \tag{178}$$

and the frequency drift rate df/dt follows then,

$$\frac{df}{dt} = -\frac{ev(sf_Q)^{-2/p}}{2\lambda_T} f^{(1+2/p)} \ . \tag{179}$$

Matching this drift rate with that of Alvarez and Haddock (1973), we obtain the following expressions for the coefficients A and α,

$$\alpha = (1 + \frac{2}{p}) \ , \qquad A = \frac{ev(sf_Q)^{-2/p}}{2\lambda_T} \ , \tag{180}$$

which determines the constants p and n_Q in our model,

$$p = \frac{2}{\alpha - 1} \ , \qquad n_Q = \left(\frac{ev}{2A\lambda_T} \right)^p (s \times 8980)^{-2} \ . \tag{181}$$

For numerical values we can use the Baumbach-Allen model for the base density n_Q and a coronal temperature of $T = 1.5$ MK, which yields then $n_Q = 4.6 \times 10^8$ cm^{-3}, $\lambda_T = 6.9 \times 10^9$ cm, $h_1 = 1.6 \times 10^{10}$ cm, and $p = 2.38$. From Eq. 181 we have a constraint for the average type III velocity,

$$v_{III} = \frac{2A\lambda_T}{e}[n_Q(s\ 8980)^2]^{1/p} . \tag{182}$$

which yields for fundamental and harmonic emission,

$$v_{III}/c = \begin{cases} 0.14 & s = 1 \quad (fundamental) \\ 0.25 & s = 2 \quad (harmonic) \end{cases} \tag{183}$$

TABLE XI

Electron Beam Speeds Inferred from Type III Exciters

Velocity v_{III}/c	Frequency Range f [MHz]	Number of analyzed events	References
0.45	45-60 MHz	8	1)
0.33	7-200 MHz	50	2)
0.15 ± 0.03	160-320 MHz	12	3)
0.16 ± 0.05	56 kHz	3	4)
0.14 ± 0.04	1 MHz	28	5)
0.18 ± 0.04	1.446 GHz	1	6)
0.15 ± 0.02	250-850 MHz	1	7)
0.30 ± 0.10	560-1240 MHz	1	7)
0.15 ± 0.04	56 kHz-1.5 GHz	all	Weighted average

References: (1) Wild, Sheridan, and Neylan, 1959; (2) Stewart, 1965; (3) Fokker, 1970; (4) Lin et al., 1981; (5) Dulk et al., 1987; (6) Aschwanden et al., 1992b; (7) Aschwanden et al. 1993b.

This model appears to be in good agreement with other measurements of type III exciter speeds, as shown in Table XI, where the mean and standard deviation of the observational values amounts to $v_{III}/c = 0.15 \pm 0.04$, consistent with our model for fundamental plasma emission. This exciter speed corresponds to a kinetic energy of $E = 6 \pm 3$ keV for the type III-emitting electrons. Note that these type III beam electrons have a lower energy than the hard X-ray producing electrons, typically observed at energies $E \gtrsim 25$ keV.

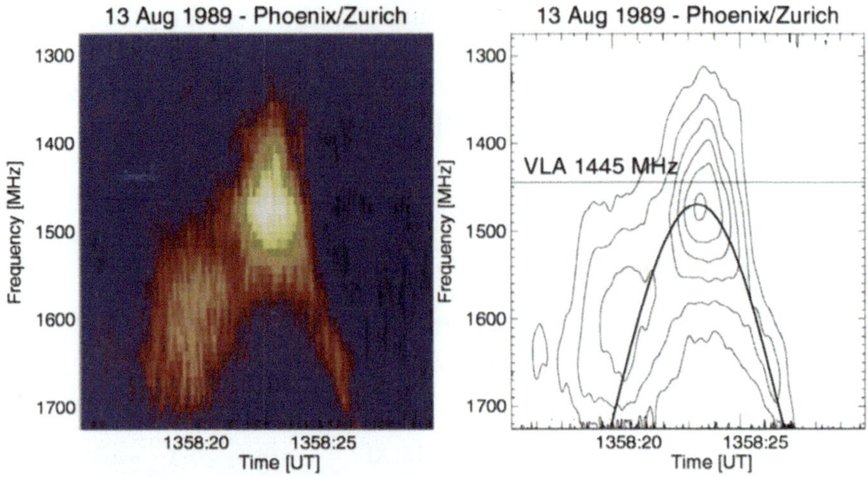

Figure 113. Radio Observations of three type U-bursts on 1989-Aug-13 with *ETH* Zurich spectrometer. The ascending branch is broader than the descending branch. The loop transit time varies around 6-9 s at 1.7 GHz for this U-burst. The same type U-burst has been imaged with the *VLA* at a frequency of 1446 MHz, see Fig. 114 [from Aschwanden et al. 1992b].

9.2.3. *Electron Beam Propagation*

Most of the electron beams are observed when they escape in upward direction from a reconnection region, because of the lower opacity and the beam focusing effect of the decreasing magnetic field. For bipolar and tripolar magnetic reconnection configurations (Fig. 2, Sections 2.1 and 2.2) we expect upward propagating electron beams, manifested as type III bursts, and for quadrupolar reconnection configurations we expect type J, U, or RS bursts, propagating along much larger loops than the primary flare loops with hard X-ray footpoints.

A well-observed case regarding spectral, temporal, and spatial resolution is a sequence of type U-bursts that occurred during 1989-Aug-13, 13:54-14:00 UT, recorded with the broadband spectrometer of *ETH* Zurich (Fig. 113) and simultaneously imaged with the *Very Large Array (VLA)* at 1446 MHz (Fig. 114) (Aschwanden et al. 1992b). It was a fortunate circumstance that the *VLA* frequency of 1446 MHz just coincided with the plasma frequency of the turnover point near the apex of the type U-burst, so that the location could uniquely be reconstructed. A potential field line extrapolation of the photospheric magnetic field yields an average height of $h \approx 130,000 \pm 15,000$ at the type U-burst looptop (Fig. 114). This yields the length of the electron beam trajectories. Together with the observed durations of the type U-bursts ($T = 6.2 - 8.8$ s), we can thus estimate an upper limit of the exciter speed of the electron beams, assuming a semi-circular trajectory,

$$\mathrm{v} < \mathrm{v}_{max} = \frac{\pi\, h}{T} \ , \qquad (184)$$

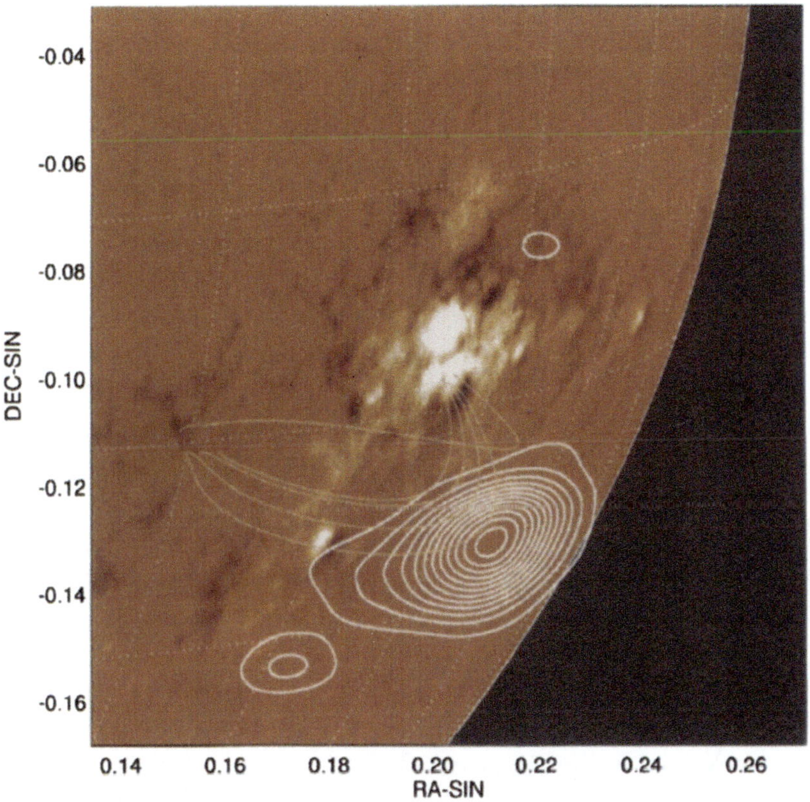

Figure 114. Photospheric magnetic field in AR 5629 from KPNO/NSO on 1989-Aug-13, 1500 UT, about 1 hour after the type U-bursts. The contours show an overlaid *VLA* map recorded at a frequency of 1446 MHz at 1355:02-1355:28 UT. The thin lines show a subset of extrapolated magnetic (potential field) field lines that intersect with the type U-burst emission, which seems to be at the top of the field lines. The apices of the field lines range from 75,000 to 142,000 km, with magnetic fields of 25 G to 8 G [from Aschwanden et al. 1992b].

which yields $v_{max} = 45,000 - 65,000$ km s^{-1} or $v_{max}/c = 0.15 - 0.22$, which is fairly consistent with electron beam speeds quoted in literature (see Table XII). In addition, using the plasma frequency as constraint for the density, we estimate looptop densities of $n_e = 2.4 \times 10^{10}$ cm^{-3} for fundamental plasma radiation, or $n_e = 6 \times 10^9$ cm^{-3} for harmonic radiation. Given the density and beam speed, we can calculate the collisional deflection time of the electrons in the loop and compare that with the loop transit time. The collisional deflection time is, expressed in terms of electron speed v_B and density n_e (e.g. Benz 1993, p.284),

$$\tau^{Defl} = 3.1 \times 10^{-20} \left(\frac{v^3}{n_e} \right) \qquad \text{cgs} - \text{units} , \tag{185}$$

which yields $\tau^{Defl} = 0.1 - 0.4$ s for fundamental emission, or $\tau^{Defl} = 0.5 - 1.4$ s for harmonic emission. Harmonic plasma emission is clearly favored, because

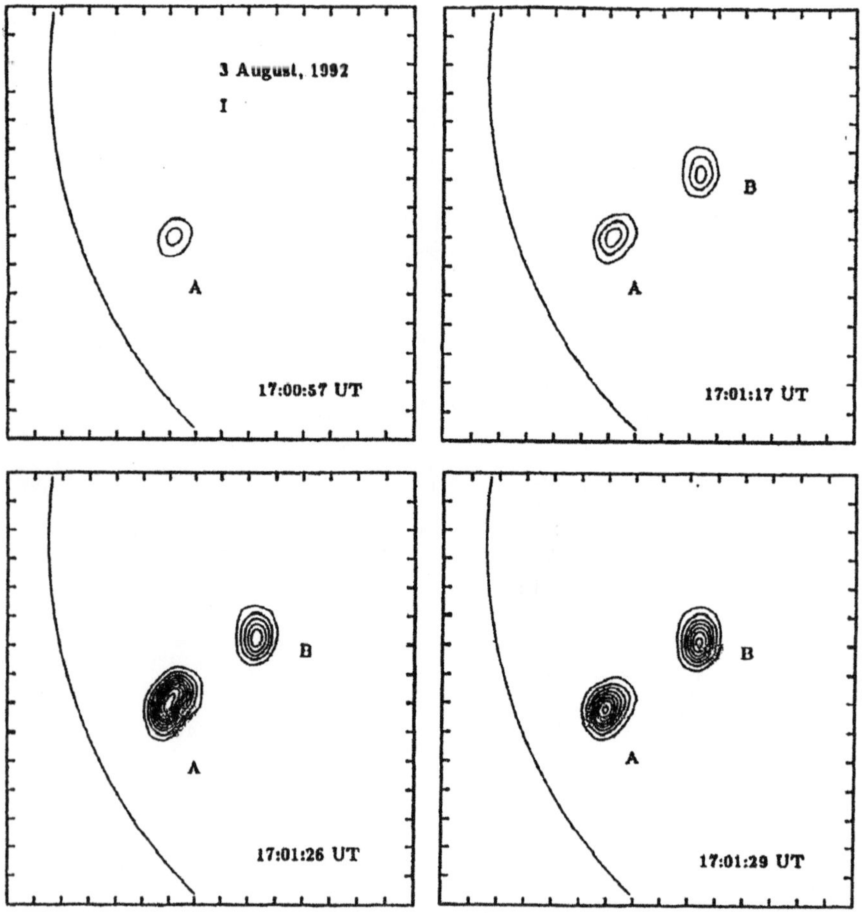

Figure 115. Radio maps recorded with the *VLA* at a frequency of 1446 MHz with a cadence of 1.7 s, showing total intensity at four times during the 1992-Aug-3, 17:00 UT, flare. The two source components A and B are separated by a distance of $\approx$ 120, 000 km. The contour levels are in units of equal brightness temperature, T_b, with an outermost contour and contour interval of $T_b = 8.1 \times 10^5$ K. The tickmarks have a spacing of 60". The curved lines demarcates the solar limb [from Willson, Lang, and Gary 1993].

of the lower collisional scattering rate as well as of the lower opacity. Although the collisional deflection time is somewhat shorter (0.5-1.4 s) than the full transit time (6.2-8.8 s), the beam may still arrive relatively focused at the other end of the loop, if the magnetic field is decreasing along the propagation direction, and this way focuses the beam additionally. The magnetic focusing effect may explain why mainly the descending branches of the U-bursts are better visible (Fig.113). Alternatively, electrons with lower energy ($\lesssim$ 5 keV) become scattered out of the beam in the ascending branch, while only higher-energy electrons ($\approx 5 - 10$ keV) survive the travel all the way to the opposite footpoint.

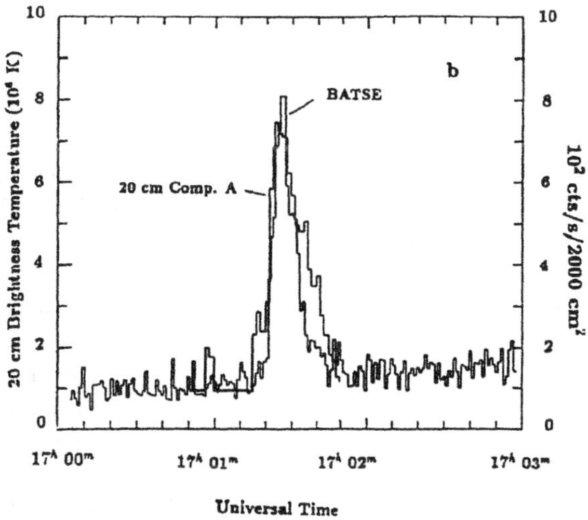

Figure 116. Time profile of the maximum brightness temperature of the 1446 MHz source mapped with the *VLA*, overlaid on the hard X-ray time profile from *BATSE/CGRO*, from the same flare shown in Fig. 115. Note the simultaneity of the rise and peak within the time resolution of 1.7 s [from Willson, Lang, and Gary 1993].

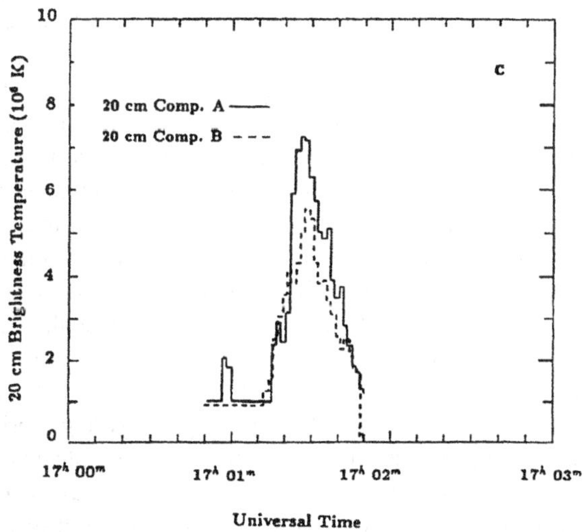

Figure 117. Time profile of the two spatial components A and B of the 1446 MHz radio source mapped with the *VLA* in Fig. 115. Note the near-simultaneity of the two components, the delay of component B relative to a is less than 1.7 s [from Willson, Lang, and Gary 1993].

A propagating electron beam may be detected twice in both the ascending and descending part of a closed field line, at altitude levels that have an electron density matching the fundamental (or harmonic) plasma frequency at the observed frequency. A nice example of such a double detection of the same electron beam was reported by Willson, Lang, and Gary (1993), using the *Very Large Array (VLA)* and the *Owens Valley Solar Array (OVRO).* A flare with hard X-ray emission detected by *BATSE/CGRO* occurred in AR 7248 (marked as location A in Fig.115). The 1446 MHz radio emission detected with the *VLA* shows a simultaneous and highly correlated time profile with the > 25 keV hard X-ray emission detected by *BATSE* (Fig.116), and thus suggests plasma emission from an electron beam accelerated in upward direction by the same acceleration process that produced also the hard X-ray emitting nonthermal electrons in downward direction. Almost simultaneously, a secondary source was detected at the same frequency of 1446 MHz in a remote active region AR 7245 (marked as location B in Fig.115) some 200,000 km away. The radio time profiles in both locations A and B are again highly correlated (Fig.117), but the component B seems to be delayed by a small amount, for which an upper limit was estimated from the used time resolution of $\Delta t = 1.7$ s. The obvious interpretation is that the same electron beam propagated over a large-scale loop from footpoint A to B. The estimated beam velocity, i.e. $v_B \gtrsim 200,000/1.7$ s = 120,000 km s^{-1} or $v_B/c \gtrsim 0.4$, is indeed typical for electron beams that excite plasma emission (Table XI). This observation clearly demonstrates that 40 keV electrons are accelerated simultaneously in upward direction (fully constrained by the double detection with the *VLA*) as well as in downward direction (as evidenced by *BATSE*), requiring a bi-directional acceleration mechanism, as provided, e.g., by the upward/downward symmetry of reconnecting current sheets. A possible flare scenario is that of quadrupolar reconnection (Section 2.3), where a small-scale loop reconnects with a (very) large-scale loop (see Fig.5 left).

A systematic timing study of the timing between hard X-ray emission at the primary flare site and nonthermal radio emission at remote secondary footpoints was performed by Hanaoka (1999). He found similar time delays of $\approx$ 500 ms between the primary hard X-ray flare site and the 17 GHz radio emission at the remote seconary flare site, indicating propagation of nonthermal electrons with speeds of $v/c \approx 0.5$. The measured distances, time delays, and inferred electron speeds are listed in Table XII for seven events analyzed by Hanaoka (1999) and others. All these cases are consistent with the flare scenario of quadrupolar magnetic reconnection (Section 2.3), where the magnetic reconnection process between a small-scale and a large-scale loop simultaneously injects accelerated electrons in upward and downward direction, which are detected in radio at both ends of the large-scale loop, with propagation delays to remote footpoints corresponding to mildly-relativistic electrons.

9.2.4. *Relative Timing of Electron Beams in Radio and Hard X-Rays*

The common origin of nonthermal electrons that produce radio type III bursts and hard X-rays has been suspected long ago, based on the general association between the emissions in the two wavelengths. In early works, the association was low, e.g. Kane (1972), Stewart (1978), and Kosugi (1981) have shown that flare-associated metric type III bursts were associated with hard X-rays only in 3% . However, statistical studies of decimetric bursts (mainly type III-like bursts) showed an increasing association rate of 48$\pm$3% with hard X-rays and 61$\pm$5% with soft X-rays

TABLE XII

Electron Beam Speeds Inferred from Double Detections [from Hanaoka, 1999]

Event	Time delay at remote source Δt [s]	Surface distance l [Mm]	Apparent velocity range v/c=l/cΔt
Hanaoka (1999)			
– 1993-Feb-6 A	0.377 ± 0.097	61	0.54
– 1993-Feb-6 B	0.493 ± 0.051	78	0.53
– 1993-Apr-10 A	0.422 ± 0.062	60	0.47
– 1993-Jun-7 A,B	0.518 ± 0.074	61	0.39
Nakajima et al. (1985)	1.7-25	150-890	0.10-0.37
Lang & Willson (1989)	3-6	260	<0.33
Willson et al. (1993)	$\leq$1.7	200	>0.40

(Aschwanden et al. 1985). The low association rate in earlier work could be due to the sensitivity limitations of earlier hard X-ray detectors and the fact that metric type III bursts originate in higher and thus in less flare-connected regions.

The ultimate proof whether radio-emitting and hard X-ray-emitting electrons are energized by the same acceleration process, however, requires a demonstration that they have the same energy spectrum and a consistent timing in both wavelengths. Because nonthermal hard X-ray bremsstrahlung is mostly produced by precipitating electrons in the chromosphere, the radio counterpart should show reverse-slope (RS) bursts, which are not observed in many flares, probably due to the higher opacity of plasma emission that downward-propagating electron beams encounter. The detailed correlation of sequences of radio type III bursts and subsecond hard X-ray pulses has been established in a number of flares (e.g. Kane, Pick and Raoult 1980; Kane et al., 1982b; Dennis et al. 1984, Aschwanden et al. 1993a, 1995d). A well-observed event is the 1992-Sep-6 flare, where a number of reverse-slope drift radio bursts with coincident hard X-ray pulses were observed (Aschwanden, Benz, and Schwartz, 1993a). The detailed relative timing is shown for 5 such bursts in Fig. 118. The cross-correlation of radio and hard X-ray time profiles is shown in Fig. 119, generally revealing a slight delay of the radio emission with respect to the hard X-ray pulses, at the start frequency as well as at the highest radio frequency, which is supposedly closest to the chromospheric hard X-ray site. The average radio delay at the start frequency was found to be $t_{R1} - t_X = 100 \pm 120$ ms, and $t_{R1} - t_X = 270 \pm 150$ ms at the highest frequency of the RS bursts, respectively. The situation of the relative timing is illustrated in Fig. 120. What is the cause for the radio delay? A likely explanation is that

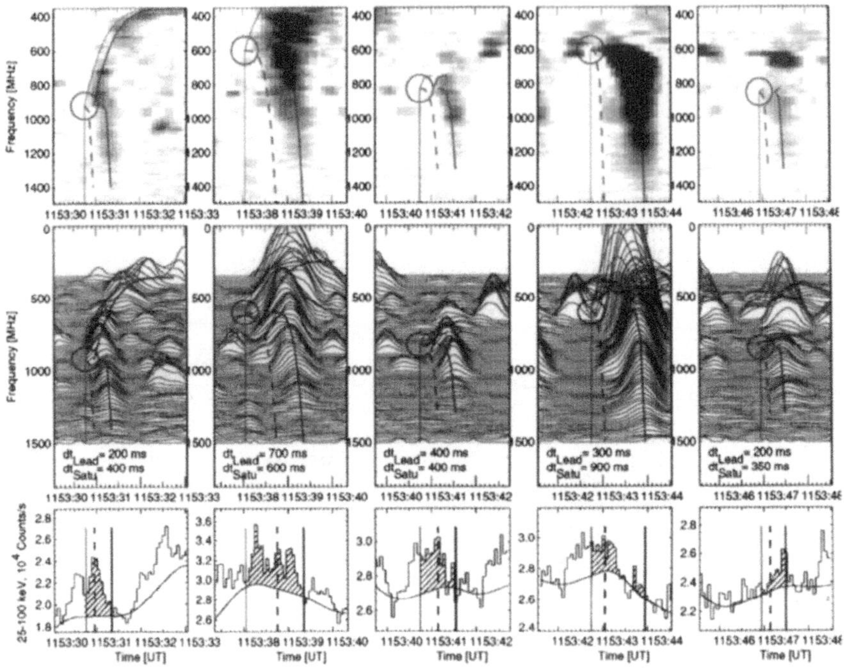

Figure 118. Five reverse-slope (RS) radio bursts observed with *ETH* Zurich (top and middle panels) and 25-100 keV hard X-ray count rate observed with *BATSE/CGRO* (bottom panels). All time intervals have a length of 3.0 s, the time resolution of the radio data is 100 ms and for hard X-rays 64 ms. The start time of the radio bursts is marked with a thin line, the leading edge with a dashed curve, and the peak time with a thick solid curve. The envelope-subtracted hard X-ray flux during the radio start and peak are hatched, consisting of a single of multiple pulses [from Aschwanden, Benz, and Schwartz, 1993a].

the radio-emitting electrons have a lower energy than the hard X-ray-emitting electrons. The minimum energy required to reach the chromosphere is 5.7 keV (for fundamental emission) or 2.3 keV (for harmonic plasma emission), based on the electron density estimated from the plasma frequency ($f < 1240$ MHz) and Eq. 130. The height difference h between the acceleration site and the precipitation site can be calculated from the radio delay

$$t_{R2} - t_{R1} = \frac{h}{\mathrm{v}} + \frac{h \cos \vartheta}{c} \tag{186}$$

which consists of the electron beam travel time and the difference of the light travel path (with an aspect angle of $\vartheta = 42^0$ for this flare). The height difference, based on the radio delay of $t_{R2} - t_{R1} = 0.2 \pm 0.1$ turns out to be $h = 8000 \pm 3000$ km. There are a number of other possible delays that could contribute to the radio delay with respect to the hard X-rays, such as the growth time for Langmuir waves, group velocity delays, radio scattering delays, wave ducting delays, or light travel path difference between the observing spacecraft in hard X-rays and the ground-

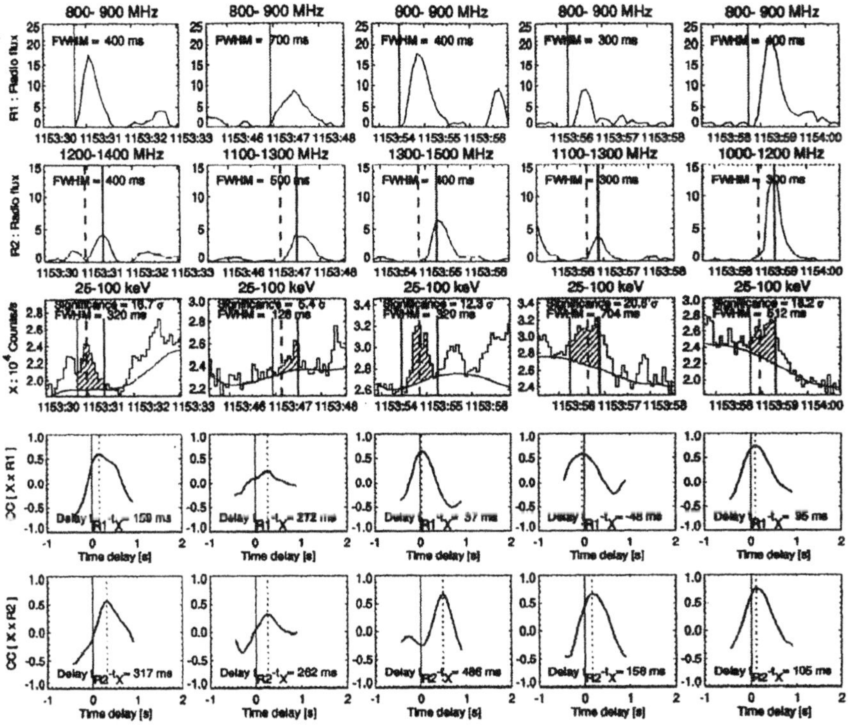

Figure 119. Cross-correlation of the time profiles of five reverse-drifting radio bursts and hard X-ray peaks. The radio flux is shown at the injection frequency f_1 (top), and at the highest frequency f_2 of the reverse-drifting bursts (second row from top). The hard X-ray flux (middle row) above the lower flux envelope is hatched between the injection time (thin line) and the radio peak time (thick line). The cross-correlation between the background-subtracted hard X-ray flux and the radio fluxes R1 and R2 are shown in the forth and fifth row. Note that the radio emission is generally delayed [from Aschwanden, Benz, and Schwartz, 1993a].

based radio observatory. These delays have all been quantified for this particular flare and are listed in Table XIII. From this it was concluded that radio emission is produced most likely at the harmonic plasma frequency and that the two largest contributors to the observed radio-hard X-ray delays ($\Delta t = 270 \pm 150$ ms) are the lower energy of radio-emitting electrons ($\Delta t < 90$ ms) and the growth time for Langmuir waves ($\Delta t < 150$ ms). This study demonstrates that the understanding of the coincident timing of radio and hard X-ray signatures requires a detailed modeling of the geometry, kinematics, and emission mechanisms.

9.2.5. *Electron Beams and Chromospheric Evaporation*

In the standard flare scenario of Kopp and Pneuman (1976), the cusp of a new-reconnected magnetic field line is connected with the outer footpoints of the soft X-ray flare loop. According to our localization of the acceleration region in the cusp we expect that nonthermal particles are accelerated in the cusp and precipitate to the footpoints, where they heat up the footpoints and drive upflows of

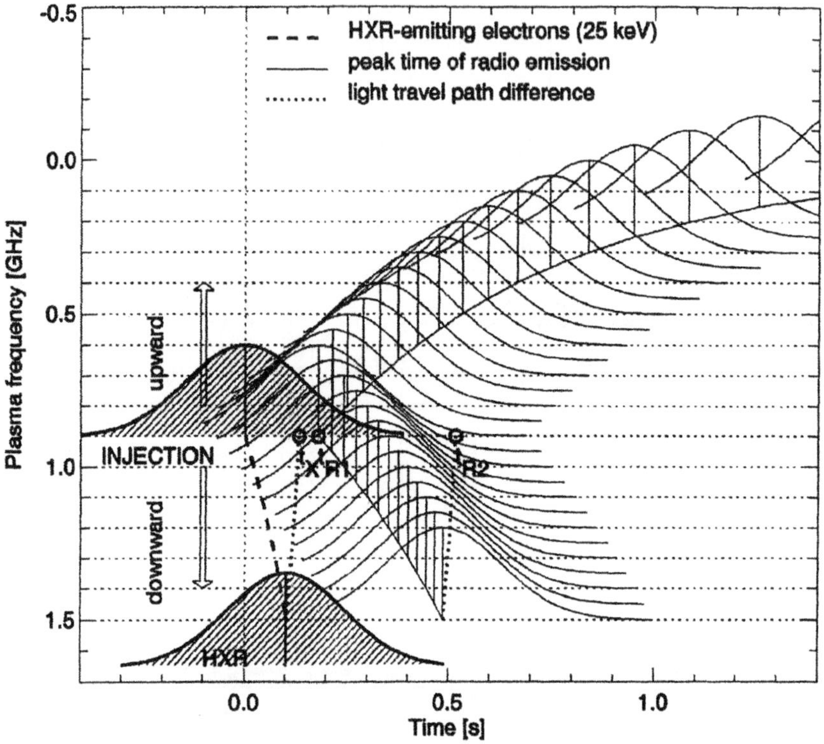

Figure 120. Frequency-dependent timing parameters of an oppositely injected pair of radio bursts. The electron density is assumed to decrease exponentially with altitude, and a constant exciter velocity is applied. The injection of particles at time $t = 0$ is represented by a Gaussian time profile, while the peak times of the upward and downward drifting bursts are delayed by t_{R1}. The observed peak times of the hard X-ray pulse (t_X) and the radio burst at the highest frequency (t_{R2}) are corrected by the light travel path difference (dotted line). The scaling of the relative delays and the duration of the pulses correspond to the observed values [from Aschwanden, Benz, and Schwartz, 1993a].

heated chromospheric plasma. The upflows eventually (typically after 20-30 s) fill the new-connected field line, which relaxes to a dipole-like field line at the outer envelope of the soft X-ray bright flare loop. This is the standard scenario for precipitation-driven chromospheric evaporation (e.g. Sturrock 1973; Canfield et al. 1980; Antonucci et al. 1982).

What are impacts of the chromospheric evaporation model on flare-produced electron beams. We expect that precipitating electron beams, which are detectable as radio type RS bursts under favorable conditions (Section 9.2.1), occasionally intercept the upward-moving chromospheric evaporation front. The electron density of upflowing chromospheric plasma has typically values of $n_e \approx 10^{10} - 10^{11}$ cm^{-3}, which corresponds to plasma frequencies of $f_p \approx 1 - 3$ GHz. Typical parameters of chromospheric upflows are, based on Doppler shift measurements in Ca XIX, $v_{CE} = 270 \pm 90$ km s^{-1}, $T_e = 16.5 \pm 2.4$ MK, $n_e = 0.8 - 1.5$ cm^{-3}, $f_p = 2.5 - 3.5$ GHz (Antonucci, Gabriel, and Dennis, 1984). We can estimate the effect of the

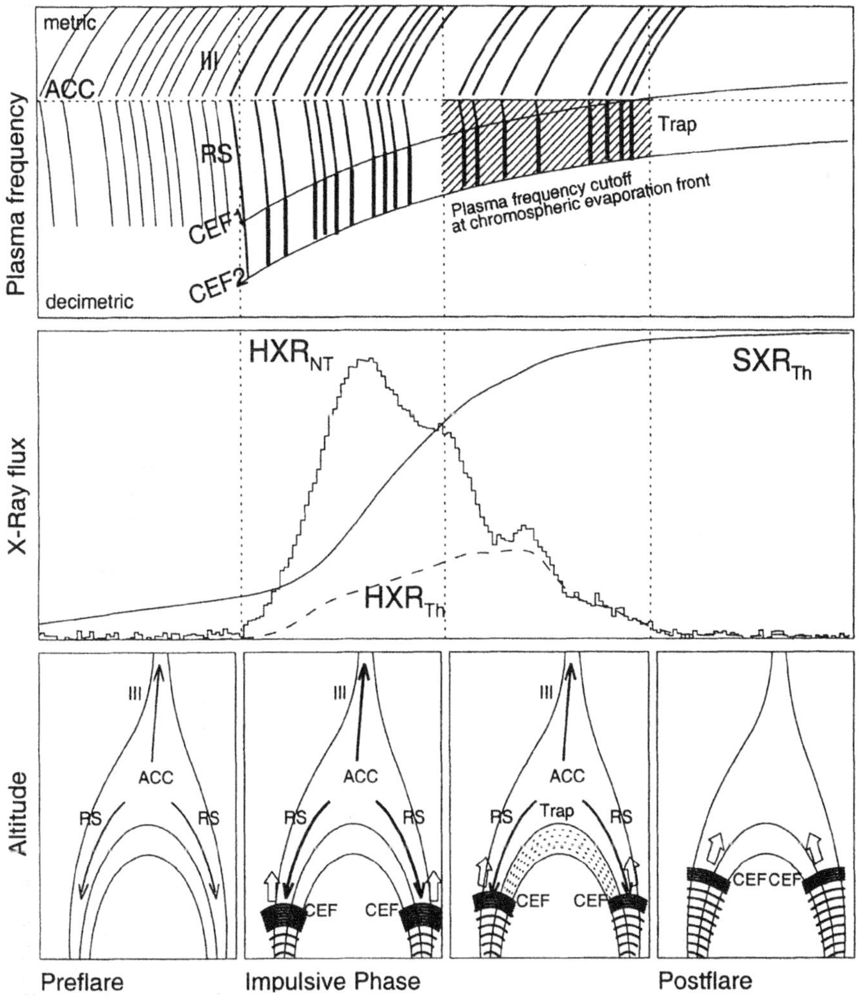

Figure 121. Flare scenario illustrating the evolution of a chromospheric evaporation front (CEF), nonthermal hard X-ray emission (HXR_{NT}), thermal hard X-ray emission (HXR_{Th}), soft X-ray emission (type III bursts, reverse-slope (RS) bursts, and plasma emission from a trap). ACC denotes the coronal acceleration region. In the radio dynamic spectrum (top) a slowly-drifting plasma frequency cutoff is expected owing to the density enhancement conveyed by the chromospheric evaporation front; CEF1 and CEF2 mark the plasma frequencies corresponding to the low-density (CEF1) and high-density edge (CEF2) at the evaporation front. The density of the upflowing chromospheric plasma is indicated with contour lines (bottom), marking a steep gradient at the evaporation front that corresponds to the jump of the plasma frequency from CEF1 to CEF2 (top). This regime CEF1-CEF2 is also the frequency range where the opacity changes for plasma emission are most dramatic, and where the drift rate of type III bursts becomes almost infinite [from Aschwanden and Benz, 1997].

TABLE XIII

Physical Parameters and Time Delays in Flare 1992-Sep-6, 1153 UT [from Aschwanden, Benz, and Schwartz, 1993a]

Parameter	Symbol	Value	
Injection frequency	f_1	$=880\pm50$ MHz	
Maximum frequency	f_2	$=1240\pm100$ MHz	
Average drift rate	df/dt	$=2350$ MHz s^{-1}	
Average drift time	t_{drift}	$=150$ ms	
Plasma emission level	s	=1 (fundamental)	=2 (harmonic)
Electron density	$n_e(f_2)$	$=1.9\times10^{10}$ cm^{-3}	$=0.5\times10^{10}$ cm^{-3}
Minimum beam velocity	v_B/c	>0.15	>0.10
Minimum beam energy	E_B	>5.7 keV	>2.3 keV
Minimum altitude	H_{min}	>6000 km	>4000 km
Maximum altitude	H_{max}	$<11{,}000$ km	$>11{,}000$ km
HXR/radio propagation delay	dt_{del}	<70 ms	<90 ms
Thermal collision time	t_{coll}	<60 ms	<20 ms
Total growth time	t_{growth}	<40 ms	<150 ms
Group delay	t_{group}	<20 ms	<1 ms
Scattering delay	$t_{scatter}$	<30 ms	<30 ms
Wave ducting delay	t_{duct}	<20 ms	<20 ms
Spacecraft orbit delay	t_{orbit}	<2 ms	<2 ms
Estimated HXR/radio delay	dt_{theo}	<190 ms	<310 ms
Observed HXR/radio delay	$t_{R2}-t_X$		$<270\pm150$ ms

chromospheric evaporation front on precipitating electron beams by investigating how the free-free opacity changes across the steep density and temperature gradient. The free-free absorption coefficient in typical coronal conditions is (e.g. Dulk 1985),

$$\kappa_{ff} \approx 0.2 n_e^2 T^{-3/2} f^{-2} \text{ cm}^{-1} \tag{187}$$

The resulting optical depth along the path r is

$$\tau_{ff}(r) \int_r^\infty \kappa_{ff} dr = 2.5 \times 10^{-9} \frac{1}{s^2} \int_r^\infty n_e(r) T(r)^{-3/2} dr. \tag{188}$$

The observed radio brightness is reduced by the absorption factor $e^{-\tau_{ff}}$. If the intrinsic brightness temperature of the radio source is characterized by the spectrum $T_{B0}(f)$, the observed spectrum is

$$T_B(f) = T_{B0}(f) \exp^{-\tau_{ff}(r[f])}, \tag{189}$$

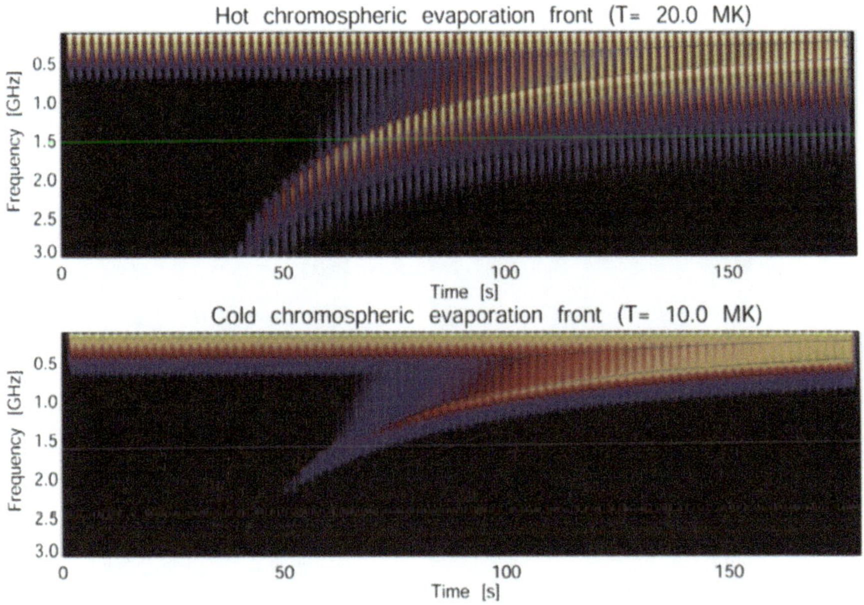

Figure 122. Simulated evolution of the radio dynamic spectrum during the first 3 minutes of a flare within the scenario outlined in Fig. 121. The radio spectrum is calculated for a periodic sequence of bursts emitting at the fundamental plasma frequency and propagating along the center of an overdense loop with a radius of 2 Mm. The curved trajectories mark the plasma frequency at the low- and high-frequency edges of a chromospheric evaporation front propagating with a velocity of 300 km s^{-1} upward into the corona, with a temperature of 20 MK (top), or 10 MK (bottom). The density enhancement in the evaporation front is assumed to be a factor of 4. Note the enhanced radio brightness and the high-frequency cutoff associated with the (slowly-drifting) passage of the chromospheric evaporation front [from Aschwanden and Benz, 1997].

where the frequency dependence of the location $r(f)$ is given by the plasma frequency relation $f(r) = s \times 8980\, n_e(r)^{-1/2}$ and the density model $n_e(r)$. Employing the two-component model described in Section 9.2.2 and adding a third component to characterize the density enhancement of the chromospheric evaporation front, as illustrated in Fig. 112, we can calculate a radio spectrum $T_B(f, t)$ (Aschwanden and Benz 1997), which is rendered in Fig. 122. What we see is that the chromospheric evaporation front produces a high-frequency cutoff for plasma emission from precipitating beams, where the radio brightness is enhanced at the evaporation front, due to the lower opacity in the much higher temperature, but the radio brightness becomes drastically suppressed upstream the evaporation front due to the much higher density. Thus, the chromospheric upflows essentially produce a slowly-drifting cutoff, which has a drift rate of

$$\left(\frac{df}{dt}\right)_{CE} = \frac{df}{dh}\frac{dh}{dt} \approx -\frac{f}{\lambda_T} v_{CE} \cos\vartheta \ , \qquad (190)$$

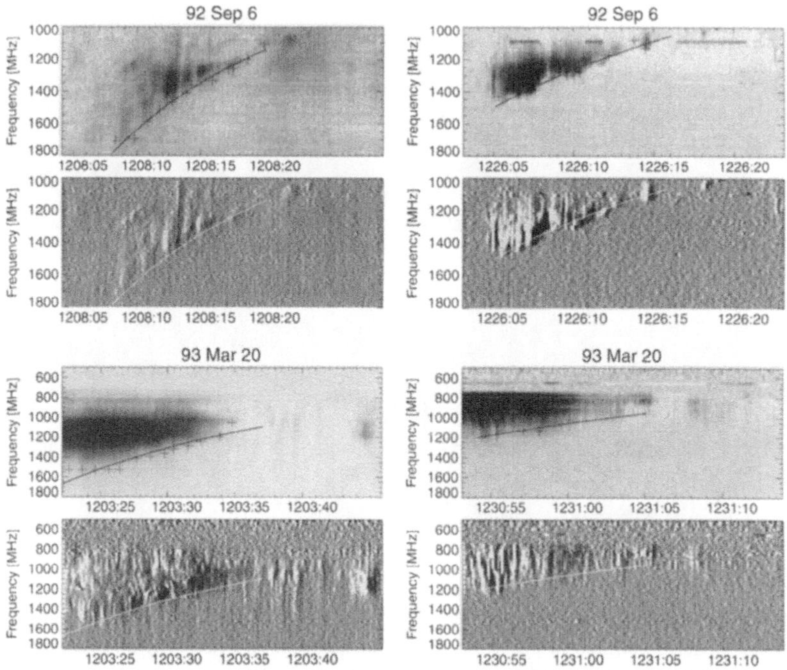

Figure 123. Four episodes with type III emission and a drifting high-frequency cutoff from the flares on 1992-Sep-6 (top rows) and 1993-Mar-20 (bottom rows). For each case the background-subtracted (upper frame) and the gradient-filtered (lower frame) radio flux is shown. The drift rates of the fitted curves at the high-frequency cutoffs are $df/dt = -82$, -53, -56, and -42 MHz s^{-1} [from Aschwanden and Benz, 1997].

yielding drift rates of $df/dt \approx -10... -100$ MHz s^{-1} for typical chromospheric upflow speeds. Such slowly-drifting high frequency cutoffs have indeed been observed for groups of type III bursts during various flare phases, e.g. see Fig. 123. These cutoffs occurred around $\approx 1-3$ GHz, as expected for densities of $n_e \approx 10^{10} - 10^{11}$ cm^{-3}. These observations corroborate the standard flare model with precipitation-driven chromospheric evaporation further and illustrate that electron beams detected as decimetric RS bursts can be used as a diagnostic of the density and upflow speed of chromospheric high-density plasma.

9.2.6. *Interplanetary Electron Beam Propagation*

A review of interplanetary electron beams is beyond the scope of this review, which is mainly focused on acceleration and propagation of electron beams in the immediate flare environment. The relative timing of propagating electron beams to the reconnection region in flare sites can be measured with sufficient accuracy in the solar corona to warrant their mutual relation, but becomes progressively less certain with increasing distance into the heliosphere, although type III trajectories have been traced far out to 35 solar radii (Stone and Fainberg 1971). For reviews on

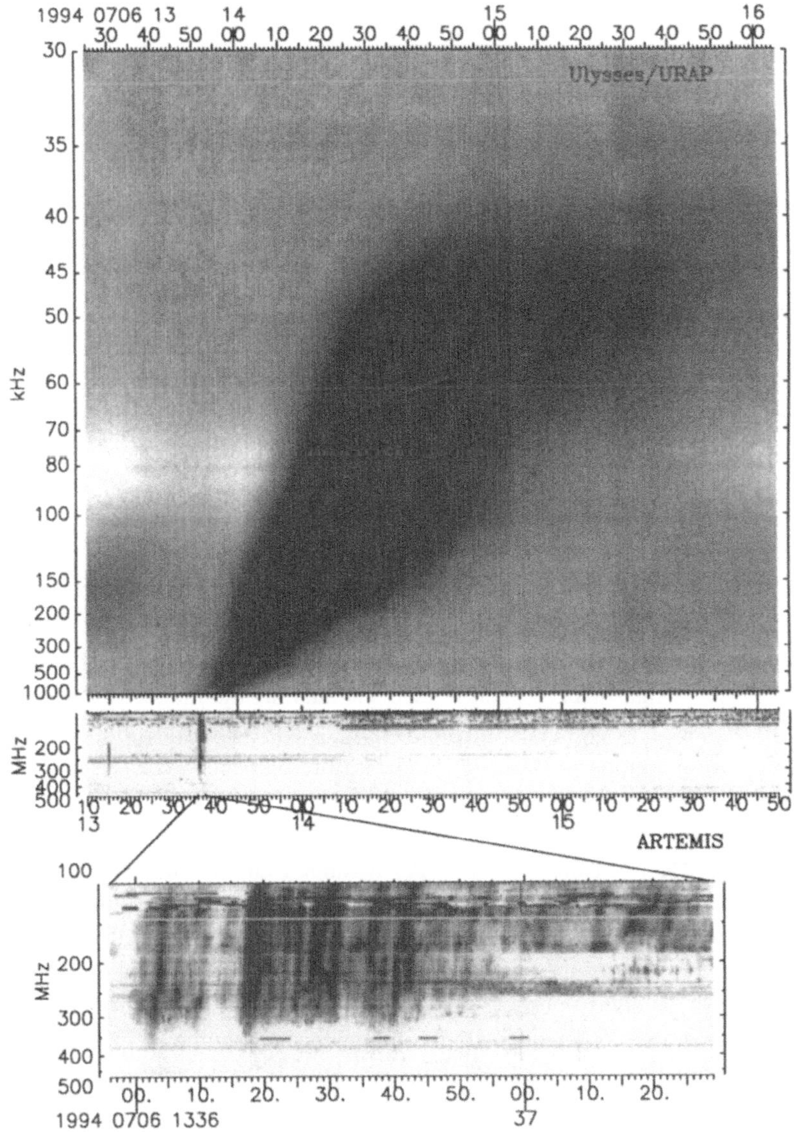

Figure 124. Artemis and Ulysses observations of a group of solar metric type III bursts and an interplanetary type III radio burst. The latter is represented with frequency decreasing down as is common for solar type III bursts. The scale is chosen in such a way that its prolongation to higher frequencies gives the starting point of the interplanetary burst in the corona. It is seen that this point coincides with the low-frequency starting point of the solar type III group. The lowest panel shows the high time resolution dynamic spectrum of the type III group identifying it as a large group consisting of many members [from Poquérusse et al. 1995].

interplanetary particle beams the reader is referred to, e.g. Lin (1974, 1993, 2000), Dulk (1990), Aschwanden and Treumann (1997), and Reames (1999, 2000). Here we want just to stress two facts from the latest developments that are most relevant for the connection between coronal and interplanetary electron beams.

First of all, the number of nonthermal electrons escaping from a flare into interplanetary space, compared with the number of nonthermal electrons that precipitate to the chromosphere and produce hard X-rays, is traditionally known to be small, in the order of $10^{-3}...10^{-2}$ (e.g. Lin, 1974). In Section 3.6 we argued that bi-directional electron beams represent a manifestation of some symmetry in the accelerator, e.g. a symmetric current sheet oriented in vertical direction, which would suggest also a symmetric number of electrons accelerated in direction to the chromosphere and toward interplanetary space. The reason why such a small number escapes into interplanetary space seems to be that most of the upward-going magnetic field lines are part of large-scale closed-field configurations. This is evident from flares that exhibit quadrupolar reconnection (Section 2.3) and from frequent observations of flare-related U-bursts (Section 9.2.3). So, given the dominance of closed-field regions over active regions, where virtually all flares are harbored, how can flare-accelerated electrons have access to interplanetary space at all? The gateway to interplanetary space seems to be faciliated by dynamic flare processes (Section 4.1) that open temporarily and locally the field, such as filament eruptions, coronal mass ejections, or some particular magnetic reconnection processes that involve open field lines, such as the magnetic breakout model (Antiochos et al. 1999). The timing between the temporary opening of the magnetic field in flaring regions and the occurrence of interplanetary type III bursts, however, has not yet been studied in detail.

What is the relation of interplanetary electron beams to coronal radio type III bursts? We cite from the review of Aschwanden and Treumann (1997): Naively, one believes that interplanetary electron beams are but the prolongation of solar type III electrons beams into interplanetary space. The connection between both interplanetary electron beams and type III bursts and coronal electron beams and solar type III bursts is not that simple. Lin (1996) has undertaken an investigation on this connection and has concluded that a coronal source of interplanetary electron streams related to solar type III bursts exists. The most sophisticated analysis of this relation is provided by Poquérusse et al. (1995), who used the newly-developed technique of the Nançay *Artemis* spectrograph to relate ground-based observations of solar type III radio bursts with *Ulysses* measurements of interplanetary type III radiation. Poquérusse et al. (1995) find that to almost every interplanetary type III burst, whose spiral field line is connected to the visible disk or limb of the Sun, a group of solar type III bursts can be related. Fig. 124 shows an example of such an *Artemis-Ulysses* reconstruction of the connection between solar and interplanetary type III bursts. Extending the slope of the leading edge of the interplanetary type III emission across the measurement gap down to the higher frequencies of the coronal band, the origin of the interplanetary type III burst is found

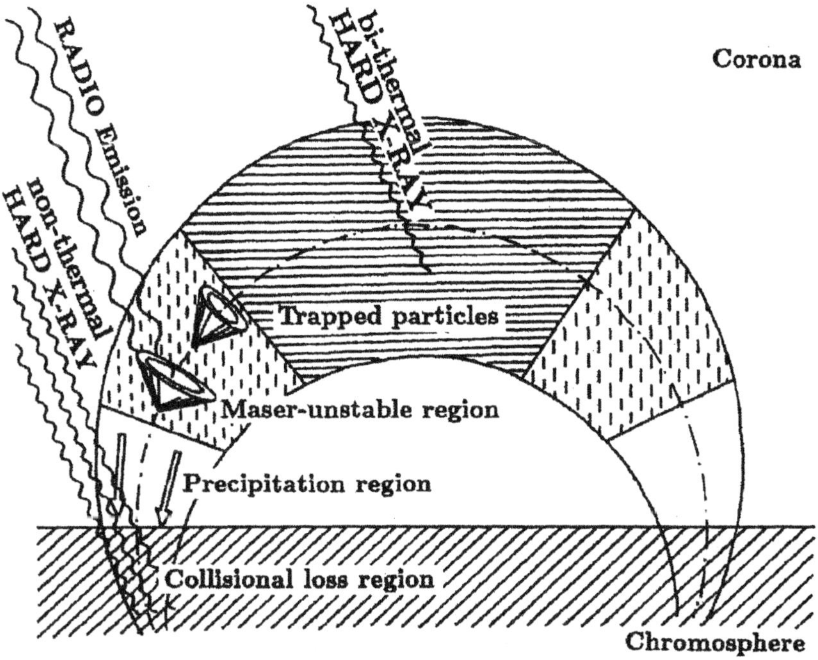

Figure 125. Schematic drawing of radio and hard X-ray sources in a magnetic loop trapping energetic electrons [from Aschwanden, Benz, and Kane, 1990].

to coincide with the end of the coronal type III group. This work must be considered as a major scientific achievement. It has a number of very important implications mentioned by Poquérusse. The first is that almost all interplanetary type III bursts are rooted in coronal type III bursts. Their electron beams originate in the solar atmosphere and propagate thus stably all the way across the corona. Inversely, not every solar type III burst, even if strong, produces an interplanetary type III burst. To become an interplanetary electron beam requires additional conditions to be satisfied, presumably open magnetic fields and possibly other conditions as well. Moreover, type III bursts giving rise to interplanetary type III bursts and electron beams usually appear in groups. These groups are not resolved at low frequencies when they merge in interplanetary space. In the solar atmosphere, each of the bursts lasts up to several seconds only. These groups of electrons obviously merge into one broad beam due to velocity dispersion, when entering interplanetary space. It should be noted, however, that the minute-scale oscillations of the electron flux in interplanetary type III electron beams reported may still memorize the origin of the beam, resulting from many bursty injections in the corona.

9.3. Radio Diagnostics of Electron Trapping

There are two different types of radio emission in trap-plus-precipitation scenarios: One component is the incoherent emission from the trapped particles itself (e.g. gyrosynchrotron and free-free emission), and the other component is coherent emission that results from the anisotropic velocity distribution of the trapped particle population, e.g. from the losscone distribution at the precipitation site (Fig. 125). The losscone instability can excite a number of waves detectable at radio wavelengths, e.g. electromagnetic waves (X, O, Z, and W magneto-ionic modes, see Fig. 54), also known as electron-cyclotron maser emission, as well as electrostatic waves (e.g. upper hybrid waves).

Radio diagnostics from trapped particles involves modeling of the particle injection function, the dynamics of particle orbits in the trapping magnetic field, the evolution of anisotropic particle distributions, the resulting wave-particle interactions, and radio emission mechanisms. This can lead to fairly complex models, from which it is difficult to invert physical parameters, given the observed time profiles and spectra that represent a convolution of all these processes. A comprehensive coverage of this subject is beyond the scope of this review, but we provide some representative literature references: Trap-plus-precipitation models have been developed in Kennel and Petschek (1966), Benz and Gold (1971), Melrose and Brown (1976), Alexander (1990), McClements (1990a, 1992), Lee, Gary, and Shibasaki (2000); The particle dynamics of trapped electrons is described in Melrose and White (1979, 1981), White, Melrose, and Dulk (1983), Craig, MacKinnon and Vilmer (1985), Vilmer, Trottet, and MacKinnon (1986), McClements (1990b), Hamilton and Petrosian (1990), Hamilton, Lu, and Petrosian (1990); Radio emission from losscone-type distributions is studied in Berney and Benz (1978), Benz (1980), Melrose and Dulk (1982), Hewitt, Melrose, and Roennmark (1982), Sharma, Vlahos, and Papadopoulos (1982), Hewitt and Melrose (1983, 1985), Hewitt, Melrose, and Dulk (1983), MacKinnon et al. (1983), Sharma, and Vlahos, (1984), Melrose, Hewitt, and Dulk (1984), Pritchett (1984, 1986), Wu (1985), Winglee (1985a,b), Winglee and Dulk (1986a,b,c), Robinson (1986, 1988, 1989, 1991a,b), Vlahos and Sprangle (1987), Vlahos (1987), Aschwanden and Benz (1988a,b), Aschwanden (1990a,b), Smith and Benz (1991), Charikov and Fleishman (1991), Fleishman and Charikov (1991), Benz (1993, chapter 8), Charikov, Mosunov, and Prokopjev, (1993), Willes and Robinson (1996), Conway and MacKinnon, (1998), Conway and Willes (2000), Fleishman and Arzner (2000), Fleishman and Melnikov (1998). There is a large body of literature on interpretations of particular radio observations in terms of losscone driven emission, which is omitted here. In the following we describe just a few selected observations that demonstrate which physical parameters of trapped particle distributions can be diagnosed with radio observations during solar flares.

9.3.1. *Trapping Times*

A first ingredient that needs to be known in trap models is the particle injection function. To a good approximation, the nonthermal $\gtrsim$25 keV hard X-ray flux $F_X(t)$ can often be used as a proxi for the electron injection function $f^{inj}(t)$. Although the hard X-ray time profile consists itself of a convolution of the electron injection function with a directly-precipitating component and a trap-plus-precipitating component, the collisional trapping time for 25 keV electrons in flare loops is fairly short, i.e. $t^{defl} \approx 0.1$ s in flare loop densities of 10^{11} cm^{-3}. The trapping time of radio-producing electrons (for which the collisional deflection time $t^{defl} \propto E^{3/2}/n_e$ is an upper limit, is generally significantly longer, either because they have a higher energy (e.g. $E \gtrsim 300$ keV for gyro-synchrotron emission) or because they are trapped in lower density regions in the cusp of flare loops. Therefore, to model the much longer trapping time of radio-emitting electrons we can approximate the injection function with the hard X-ray flux, i.e. $f^{inj}(t) \approx f_X(t)$. The instantaneous number of electrons in the trap, $n^{trap}(t)$, is then the convolution of the injection function $f^{inj}(t)$ with an e-folding trapping time t^{trap}, where only particles with a pitch-angle larger than the losscone angle are trapped ($\alpha > \alpha_0$),

$$n^{trap}(t) = \frac{1}{t^{trap}} \int_0^t f^{inj}(t', \alpha > \alpha_0) \exp[-\frac{(t-t')}{t^{trap}}] \, dt' \,. \tag{191}$$

This trap number density $n^{trap}(t)$ is relevant for radio emission from trapped particles, e.g. for gyrosynchrotron emission from the trap. The gyrosynchrotron emissivity depends also on the magnetic field $B(\mathbf{x})$, the aspect angle $\theta(\mathbf{x})$, and the energy injection spectrum $f^{inj}(E)$, which vary for every location inside the trap, and has to be modeled to interpret imaging data (e.g. Kucera et al. 1993). For total flux measurements, Eq. 191 can be used to determine an average trapping time t^{trap}. If the trapping time is energy-dependent, as it is for the collisional deflection time (Eq. 104), the average trapping time is weighted by those electron energies that contribute most to the radio emissivity, as it can be seen from relations between electron spectrum and peak frequency of a gyrosynchrotron spectrum (e.g. Dulk 1985).

The second component of radio emission from trapped particles, the loss conedriven coherent emission, depends on the number of electrons that form a losscone distribution. This number density is a mixture of trapped and precipitating electrons and requires detailed modeling of the particle dynamics (e.g. see Hamilton et al. 1990). In first order, the number of electrons scattered into the losscone is proportional to the number of trapped electrons,

$$\left(\frac{dn^{loss}(t)}{dt}\right) = \frac{dn^{trap}(t)}{d\alpha}\frac{d\alpha}{dt} \approx \frac{n^{trap}(t)}{\alpha_0} < \frac{d\alpha}{dt} > , \tag{192}$$

so that we obtain with Eq. 191,

$$\left(\frac{dn^{loss}(t)}{dt}\right) = \frac{< d\alpha/dt >}{\alpha_0 \, t^{trap}} \int_0^t f^{inj}(t', \alpha > \alpha_0) \exp[-\frac{(t-t')}{t^{trap}}] \, dt' \,. \tag{193}$$

Radio emission generated by a losscone is in first order proportional to the number of precipitating electrons, which consists of directly-precipitating electrons (with $\alpha < \alpha_0$) plus the trapped-precipitating electrons (with initial $\alpha < \alpha_0$),

$$S(t) \propto f^{inj}(t, \alpha < \alpha_0) + \frac{<d\alpha/dt>}{\alpha_0\, t^{trap}} \int_0^t f^{inj}(t', \alpha > \alpha_0) \exp[-\frac{(t-t')}{t^{trap}}]\, dt' \, . \tag{194}$$

The latter equation can be used to deconvolve the average trapping time t^{trap}, if the injection function $f^{inj}(t)$ is known, e.g. in form of the hard X-ray profile $f_X(t)$ as proxi. For an isotropic injection function, Eq. 194 can we expressed in terms of a parameter q^{prec} that quantifies the fraction of directly-precipitating particles,

$$S(t) \propto q^{prec} f^{inj}(t) + (1 - q^{prec}) n^{trap}(t) \, , \tag{195}$$

or the fraction $q^{trap} = 1 - q^{prec}$ of trapped particles, respectively.

An example of decimetric radio emission that is highly correlated with hard X-rays but is delayed by some amount, is the burst type of *decimetric millisecond spikes*. Four cases with simultaneous hard X-ray and radio data are shown in Fig. 126. Using the hard X-ray data as proxi for the electron injection function, a convolution with an exponential decay time (Eq. 194) was performed to fit the radio data. The four cases shown in Fig. 126 exhibit radio delays of 4.2 s, 2.4 s, 4.0 s, 8.8 s, for those structures that appear to be correlated with hard X-ray enhancements. Statistics of some 27 such events yielded a typical delays in the range of $t^{delay} \approx 2 - 5$ s (Aschwanden and Güdel, 1992). *Decimetric millisecond spikes* are commonly interpreted in terms of losscone driven electron-cyclotron maser emission, because of their high brightness temperature, short time scales, and narrow bandwidth (e.g. Holman, Eichler, and Kundu 1980; Melrose and Dulk 1982; Benz 1986; Aschwanden and Benz 1988b; Aschwanden 1990a). The observed delay of the decimetric millisecond spikes with respect to the hard X-ray emission could be interpreted as trapping delay within the model outlined in Eqs. 192–194. The question is now what the physical meaning of this trapping delay is. The maser growth rate is as fast as a fraction of a microsecond for the magneto-ionic X- and O-mode (Melrose and Dulk, 1982), and thus a losscone would be transformed into a relaxed plateau distribution within a few growth times by quasi-linear diffusion of the gyro-resonant waves. Positive growth of electron-cyclotron maser emission occurs only in an undamped regime of the velocity distribution where the resonance condition is not affected by thermal suppression (Fig. 127), say above a critical energy E_{crit}. Electrons with large initial pitch angles, which are populated outside the losscone, $\alpha > \alpha_0$, mirror forth and back in the trap until they become scattered towards the losscone edge $\alpha \approx \alpha_0$ by another pitch-angle scattering mechanism, e.g. by Coulomb collisional scattering. In this scenario, every losscone driven emission of particles with energies $E > E_{crit}$ is delayed by the scattering delay from large pitch-angles to small pitch-angles. Consequently, we can interpret our measured trap delay of $t^{trap} \approx 2 - 5$ s as collisional trapping time of the

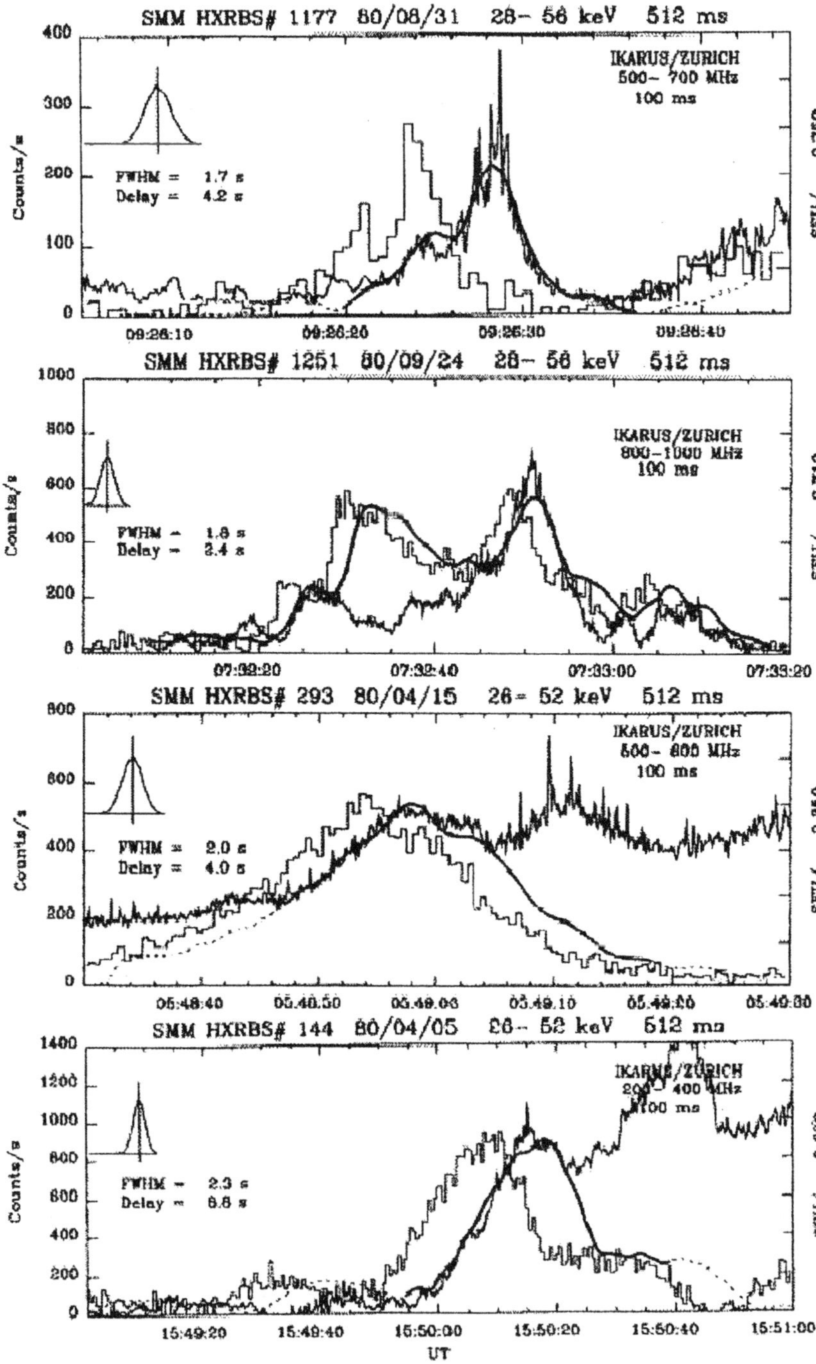

Figure 126. Convolution fits (thick solid curve) of the hard X-ray emission (histogrammed) to the radio flux (fluctuating curve) of four *decimetric millisecond spike* events. All time profiles are background-subtracted. The Gaussian kernel function, the FWHM, and the delays are indicated on the left-hand side of each frame. Note that not every hard X-ray peak has a radio counterpart, but those which have can be fitted well with a convolution function. The radio emission is delayed by 2.4-8.8 s [from Aschwanden and Güdel, 1992].

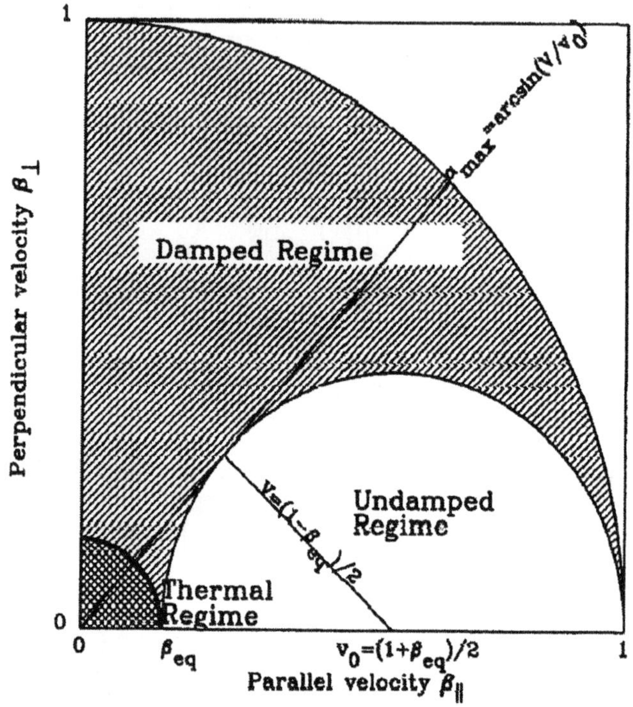

Figure 127. The undamped regime of positive growth rates for electron cyclotron maser emission in velocity space. Approximating the resonance ellipses by circles, the undamped regime is confined by a circle with radius $V = (1 - \beta_{eq})/2$. This is the envelope of all undamped resonance circles. The geometrical sketch shows which part of the thermal tail can not be affected by quasi-linear maser diffusion, mainly low energetic particles with higher pitch angles [from Aschwanden 1990b].

large pitch-angle electrons with energies $E > E_{crit}$. For typical flare loop densities of $n_e \gtrsim 10^{10}$ cm^{-3}, a collisional deflection time of $t^{Defl} = 2 - 5$ s required electrons with energies $E_{crit} > 35 - 65$ keV, which corresponds to relativistic velocities of $\beta > 0.35 - 0.45$. This value seems to fit a large range of losscone angles according to the diagram of the undamped regime shown in Fig. 127. The measurement of trapping delays can this way constrain the energy thresholds of losscone instabilities.

9.3.2. *Ratio of Direct-Precipitating to Trap-Precipitating Electrons*

Microwave emission from solar flares often consists of (1) impulsive peaks that are highly correlated with near-simultaneous hard X-ray peaks, and (2) slowly-decaying tails in their time profiles. An example of a flare with a simple spiky time structure is shown in Fig. 128, observed with *Yohkoh/HXT* and the Nobeyama radio telescope at 17 and 34 GHz (Kundu et al. 2001). This simple time profile clearly shows the two components of an impulsive peak and a gradual decay, which can be naturally modeled with the simple trap model as outlined in Eqs. 192–194. The fitting of this trap model reveals a trapping time of $\tau = 12.9$ s at 17 GHz and

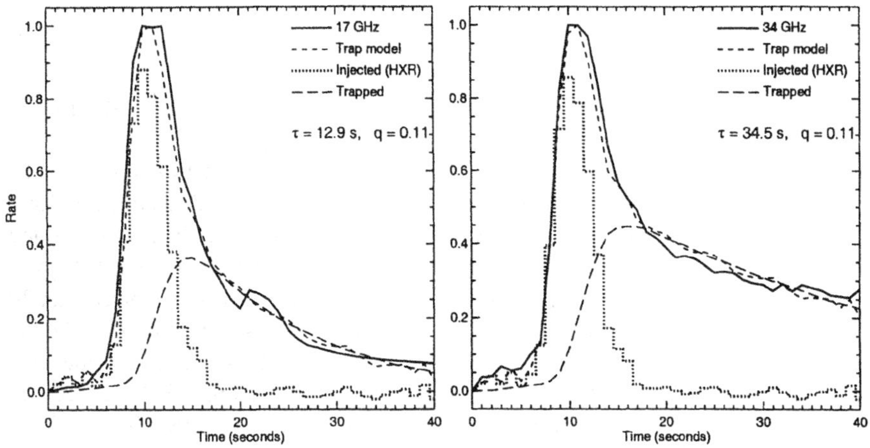

Figure 128. Comparison of the radio time profiles for the event on 1998-Jun-13 for 17 GHz (left) and 34 GHz (right) with a trap model (short-dashed line) derived using the *Yohkoh/HXT* 53-93 keV hard X-ray time profile (dotted histogram) as an injection function. The radio time profile is modeled as the sum of a component identical to the hard X-ray time profile (the injection function) and a trapped component (long-dashed line) derived by integrating over the injection function convolved with an exponential decay term (time constant τ). All time profiles are normalized to a peak of unity, and the parameter q specifies the relative contributions of the injected and trapped components [from Kundu et al. 2001].

$\tau = 34.5$ s at 34 GHz. These time scales are compatible with collisional deflection time scales of $E \approx 100 - 250$ keV electrons in densities of $n_e \approx 10^{10}$ cm^{-3}, and thus may explain the slowly-decaying tail of the emission in terms of trapping of gyro-synchrotron emitting electrons. However, the difference in trapping times between the two frequencies of 17 and 34 GHz is not straightforward to understand, because more or less the same energy range of the electron injection spectrum is expected to contribute to the optically thin frequencies. In the same study, also the relative ratio of the radio flux produced by direct-precipitating and trap-precipitating electrons was determined, according to Eq. 195, yielding a ratio of $q = 0.11$. While this ratio is directly proportional to the electron numbers of the two components for hard X-ray bremsstrahlung (Eq. 100–101), the situation is somewhat more complicated for gyrosynchrotron emission. Gyro-synchrotron is not simply proportional to the number of energetic electrons, but also highly sensitive to the pitch angle distribution and the magnetic field. For a powerlaw electron spectrum (with slope δ and electron density N) the gyrosynchrotron emissivity η varies as function of the angle θ to the magnetic field B (or gyrofrequency f_B) approximately as (Dulk, 1985),

$$\frac{\eta(\delta, \theta)}{BN} \approx 3.3 \times 10^{-24} 10^{-0.52\delta} \sin\theta^{-0.43+0.65\delta} \left(\frac{f}{f_B} \right)^{1.22-0.90\delta} , \qquad (196)$$

Therefore, the radio flux ratio is additionally weighted by the gyrosynchrotron emissivity function $\eta(\theta, \delta)$, which differs in the trapping region near the looptop

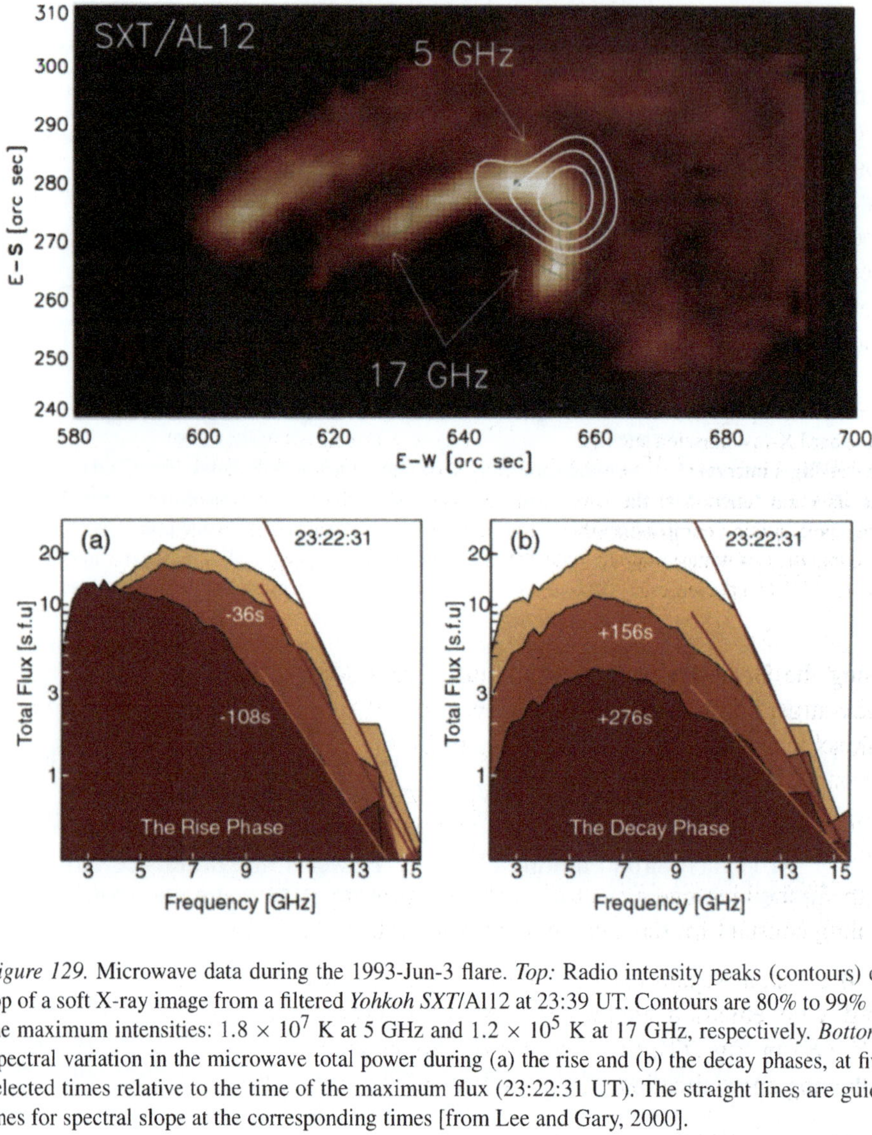

Figure 129. Microwave data during the 1993-Jun-3 flare. *Top:* Radio intensity peaks (contours) on top of a soft X-ray image from a filtered *Yohkoh SXT*/Al12 at 23:39 UT. Contours are 80% to 99% of the maximum intensities: 1.8×10^7 K at 5 GHz and 1.2×10^5 K at 17 GHz, respectively. *Bottom:* Spectral variation in the microwave total power during (a) the rise and (b) the decay phases, at five selected times relative to the time of the maximum flux (23:22:31 UT). The straight lines are guide lines for spectral slope at the corresponding times [from Lee and Gary, 2000].

(θ_T) and in the precipitation sites near the footpoints (θ_P). A more general formulation of Eq. 195 thus needs to include the dependence of the gyrosynchrotron emissivity (Kundu et al. 2001),

$$\begin{aligned} S(t) = & \; Q \int dE \; \eta(E, \theta_I) f_I(E,t) \tfrac{L}{2v} \\ & + (1-Q) \int dE \; \eta(E, \theta_T) n^{trap}(E, \tau_E, t) \; . \end{aligned} \qquad (197)$$

9.3.3. *Pitch-Angle Distributions*

Some microwave observations even allow to make conclusions about the pitch-angle distribution of the particles injected into the trap region. An example of such an observation and analysis is given in Lee and Gary (2000), illustrated in Fig. 129. A soft X-ray flare loop is identified in a *Yohkoh/SXT* image, where the microwave emission at 5 GHz is located at the looptop, and the 11-17 GHz emission near both footpoints (Fig. 129). The magnetic field is found to be asymmetric for this flare loops, with the weaker field at the east side, where a higher direct-precipitation rate is expected, while the west side has a higher trapping efficiency. The microwave spectrum is observed to harden in the decay phase (Fig. 129 bottom right), as expected for trapping in the weak-diffusion limit, because the trapping or collisional deflection time has an energy dependence of $\tau^{defl} \propto E^{3/2}$ (Eq. 137). The authors model the injected electron population with a Gaussian pitch angle distribution,

$$P_E(\mu) = \frac{1}{\sqrt{\pi}\Delta\mu} \exp\left[-4\frac{(\mu-\mu_0)^2}{\Delta\mu^2}\right] , \qquad (198)$$

where $\mu = \cos\alpha$ is the cosine of the electron pitch angle. Interestingly, fitting the microwave time profiles at multiple frequencies, using the model with gyrosynchrotron emission produced by trapped electrons within the given spatial geometry and magnetic field configuration, leads to an anisotropic injection function. They find that the data are best fit by an injection with pitch angles confined to a narrow range of $\Delta\mu(t \approx 0) \approx 0.26$ ($\Delta\theta(t \approx 0) \approx 30^0$) and a short injection time of ≈ 32 s, into a magnetic loop with a relatively low density of $n_e \approx 4.1 \times 10^9$ cm^{-3}. This result provides an interesting constraint for acceleration mechanisms. Direct DC electric field acceleration as well as stochastic acceleration tend to produce strongly beamed distributions with small pitch angles. This analysis demonstrates two effects that make microwave observations useful for diagnosing the injection pitch-angle distribution: (1) the high sensitivity of gyrosynchrotron emissivity on the pitch angle (Eq. 196), and (2) microwaves are produced by the trapped electron population where information on the injection function is preserved.

10. Conclusions

In this review we focused on the kinematic aspects of particle acceleration and propagation, which entail the basic physics of nonthermal particles that is needed to interpret many flare phenomena observed in gamma rays, hard X-rays, and radio wavelengths. The exploration of this research field saw a vigorous development over the last five years, thanks to new observational capabilities such as high-precision timing measurements with the *Compton Gamma Ray Observatory (CGRO)*, with high-resolution imaging with *Yohkoh* in soft and hard X-rays, and with the *Transition Region and Coronal Explorer (TRACE)* in EUV. These new instrument capabilities ideally complement previous observations we traditionally obtained from non-imaging hard X-ray and radio observations over the last 20 years. Significant advances have also been made on the theoretical side and with numerical simulations and modeling, especially those modeling attempts that could be fully constrained with measured observables. In the following we summarize the major lessons we learned on the subject, organized in the same order as the review sections:

1. Magnetic Topology of Acceleration Region : *Yohkoh* and *TRACE* images of solar flares clearly provide compelling evidence for ubiquitous magnetic reconnection processes, which can be subdivided into *X-type reconnection geometries* (2D as well as 3D dipolar, tripolar, and quadrupolar) and *3D-nullpoint reconnection geometries* (fan, spine, and separator reconnection). Each of these basic magnetic configurations can manifest itself in single loops or in multi-loop arcades (the latter also termed double-ribbon flares). Magnetic reconnection processes can involve closed as well as open magnetic field lines, and their dynamic evolution can open the field temporarily during eruptive flare phases, which then tends to close in the post-flare phase, so that we observe mostly relaxed, dipole-like postflare loop systems in soft X-rays and EUV. The dynamic phase, which is manifested in filament eruptions at the onset of flares with subsequent launch of coronal mass ejections, can best be observed in EUV. 3D reconstruction of the topology of magnetic reconnection regions and modeling of its evolution is still in its infancy stage, but provides the most fundamental basis to understand the physics of the electrodynamics that controls particle acceleration and propagation.
2. Geometry of Acceleration Region : The location of particle acceleration regions could be detected by direct imaging of coronal *above-the-loop-top* hard X-ray sources as discovered by Masuda et al., and indirectly inferred by electron time-of-flight measurements, conjugate footpoint constraints, and remote footpoint timing. All methods consistently reveal a location of the acceleration region above the soft X-ray-bright flare loops, coincident with the cusp region in the downward outflow region of magnetic reconnection points. Radio observations of bi-directional electron beams moreover confirm the symmetrical accelerating fields in upward and downward direction, and the plasma frequency at the starting point of radio-emitting electron beams is consistent with a significantly lower electron density in the reconnection region than the high-density loops seen in soft X-rays, which are filled by heated chromospheric upflows.
3. Dynamics of Acceleration Region : Given the omnipresence of fast (subsecond) time structures observed in hard X-rays and in radio bursts, the spectral fragmentation of radio bursts, and the multi-loop structure of EUV postflare arcades, it is clear that the the acceleration region consists of many small-scale elements and operates intermittently in time. A spatio-temporal correlation found between wavelet-based time scales and flare loop sizes indicates a scale-invariant geometry of magnetic reconnection regions and its associated magnetic islands. The high shear observed at the onset of double-ribbon flares strongly suggests a shear-driven bursty reconnection regime, where tearing mode instabilities and coalescence of magnetic islands interleave, producing intermittent episodes of acceleration episodes, a process that was also

reproduced with numerical MHD simulations. Steady Sweet-Parker or steady Petschek reconnection cannot explain the bursty time structures observed in hard X-rays and radio.

4. *Accelerating Electromagnetic Fields and Waves :* Asking the question which acceleration mechanism is operating in solar flares, the answer is that all three known major categories are possible: electric DC-field acceleration, stochastic acceleration, or shocks, as long as they operate on small scales compared with the flare loop size or height of the reconnection region. All we can rule out at this point are large-scale electric DC fields as large as the height of the reconnection region, based on the observational constraints of electron time-of-flight measurements. Acceleration in electric DC fields is possible near X-points in filamentary current sheets or in convective electric fields during coalescence of magnetic islands. Stochastic acceleration is possible in turbulent reconnection outflows. Shock acceleration can happen by the first-order Fermi mechanism in fast shock regions between downward reconnection outflows and the underlying high-density postflare loops, or in mirror trap regions in the cusp of reconnection outflows, as well as in upward propagating shock waves associated with CME fronts.
5. *Particle Kinematics :* Kinematics of relativistic electrons became an important tool to analyze energy-dependent time delays in hard X-rays. The nonthermal hard X-ray emission observed in flares, mainly produced by bremsstrahlung of relativistic electrons in chromospheric thick-target regions, could be disentangled into two components with different energy-dependent timing. The spiky component of hard X-ray time profiles reveal energy-dependent time delays ($\tau \propto E^{-1/2}$) that are consistent with electron time-of-flight dispersion between the coronal acceleration site and the chromospheric energy loss site. The inferred time-of-flight distances scale with the flare loop size and locate the acceleration region in the cusp region of reconnection points. The slowly-varying hard X-ray component reveals energy-dependent time delays that are consistent with Coulomb collisional scattering ($\tau \propto E^{3/2}$), and the inferred electron densities are consistent with flare loop densities measured in soft X-rays. The two timing components confirm the theoretically expected bifurcation of accelerated electrons: electrons with small pitch angles precipitate directly, while electrons injected with large pitch angles are temporarily trapped and precipitate after one collisional time scale. The agreement of the measured delays with the expected collisional time scales indicates trapping in the *weak-diffusion limit*, while there is no evidence found for the *strong-diffusion limit* aided by wave turbulence. A new aspect is the duality of two different trapping regions: a primary trap is given by the cusp geometry surrounding the acceleration region, while a secondary trapping region is given by the traditionally known mirror points near the footpoints of flare loops.
6. *Gamma-Ray Emission :* The small time delay ($\approx 1-2$ s) of prompt nuclear deexcitation line emission in gamma rays with respect to hard X-ray brems-

strahlung contains information on the relative time-of-flight delays between protons and electrons on their propagation from the coronal acceleration site to the chromosphere. For equipartition of the kinetic energies of protons and electrons, one expects a delay of $\approx 4-8$ s between hard X-ray and gamma-ray pulses, for typical flare loop sizes, but it can be as small as $\approx 1-2$ s for small flare loops. Equipartition of kinetic energy in accelerated ions and electrons is also called for from the latest analysis of the 1.634 MeV ^{20}Ne gamma-ray line. By similar time-of-flight arguments it can also be ruled out that the $\gtrsim$ 20 keV bremsstrahlung could be produced by protons or ions, corroborating earlier arguments based on inconsistent hard X-ray/γ-ray flux ratios. Long-term trapping of gamma-ray producing protons is possible for large-scale, low-density loops with sufficient twist. Large-scale loops connected to flare sites are also required to explain gamma-ray emission from behind-the-limb flares.

7. *Hard X-Ray Emission :* Energy-dependent time delays in the order of $\approx 10-100$ ms can now reliably be used to determine electron time-of-flight distances in the framework of the thick-target bremsstrahlung model. This method allows to determine distances between coronal acceleration sites and the chromosphere with an accuracy down to $\lesssim$ 10%, during large flares with high photon count rates and using large-area detectors, such as from *BATSE/CGRO*. The accuracy of the method scales with the photon statistics and detector area. Furthermore, measurements of hard X-ray delays provide also trapping times, but they have to be deconvolved from time-of-flight delays. - A recent discovery is the detection of double ribbons in hard X-rays, which match the double ribbons in EUV spatially as well as regarding time evolution. Previous lack of hard X-ray ribbon detections seems to be rooted in the limitations of Fourier imaging. The discovery of hard X-ray ribbons and their close relation to EUV double ribbons corroborates the standard magnetic reconnection model of Kopp-Pneuman to a large extent.
8. *Radio Emission :* Imaging radio observations map the locations of propagating electron beams at a particular density given by the local plasma frequency. Such radio imaging of electron beams, although rare with not solar-dedicated instruments such as the *VLA*, have confirmed the propagation of electron beams in large-scale secondary flare loops that are interacting with the small-scale primary flare loops seen in hard X-rays. Besides the plasma emission from electron beams, gyrosynchrotron emission from trapped relativistic electron beams has been imaged for many years, which outlines the trapping regions in double-ribbon arcades and large-scale secondary flare loops. - Non-imaging radio observations with high time and spectral resolution provide excellent diagnostic of plasma densities in the acceleration region, the intermittent temporal pattern of particle acceleration, electron beam speeds, the access of accelerated electrons to open and/or closed field lines, the directivity and symmetry of upward and downward acceleration, the interaction of downward propagating electron beams with chromospheric evaporation upflows, trapping times, and ratios of

precipitating and trapped electrons. Combining imaging and spectral information radio data can even be used to constrain the pitch-angle distribution of accelerated electrons.

This review summarizes the achievements of the pre-HESSI era. A next breakthrough in the study of particle acceleration and kinematics is expected from the *High Energy Solar Spectroscopic Imager (HESSI)* mission, which was launched on 2002-Feb-5. *HESSI* will for the first time provide imaging of hard X-ray energies $\gtrsim$ 100 keV and of gamma-rays, will for the first time spectrally resolve many gamma-ray lines thanks to the cooled-germanium detector technology, and thus will allow for unprecedented spectral deconvolution and modeling. We are looking to an exciting future of modeling particle kinematics in solar flares.

Acknowledgements

This review is dedicated to the memory of Drs. Reuven Ramaty and Natalie Mandzhavidze, who passed away on April 8 and 9, 2001. Dr. Ramaty was a renowned expert in cosmic rays, gamma rays and nuclear astrophysics and a leading theorist at NASA's Goddard Space Flight Center, Greenbelt, Maryland. - The author is particularly grateful to Drs. H.S. Hudson and B. Kliem for careful reading, thoughtful commentaries, and corrections of the manuscript. This review benefited from continuous support of flare studies by a number of NASA missions and Guest Investigator Programs (*SMM, CGRO, Yohkoh, TRACE*). Also I like to thank many individuals for insightful discussions, scientific collaborations, provision of image material, and reviewing of this manuscript: (in alphabetical order) David Alexander, Guillaume Aulanier, Tim Bastian, Arnold Benz, John Brown, Brian Dennis, Gordon Emslie, Lyndsay Fletcher, Terry Forbes, Gordon Holman, Hugh Hudson, Judith Karpen, Alan Kiplinger, Bernhard Kliem, Takeo Kosugi, Mukul Kundu, JeongWoo Lee, Jim Lemen, Bob Lin, Satoshi Masuda, Tom Metcalf, Jim Miller, Ron Murphy, Masanori Nishio, Nariaki Nitta, Michel Poquérusse, Richard Schwartz, Boris Somov, Toshi Tajima, Saku Tsuneta, Harry Warren, Stephen White, and Rob Willson. The *CGRO* data used in this review have been provided through the *Compton Observatory Science Support* Center GOF account provided by the NASA/Goddard Space Flight Center. We thank the *BATSE/CGRO* principal investigator team of G.Fishman, W.Paciesas, R.Wilson, and C.Meegan at MSFC, T. McGlynn at the *CGRO* science support center *(GROSSC)*, and the *SDAC* at *GSFC* for access and archival services of *BATSE* data. The *Yohkoh* data used in this review are taken by the *Yohkoh* mission of *ISAS*, Japan, which was prepared and is operated by the international collaboration of Japanese, US, and UK scientists under the support of *ISAS, NASA*, and *SERC*, respectively. The *TRACE* team includes scientists from Lockheed Martin Advanced Technology Center, Stanford University, NASA Goddard Space Flight Center, the University of Chicago, Montana State University, and the Harvard-Smithsonian Center for Astrophysics. This work was supported by NASA contract NAS8-40108 (*Yohkoh*) and NAS5-38099 (*TRACE*).

REFERENCES

Achterberg, A. 1979, *Astron. Astronomy* **76**, 276.

Achterberg, A. and Norman,C.A. 1980, *Astron. Astronomy* **89**, 353.

Alexander, D.1990, *Astron. Astronomy* **235**, 431.

Alexander, D. and Metcalf, T.R. 1997, *Astrophys. J.* **489**, 442.

Alexander, D. and Fletcher, L. 1999, *Solar Phys.* **190**, 167.

Alvarez, H. and Haddock, F.T. 1973, *Solar Phys.* **29**, 197.

Ambrosiano, J., Matthaeus, W.H., Goldstein, M.L., and Plante, D. 1988, *J.Geophys.Res.* **93**, 14383.

Antiochos, S.K. 1998, *Astrophys. J.* **502**, L181

Antiochos, S.K., DeVore, C.R., and Klimchuk, J.A. 1999, *Astrophys. J.* **510**, 485.

Antonucci, E., Gabriel, A.H., Acton, L.W., Culhane, J.L., Doyle, J.G., Leibacher, J.W., Machado, M.E., Orwig, L.E., Rapley, C.G. 1982, *Solar Phys.* **78**, 107.

Antonucci, E., Gabriel, A.H., Dennis, B.R. 1984, *Astrophys. J.* **287**, 917.

Aschwanden, M.J., Wiehl, H.J., Benz, A.O., and Kane, S.R. 1985, *Solar Phys.* **97**, 159.

Aschwanden, M.J.1987, *Solar Phys.* **111**, 113.

Aschwanden, M.J. and Benz, A.O. 1988a, *Astrophys. J.* **332**, 447.

Aschwanden, M.J. and Benz, A.O. 1988b, *Astrophys. J.* **332**, 466.

Aschwanden, M.J., Benz, A.O., Schwartz, R.A., Lin, P.R., Pelling, R.M., Stehling, W. 1990, *Solar Phys.* **130**, 39.

Aschwanden, M.J. 1990a, *Astron. Astronomy* **237**, 512.

Aschwanden, M.J. 1990b, *Astron. Astronomy Suppl. Ser.* **85**, 1141.

Aschwanden, M.J., Benz, A.O., and Kane, S.R. 1990, *Astron. Astronomy* **229**, 206.

Aschwanden, M.J. and Güdel, M. 1992, *Astrophys. J.* **401**, 736.

Aschwanden, M.J., Bastian, T.S., Benz, A.O., and Brosius, J.W. 1992b, *Astrophys. J.* **391**, 380.

Aschwanden, M.J., Benz, A.O. and Schwartz, R.A. 1993a, *Astrophys. J.* **417**, 790.

Aschwanden, M.J., Benz, A.O., Dennis, B.R. and Gaizauskas, V. 1993b, *Astrophys. J.* **416**, 857.

Aschwanden, M.J., Benz, A.O., and Montello, M. 1994a, *Astrophys. J.* **431**, 432.

Aschwanden, M.J., Benz, A.O., Dennis, B.R., and Kundu, M.R. 1994b, *Astrophys. J. Suppl. Ser.* **90**, 631.

Aschwanden, M.J., Schwartz, R.A., and Alt, D.M. 1995a, *Astrophys. J.* **447**, 923.

Aschwanden, M.J. and Schwartz, R.A. 1995b, *Astrophys. J.* **455**, 699.

Aschwanden, M.J., Benz, A.O., Dennis, B.R., and Schwartz, R.A. 1995c, *Astrophys. J.* **455**, 347.

Aschwanden, M.J., Montello, M.L., Dennis, B.R., and Benz, A.O. 1995d, *Astrophys. J.* **440**, 394.

Aschwanden, M.J., and Schwartz, R.A. 1996a, *Astrophys. J.* **464**, 974.

Aschwanden, M.J., Hudson, H.S., Kosugi, T., and Schwartz,R.A. 1996b, *Astrophys. J.* **464**, 985.

Aschwanden, M.J., Wills,M.J., Hudson,H.S., Kosugi,T., and Schwartz, R.A. 1996c, *Astrophys. J.* **468**, 398.

Aschwanden, M.J., Kosugi, T., Hudson, H.S., Wills, M.J., and Schwartz, R.A. 1996d, *Astrophys. J.* **470**, 1198.

Aschwanden, M.J. 1996e, *Astrophys. J.* **470**, L69.

Aschwanden, M.J. 1996f, in Proc. of Workshop on "High Energy Solar Physics", NASA/GSFC, Greenbelt, Aug.16-18, 1995, AIP Conf. Proc. **374**, (eds. R. Ramaty, Mandzhavidze, N., and Hua, X.-M.), p.300.

Aschwanden, M.J. and Treumann, R.A. 1997, *Lecture Notes in Physics* **483**, 108.

Aschwanden, M.J., Bynum, R.M., Kosugi, T., Hudson, H.S., and Schwartz, R.A. 1997a, *Astrophys. J.* **487**, 936.

Aschwanden, M.J. and Benz, A.O. 1997, *Astrophys. J.* **480**, 825.

Aschwanden, M.J. 1997c, in Proc. Yohkoh 5th Anniversary Symposium on Observational Plasma Astrophysics: Five Years of Yohkoh and Beyond, Yoyogi, Japan, (eds. T. Watanabe, T. Kosugi, and A.C. Sterling), Dordrecht: Kluwer

Aschwanden, M.J. 1998a, *Astrophys. J.* **502**, 455.

Aschwanden, M.J., Schwartz, R.A., and Dennis, B.R. 1998b, *Astrophys. J.* **502**, 468.

Aschwanden, M.J., Kliem, B., Schwarz, U., Kurths, J., Dennis, B.R., and Schwartz, R.A. 1998c, *Astrophys. J.* **505**, 941.

Aschwanden, M.J. 1998d, in "Observational Plasma Astrophysics: Five Years of Yohkoh and Beyond", (Watanabe, T., Kosugi, T. and Sterling, A.C. eds.), Kluwer Academic Publishers, Dordrecht, p.285.
Aschwanden, M.J., Dennis, B.R., and Benz, A.O. 1998, *Astrophys. J.* **497**, 972.
Aschwanden, M.J. 1999, in The Many Faces of the Sun, A Summary of the Results from NASA's Solar Maximum Mission, (eds. K.T. Strong, J.L.R. Saba, B.M. Haisch, and J.T. Schmelz), Springer, p.273.
Aschwanden, M.J., Fletcher, L., Sakao, T., Kosugi, T., and Hudson, H. 1999a, *Astrophys. J.* **517**, 977.
Aschwanden, M.J., Kosugi, T., Hanaoka, Y., Nishio, M., and Melrose, D.B. 1999b, *Astrophys. J.* **526**, 1026.
Aschwanden, M.J., Fletcher, L., Schrijver, C., and Alexander, D., 1999c, *Astrophys. J.* **520**, 880.
Aschwanden, M.J. 2000a, in "High Energy Solar Physics: Anticipating HESSI", Astron.Soc.Pac. Conf.Ser., Vol.206, R. Ramaty and N. Mandzhavidze (eds.), p.197.
Aschwanden, M.J. 2000b, in Encyclopedia of Astronomy and Astrophysics, (ed. Paul Murdin), Institute of Physics Publishing, Grove's Dictionaries, Inc., New York, Vol.**3**, p.2543.
Aschwanden, M.J. and Alexander, D. 2001, *Solar Phys.* **204**, 91.
Aulanier, G., DeLuca, E.E., Antiochos, S.K., McMullen, R.A. and Golub, L. 2000, *Astrophys. J.* **540**, 1126.
Aurass, H., Klein, K.L., and Martens, P.C.H. 1995, *Solar Phys.* **155**, 203
Bai, T. and Ramaty, R. 1978, *Astrophys. J.* **219**, 705.
Bai, T. 1982, AIP Conf.Ser. **77**, 409.
Bai, T., Hudson, H.S., Pelling, R.M., Lin, R.P., Schwartz, R.A., and VonRosenvinge, T.T. 1983, *Astrophys. J.* **267**, 433.
Bai, T. and Sturrock, P. 1989, *Annu. Rev. Astron. Astrophys.* **27**, 421.
Barat, C., Trottet, G., Vilmer, N., Dezalay, J.P., Talon, R., Sunyaev, R., Terekhov, O., Kuznetsov, A. 1994, *Astrophys. J.* **425**, L109.
Barbosa, D.D. 1979, *Astrophys. J.* **233**, 383.
Bastian, T.S., Benz, A.O., and Gary, D.E. 1998, *Annu. Rev. Astron. Astrophys.* **36**, 131.
Bastian, T.S. 2000 in Encyclopedia of Astronomy and Astrophysics, (ed. Paul Murdin), Institute of Physics Publishing, Grove's Dictionaries, Inc., New York, Vol.3, p.2553.
Baum, P.J. and Bratenahl, A. 1980, *Solar Phys.* **67**, 245
Benka, S.G. and Holman, G.D. 1994, *Astrophys. J.* **435**, 469.
Benz, A.O. and Gold, T. 1971 *Solar Phys.* **21**, 157.
Benz, A.O. 1980 *Astrophys. J.* **240**, 892.
Benz, A.O. 1985, *Solar Phys.* **96**, 357.
Benz, A.O. 1986, *Solar Phys.* **104**, 99.
Benz, A.O., Magun, A., Stehling, W., and Su, H. 1992, *Solar Phys.* **141**, 335.
Benz, A.O. 1993, *Plasma Astrophysics. Kinetic Processes in Solar and Stellar Coronae*, Dordrecht: Kluwer Academic Publishers.
Benz, A.O., Csillaghy, A., and Aschwanden, M.J. 1996, *Astron. Astronomy* **309**, 291.
Benz, A.O. and Smith, D.F. 1987, *Solar Phys.* **107**, 299.
Benz, A.O. 2000 in Encyclopedia of Astronomy and Astrophysics, (ed. Paul Murdin), Institute of Physics Publishing, Grove's Dictionaries, Inc., New York, Vol.**3**, p.2529.
Berney, M. and Benz, A.O.1978, *Astron. Astronomy* **65**, 369.
Biskamp, D. and Welter, H. 1979, *Phys.Rev.Lett.* **44**, 1069.
Biskamp, D. and Welter,H. 1989, *Phys.Fluids* B**1**, 1964.
Blackman, E.G. 1997, *Astrophys. J.* **484**, L79.
Block, L.P. 1978, *Astrophys. Space Sci.* **55**, 59.
Brown, J.C. 1971, *Solar Phys.* **18**, 489.
Brown, J.C. 1972, *Solar Phys.* **26**, 441.
Brown, J.C. and Melrose, D.B. 1977, *Solar Phys.* **52**, 117.
Brown, J.C. and Hoyng, P. 1975, *Astrophys. J.* **200**, 734.
Brown, J.C., Karlický, M., MacKinnon, A.L., and Van den Oord, G.H.J. 1990, *Astrophys. J. Suppl. Ser.* **73**, 343.
Brown, J.C. 1991, in The Physics of Solar Flares, Proc. A Royal Society Discussion Meeting, London, (eds. Culhane, J.L. and Jordan, C.), Cambridge: University Press.
Brown, J.C., Conway, A.J. and Aschwanden, M.J. 1998, *Astrophys. J.* **509**, 911.
Cargill, P.J., Goodrich, C.C., and Vlahos, L. 1988, *Astron. Astronomy* **189**, 254.
Canfield, R.C., Brown, J.C., Brueckner, G.E., Cook, J.W., Craig, I.J.D., Doschek, G.A., Emslie, A.G., Henoux, J.C., Lites, B.W., Machado, M.E., Underwood, J.H. 1980, in Solar Flares,

Skylab Workshop II, (ed. P.A. Sturrock), Boulder, Colorado: Colorado Associated University Press, p.231.
Canfield, R.C., Blais, K.A., Reardon, K.P., Acton, L., and Kurokawa, H. 1994, in Solar Active Region Evolution: Comparing Models with Observations, (K.S. Balasubramaniam and G.W. Simon, eds), ASP Conf.Ser. **68**, 411.
Canfield, R.C. and Reardon, K.P. 1998, *Solar Phys.* **182**, 145.
Canfield, R.C., Hudson, H.S., and McKenzie, D.E. 1999, *Geophys. Res. Lett.* **26**/6, 627.
Chae, J. 1999, in "High Resolution Solar Physics: Theory, Observations, and Techniques", ASP Conference Series #183, (eds. T.R. Rimmele, K.S. Balasubramaniam, and R.R. Radick), 375.
Chae, J., Qiu, J., Wang, H., and Goode, P.R. 1999a, *Astrophys. J.* **513**, L75.
Charikov, Y.E. and Fleishman, G.D. 1991, *Solar Phys.* **139**, 387.
Charikov, Y.E., Mosunov, A.N., and Prokopjev, A.V. 1993, *Solar Phys.* **147**, 157.
Chen, F. 1974, Introduction to Plasma Physics, New York:Plenum Press.
Cheng, C.C. 1977, *Astrophys. J.* **213**, 558.
Chupp, E.L. 1996, in Proc. of Workshop on "High Energy Solar Physics", NASA/GSFC, Greenbelt, Aug.16-18, 1995, AIP Conf. Proc. **374**, (eds. R. Ramaty, N. Mandzhavidze, and Hua, X.-M.), p.3.
Conway, A.J., and MacKinnon, A.L. 1998, *Astron. Astronomy* **339**, 298.
Conway, A.J., and Willes, A.J. 2000, *Astron. Astronomy* **355**, 751.
Craig, I.J.D., MacKinnon, A.L., and Vilmer, N. 1985, *Astron. Astronomy Suppl. Ser.* **116**, 377.
Culhane, J.L. and Jordan, C. (eds.) 1991, The Physics of Solar Flares, Proc. A Royal Society Discussion Meeting, London, Cambridge: University Press.
Czaykowska, A., DePontieu, B., Alexander, D., Rank, G. 1999, *Astrophys. J.* **521**, L75.
Czaykowska, A., Alexander, D., and DePontieu, B. 2001, *Astrophys. J.* **552**, 849.
Decker, R.B. and Vlahos, L. 1986, *Astrophys. J.* **306**, 710.
Dennis, B.R., Benz, A.O., Ranieri, M., and Simnett, G.M. 1984, *Solar Phys.* **90**, 383.
Dennis, B.R. 1985, *Solar Phys.* **100**, 465.
Dennis, B.R., Orwig, L.E. and Kiplinger, A.L.(eds.) 1987, NASA CP **2449**, 478p.
Dennis, B.R. 1988, *Solar Phys.* **118**, 49.
Dennis, B.R. and Schwartz, R.A. 1989, *Solar Phys.* **121**, 75.
De Jager, C. and De Jonge, G. 1978, *Solar Phys.* **58**, 127.
De Jager, C. 1986, *Space Sci.Rev.* **44**, 43.
Ding, M.D., Qiu, J., Wang, H., and Goode, P.R. 2001, *Astrophys. J.* **552**, 340.
Drake, J.F., Biskamp, D., and Zeiler, A. 1997, *Geophys. Res. Lett.* **24**/22, 2921.
Dreicer, H. 1959, *Phys.Rev.* **115**, 238.
Dungey, J.W. 1953, *Phil.Mag.* **44**, 725.
Dulk, G.A. 1985, *Annu. Rev. Astron. Astrophys.* **23**, 169.
Dulk, G.A., Steinberg, J.L., Hoang, S., and Goldman, M.V. 1987, *Astron. Astronomy* **173**, 366.
Dulk, G.A. 1990, *Solar Phys.* **130**, 139.
Dulk, G.A., Kiplinger, A.L., and Winglee, R.M. 1992, *Astrophys. J.* **389**, 756.
Eichler, D. 1979, *Astrophys. J.* **229**, 413.
Elwert, G. 1939, *Ann.Physik* **34**, 178.
Emslie, A.G. and Rust, D.M. 1980, *Solar Phys.* **65**, 271.
Emslie, A.G. 1983, *Astrophys. J.* 271, 367.
Emslie, A.G. and Brown, J.C. 1985, *Astrophys. J.* **295**, 648.
Emslie, A.G. and Henoux, J.-C. 1995, *Astrophys. J.* **446**, 371.
Emslie, A.G., Brown, J.C., and MacKinnon, A.L. 1997, *Astrophys. J.* **485**, 430.
Ellison, D.C. and Ramaty, R. 1985, *Astrophys. J.* **298**, 400.
Fan, Y., Zweibel, E.G., Linton, M.G. and Fisher, G.H. 1999, *Astrophys. J.* **521**, 460.
Fermi, E. 1949, *Phys.Rev.* **75**, 1169.
Fermi, E. 1954, *Astrophys. J.* **119**, 1.
Filippov, B. 1999, *Solar Phys.* **185**, 297.
Fisher, G.H., McClymont, A.N., and Chou, D.Y. 1991, *Astrophys. J.* **374**, 766.
Fisk, L.A. 1976, *J.Geophys.Res.* **81**, 4633.
Fleishman, G.D. and Charikov, Y.E. 1991, *Sov.Astron.* **35**/4, 354.
Fleishman, G.D. and Arzner, K.J. 2000, *Astron. Astronomy* **358**, 776.
Fleishman, G.D. and Mel'nikov, V.F. 1998, *Physics - Uspekhi* **41** (12), 1157.
Fletcher, L. 1995, *Astron. Astronomy* **303**, L9.
Fletcher, L. and Martens, P.C.H. 1998, *Astrophys. J.* **505**, 418.
Fletcher, L., Metcalf, T.R., Alexander, D., Ryder, L.A., and Brown, D.S. 2001, *Astrophys. J.* **554**, 451.

Fletcher, L., and Hudson, H.S. 2001, *Solar Phys.* **204**, 69.
Fokker, A.D. 1970, *Solar Phys.* **11**, 92.
Forbes, T.G. and Priest, E.R. 1995, *Astrophys. J.* **446**, 377.
Forbes, T.G. and Acton, L.W. 1996, *Astrophys. J.* **459**, 330.
Forrest, D.J. and Chupp, E.L. 1983, *Nature* **305**, 291.
Furth, H.P, Killeen, J., and Rosenbluth, M.N. 1963, *Phys.Fluids* **6**, 459.
Galsgaard, K., Priest, E.R., and Nordlund, *Astron. Astronomy* 2001, *Solar Phys.* **193**, 1.
Gary, A. 2001, *Solar Phys.* **203**, 71
Gayley, K.G. and Canfield, R.C. 1991, *Astrophys. J.* **380**, 660.
Gold, T. and Hoyle, F. 1960, *Month. Not. Royal Astron.Soc.* **120**/2, 89.
Goldman, M.V. and Smith, D.F.1986 in "Physics of the Sun, Vol.II", Reidel, Dordrecht, p.325.
Golub, L. and 11 co-authors 1999, *Physics of Plasmas* **6/5**, 2205.
Gorbachev, V.S. and Somov, B.V. 1989, *Soviet Astronomy* **33**, 57.
Güdel, M., Aschwanden, M.J., and Benz, A.O. 1991, *Astron. Astronomy* **251**, 285.
Hagyard, M.J., Smith, J.B.Jr., Teuber, D., and West, E.A. 1984, *Solar Phys.* **91**, 115.
Hagyard, M.J., Venkatakrishnan, P., and Smith, J.B.Jr. 1990, *Astrophys. J. Suppl. Ser.* **73**, 159.
Hamilton, R.J. and Petrosian, V. 1990, *Astrophys. J.* **365**, 778.
Hamilton, R.J., Lu, E.T., and Petrosian, V. 1990, *Astrophys. J.* **354**, 726.
Hamilton, R.J. and Petrosian, V. 1992, *Astrophys. J.* **398**, 350.
Hanaoka, Y. 1996, *Solar Phys.* **165**, 275.
Hanaoka, Y. 1999, *Proc. Astr. Soc. Japan* **51**, 483.
Haerendel, G. 1994, *Astrophys. J. Suppl. Ser.* **90**, 765.
Haruki, T. and Sakai, J.I. 2001a, *Phys.Plasmas* **8**(5), 1538.
Haruki, T. and Sakai, J.I. 2001b, *Astrophys. J.* **552**, L175.
Heristchi, D. 1986, *Astrophys. J.* **311**, 474.
Heristchi, D. 1987, *Astrophys. J.* **323**, 391.
Hewitt, R.G., Melrose, D.B., and Roennmark, K.G. 1982, *Aust.J.Phys.* **35**, 447.
Hewitt, R.G. and Melrose, D.B. 1983, *Aust.J.Phys.* **36**, 725.
Hewitt, R.G., Melrose, D.B., and Dulk, G.A. 1983, *J.Geophys.Res.* **88**, 10065.
Hewitt, R.G. and Melrose, D.B. 1985, *Solar Phys.* **96**, 157.1
Heyvaerts, J., Priest, E.R., and Rust, D.M. 1977, *Astrophys. J.* **216**, 123.
Hirayama, T. 1974, *Solar Phys.* **34**, 323.
Hirose, S., Uchida, Y., Uemura, S., Yamaguchi, T., and Cable, S.B. 2001, *Astrophys. J.* **551**, 586.
Holman, G.D., Eichler, D., and Kundu, M.R. 1980, in Proc. IAU Symp. 86, "Radio Physics of the Sun, eds. M.R. Kundu, and T.E. Gergely, Dordrecht:Reidel, 457.
Holman, G.D. 1985, *Astrophys. J.* **293**, 584.
Holman, G.D. and Pesses, M.E. 1983, *Astrophys. J.* **267**, 837.
Holman, G.D. 1996, in Proc. High Energy Solar Physics Workshop, (eds. R. Ramaty, N. Mandzhavidze, and Hua, X.-M.), AIP **374**, 479.
Holt, S.S. and Cline, T.L. 1968, *Astrophys. J.* **154**, 1027.
Hood, A.W. and Priest, E.R. 1979, *Solar Phys.* **64**, 303.
Hudson, H.S. 1972, *Solar Phys.* **24**, 414.
Hudson, H.S. 1973, in Ramaty. R. and Stone. R. (eds.), NASA SP **342**, 207.
Hudson, H.S. 1978, *Astrophys. J.* **224**, 235.
Hudson, H.S., Bai, T., Gruber, D.E., Matteson, J.L., Nolan, P.L., and Petterson, L.E. 1980, *Astrophys. J.* **236**, L91.
Hudson, H.S., Strong, K.T., Dennis, B.R., Zarro, D., Inda, M., Kosugi, T., and Sakao,T. 1994, *Astrophys. J.* **422**, L25.
Hudson, H.S. and Ryan, J. 1995, *Annu. Rev. Astron. Astrophys.* **33**, 239.
Hudson, H.S. 2000, *Astrophys. J.* **531**, L75.
Isliker, H. 1996, *Astron. Astronomy* **310**, 672
Jackson, J.D. 1962, Classical Electrodynamics, New York: John Wiley and Sons, Inc.
Jakimiec, J., Tomczak, M., Falewicz, R., Phillips, K.J.H., and Fludra, A. 1998, *Astron. Astronomy* **334**, 1112
Jakimiec, J. 1999, in Proc. 9th European Meeting on Solar Physics, "Magnetic Fields and Solar Processes", Florence, Italy, ESA SP-**448**, p.729.
Jiang, Y. and Wang, J. 2001, *Astron. Astronomy* **367**, 1022.
Johns, C.M. and Lin, R.P. 1992, *Solar Phys.* **137**, 121.
Jokipii, J.R. 1966, *Astrophys. J.* **143**, 961.
Jones, T.W. 1994, *Astrophys. J. Suppl. Ser.* **90**, 969.

Kanbach, G., Bertsch, D.L., Fichtel, C.E., Hartman, R.C., Hunter, S.D., et al. 1993, *Astron. Astronomy Suppl. Ser.* **97**, 349.
Kane, S.R. 1972, *Solar Phys.* **27**, 174.
Kane, S.R., Anderson, K.A., Evans, W.D., Klebesadel, R.W., and Laros, J. 1979 *Astrophys. J.* **233**, L151.
Kane, S.R., Pick, M., and Raoult, A. 1980, *Astrophys. J.* **241**, L113.
Kane, S.R., Fenimore, F.E., Klebesadel, R.W., and Laros, J.G. 1982a, *Astrophys. J.* **254**, L53.
Kane, S.R., Benz, A.O., and Treumann, R.A. 1982b, *Astrophys. J.* **263**, 423.
Kane, S.R. 1983, *Solar Phys.* **86**, 355.
Kane, S.R., Chupp, E.L., Forrest, D.J., Share, G.H., and Rieger, E. 1986, *Astrophys. J.* **300**, L95.
Kane, S.R., McTiernan, J., Loran, J., Fenimore, E.E., Klebesadel, R.W., and Laros, J.G. 1992, *Astrophys. J.* **390**, 687.
Karimabadi, H., Menyuk, C.R., Sprangle, P., and Vlahos, L. 1987, ApJ **316**, 462.
Karpen, J.T. 1982, *Solar Phys.* **77**, 205.
Karpen, J.T., Antiochos, S.K., and DeVore, C.R. 1991, *Astrophys. J.* **382**, 327.
Karpen, J.T., Antiochos, S,K., and DeVore, C.R. 1995, *Astrophys. J.* **450**, 422.
Karpen, J.T., Antiochos, S.K., DeVore, C.R., and Golub, L. 1998, *Astrophys. J.* **495**, 491.
Kennel, C.F. and Petschek, H.E. 1966, *J.Geophys.Res.* **71**(1), 1.
Kiplinger, A.L., Dennis, B.R., Emslie, A.G., Frost, K.J., Orwig, L.E. 1983, *Astrophys. J.* **265**, L99.
Kiplinger, A.L., Dennis, B.R., Frost, K.J., and Orwig, L.E. 1984, *Astrophys. J.* **287**, L105.
Kirk, J.G., Melrose, D.B., and Priest, E.R. 1994, Lecture Notes, Saas-Fee Advanced Course **24**, Berlin: Springer-Verlag.
Kivelson, M.G. and Russell, C.T. 1995, *Introduction to Space Physics*, Cambridge: Cambridge University Press.
Kliem, B. 1988, ESA SP-**285**, 117.
Kliem, B. 1990, *Astronomische Nachrichten* **311**, 399.
Kliem, B. 1994, *Astrophys. J. Suppl. Ser.* **90**, 719.
Kliem, B. 1995, *Lecture Notes in Physics* **444**, 93.
Kliem, B., Schumacher, J., Shklyar, D.R. 1998, *Adv.Space Res.* **21**(4), 563.
Kliem, B., Karlický, M., and Benz, A.O. 2000, *Astron. Astronomy* **360**, 715.
Koch, H.W. and Motz, J.W. 1959, *Rev. Mod. Phys.* **31**, 920.
Kopp, R.A. and Pneuman, G.W. 1976, *Solar Phys.* **50**, 85.
Kosugi, T. 1981, *Solar Phys.* **71**, 91.
Krauss-Varban, D., Burgess, D., and Wu, C.S. 1989, *J.Geophys.Res.* **94**, 15089.
Krauss-Varban, D. and Burgess, D. 1991, *J.Geophys.Res.* **96**, 143.
Krucker, S., Benz, A.O., and Aschwanden, M.J. 1997, *Astron. Astronomy* **317**, 569.
Krüger, A. 1979, Introduction to Solar Radio Astronomy and Radio Physics, Dordrecht: Reidel.
Kucera, T.A., Dulk, G.A., Kiplinger, A.L., Winglee, R.M., Bastian, T.S., and Graeter, M. 1993, *Astrophys. J.* **412**, 853.
Kundu, M.R. and Vlahos, L.1982 *Space Sci. Rev.* **32**, 405.
Kundu, M.R. 1985, *Solar Phys.* **100**, 491.
Kundu, M.R., Raulin, J.P., Pick, M., and Strong, K.T. 1995, *Astrophys. J.* **444**, 922.
Kundu, M.R., White, S.M., Shibasaki, K., Sakurai, T., and Grechnev, V.V. 2001, *Astrophys. J.* **547**, 1090.
Kurokawa, H. 1987, *Solar Phys.* **113**, 259.
Kurokawa, H. 1991, *Lecture Notes in Physics* **387**, 39.
Lampe, M. and Papadopoulos, K. 1977, *Astrophys. J.* **212**, 886.
Lang, K.R. 1980, Astrophysical Formulae, 2nd edition, Berlin:Springer.
Lang, K.R. and Willson, R.F. 1989, *Astrophys. J.* **344**, L77.
LaRosa, T.N. and Moore, R.L. 1993, *Astrophys. J.* **418**, 912.
LaRosa, T.N., Moore, R.N., and Shore, S.N. 1994, *Astrophys. J.* **425**, 856.
Lau, Y.T., Northrop, T.G., and Finn, J.M. 1993, *Astrophys. J.* **414**, 908.
Leboef, J.N., Tajima, T., Dawson, J.M. 1982, *Phys.Fluids* **25**, 784.
Lee, J.W., Gary, D.E., and Shibasaki, K. 2000, *Astrophys. J.* **531**, 1109.
Lee, J.W., and Gary, D.E. 2000, *Astrophys. J.* **543**, 457.
Lee, M.A. and Völk, H.J. 1975, *Astrophys. J.* **198**, 485.
Lee, M.A. and Völk, H.J. 1973, *Space Sci.* **24**, 31.
Leka, K.D., Canfield, R.C., McClymont, A.N., and Van Driel-Gesztelyi, L. 1996, *Astrophys. J.* **462**, 547.
Lemberge, B. 1995, *Lecture Notes in Physics* **444**, 201.
Li, J., Metcalf, T.R., Canfield, R.C., Wülser, J.P., and Kosugi, T. 1997, *Astrophys. J.* **482**, 490.
Li, P., Hurley, K., Barat, C., Niel, M., Talon, R., and Kurt, V. 1994, *Astrophys. J.* **426**, 758.

Li, P. 1995, *Astrophys. J.* **443**, 855.
Lin, R.P. 1974, *Space Science Reviews* **16**, 189.
Lin, R.P., Potter, D.W., Gurnett, D.A., and Scarf, F.L. 1981, *Astrophys. J.* **251**, 364.
Lin, R.P., Schwartz, R.A., Pelling, R.M., and Hurley, K.C. 1981, *Astrophys. J.* **251**, L109.
Lin, R.P. 1993, *Adv.Space Res.* **13**/9, 265.
Lin, R.P. and Johns,C.M. 1993, *Astrophys. J.* **417**, L53.
Lin, R.P. 1996, *Lecture Notes in Physics* **483**, 93.
Lin, R.P. 1998, *Proc. SPIE* **3442**, 2.
Lin, R.P. 2000, IAU Symp. 195: Highly Energetic Physical Processes and Mechanisms for Emission from Astrophysical Plasmas, **195**, 15.
Litvinenko, Y.E. 1996, *Astrophys. J.* **462**, 997.
Litvinenko, Y.E. and Somov, B.V. 1995, *Solar Phys.* **158**, 317.
Machado, M.E., Moore, R.L., Hernandez, A.M., Rovira, M.G., Hagyard, M.J. and Smith, J.B.Jr. 1988, *Astrophys. J.* **326**, 425.
MacKinnon, A.L., Brown, J.C., Trottet, G., and Vilmer, N.1983, *Astron. Astronomy* **119**, 297.
Malara, F., Veltri, P., Carbone, V. 1992, *Phys.Fluids* B**4**, 3070.
Mariska, J.T., Sakao, T., and Bentley, R.D. 1996, *Astrophys. J.* **459**, 815.
Marsh, K.A., and Hurford, G.J.1982, *Annu. Rev. Astron. Astrophys.* **20**, 497.
Masuda, S., Kosugi, T., Hara, H., Tsuneta, S., and Ogawara, Y. 1994a, *Nature* 371, No. **6497**, 495.
Masuda, S. 1994b, PhD Thesis, "Hard X-Ray Sources and the Primary Energy Release Site in Solar Flares", Natl. Astronom. Obs., Mitaka, Tokyo.
Masuda, S. 2000, in Highly Energetic Physical Processes and Mechanisms for Emission from Astrophysical Plasmas, IAU Symp. **195**, 2000, (eds. P.C.H. Martens S.Tsuruta, and M.A.Weber), 413.
Masuda, S., Kosugi, T., and Hudson, H.S. 2001, *Solar Phys.* , **204**, 55.
Matsushita, K., Masuda, S., Kosugi, T., Inda, M., and Yaji, K. 1992, *Proc. Astr. Soc. Japan* **44**, L89.
Mazur, J.E., Mason, G.M., Klecker, B., and McGuire, R.E. 1992, *Astrophys. J.* **401**, 398.
McClements, K.G. 1990a, *Astron. Astronomy* **230**, 213.
McClements, K.G.1990b, *Astron. Astronomy* **234**, 487.
McClements, K.G.1992, *Astron. Astronomy* **258**, 542.
McClements, K.G., Su, J.J., Bingham, R., Dawson, J.M., and Spicer, D.S. 1990, *Solar Phys.* **130**, 229.
McConnell, M.L., Forrest, D., Vestrand, W.T., and Finger, M. 1996, in High Energy Solar Physics (eds. R. Ramaty, N. Mandzhavidze, X.-M.Hua), AIP **374**, 368.
McKenzie, D.E. 2000, *Solar Phys.* **195**, 381.
McKenzie, D.E. and Hudson, H.S. 1999, *Astrophys. J.* **519**, L93.
McLean, D.J. and Labrum, N.R. 1985, Solar Radiophysics, Cambridge: Cambridge University Press.
McTiernan, J.M., Kane, S.R., Loran, J.M., Lemen, J.R., Acton, L.W., Hara, H., Tsuneta, S., and Kosugi, T. 1993, *Astrophys. J.* **416**, L91.
Melendez, J.L. 1998, Masters Thesis, INPE-6832-TDI/601, Instituto de Pesquisas Espaciais.
Melendez, J.L., Sawant, H.S., Fernandes, F.C.R., and Benz,A.O. 1999, *Solar Phys.* **187**, 77.
Melrose, D.B. 1974, *Solar Phys.* **37**, 353.
Melrose, D.B. and Brown, J.C.1976, *Month. Not. Royal Astron.Soc.* **176**, 15.
Melrose, D.B. and White, S.M.1979, *Proc.ASA* **3**(6), 369.
Melrose, D.B. 1980a, Plasma Astrophysics: Nonthermal Processes in Diffuse Magnetized Plasmas, Volume I, New York: Gordon and Breach.
Melrose, D.B. 1980b, Plasma Astrophysics: Nonthermal Processes in Diffuse Magnetized Plasmas, Volume II, New York: Gordon and Breach.
Melrose, D.B. and White, S.M.1981 *J.Geophys.Res.* **86**/A4, 2183.
Melrose, D.B., and Dulk, G.A.1982 *Astrophys. J.* **259**, L41.
Melrose, D.B., Hewitt, R.G., and Dulk, G.A. 1984, *J.Geophys.Res.* **89**(A2), 897.
Melrose, D.B. 1997, *Astrophys. J.* **486**, 521.
Mikić, Z., Schnack, D.D., and Van Hoven, G. 1990, *Astrophys. J.* **361**, 690.
Miller, J.A. 1991, *Astrophys. J.* **376**, 342.
Miller, J.A., Cargill, P.J., Emslie,A.G., Holman, G.D., Dennis, B.R., LaRosa, T.N., Winglee, R.M., Benka, S.G., and Tsuneta, S. 1997, *J.Geophys.Res.* **102**, No.A7, 14,631
Miller, J.A., Guessoum, N., and Ramaty, R. 1990, *Astrophys. J.* **361**, 701.
Miller, J.A., LaRosa, T.N., and Moore, R.L. 1996, *Astrophys. J.* **461**, 445.
Miller, J.A. and Ramaty, R. 1987, *Solar Phys.* **113**, 195.
Miller, J.A. and Ramaty, R. 1992, in *Particle Acceleration in Cosmic Plasmas*, (eds. G.P. Zank and T.K. Gaisser), p.223, New York: AIP Press.

Miller, J.A. and Reames, D.V. 1996, in *High Energy Solar Physics*, eds. R. Ramaty, N. Mandzhavidze, and X.-M. Hua, p.450, New York: AIP Press.

Miller, J.A. and Roberts, D.A. 1995, *Astrophys. J.* **452**, 912.

Mori, K.I., Sakai, J.I., and Zhai, J. 1998, *Astrophys. J.* **494**, 430.

Moore, R.L. and Roumeliotis, G. 1992, in Eruptive Solar Flares, (Z. Svestka, B.V. Jackson, M.E. Machado, eds.), *Lecture Notes in Physics* **399**, 69.

Moore, R.L., Sterling, A.C., Hudson, H.S., and Lemen, J.R. 2001, *Astrophys. J.* **552**, 833.

Murphy, R.J., Kozlovsky, B., and Ramaty, R. 1988, *Astrophys. J.* **331**, 1029.

Murphy, R.J., Ramaty, R., Kozlovsky, B., and Reames, D.V. 1991, *Astrophys. J.* **371**, 793.

Murphy, R.J., Share, G.H., Grove, J.E., Johnson, W.N., Kinzer, R.L., Kurfess, J.D., Strickman, M.S., and Jung, G.V. 1997, *Astrophys. J.* **490**, 883.

Murphy, R.J., Share, G.H., DelSignore, K.W., Hua, X.M. 1999, *Astrophys. J.* **510**, 1011.

Nakajima, H., Dennis, B.R., Hoyng, P., Nelson, G., Kosugi, T. and Kai, K. 1985, *Astrophys. J.* **288**, 806.

Nishio, M., Yaji, K., Kosugi, T., Nakajima, H., Sakurai, T. 1997, *Astrophys. J.* **489**, 976.

Nitta, N., Van Driel-Gesztelyi, L., Leka, K.D., Mickey, D.L., Metcalf, T.R., Wülser, J., Ichimoto, K., Sakurai, T., and Shibata, K. 1994, in Proc. Kofu Symp. on "New Look at the Sun with Emphasis on Advanced Observations of Coronal Dynamics and Flares", (eds. S. Enome and T. Hirayama), NRO Rep. No. **360**, p.385.

Øieroset, M., Phan, T.D., Fujimoto, M., and Lepping, R.P. 2001, *Nature* **412**, 6845.

Paesold, G., Benz, A.O., Klein, K.L., and Vilmer, N. 2001, *Astron. Astronomy* **371**, 333.

Papadopoulos, K. 1979, in *Dynamics of the Magnetosphere*, (eds. S.I.Akasofu), Norwell, Mass.: Kluwer, p.289,

Parker, E.N. 1957, *J.Geophys.Res.* **62**, 509.

Parker, E.N. 1963, *Astrophys. J. Suppl. Ser.* **8**, 177.

Parker, E.N. 1965, *Planet. Space Sci.* **13**, 9.

Parker, E.N. 1988, *Astrophys. J.* **330**, 474.

Petschek, H.E. 1964, in Physics of Solar Flares, (ed. W.N. Hess), NASA SP-**50**, Washington, DC, 425.

Poquérusse, M., Hoang, S., Bougeret, J.L., and Moncuquet, M. 1995, Solar Wind 8 Proc., Dana Point/California, June 1995, p.62.

Priest, E.R. 1982, Solar Magnetohydrodynamics, Dordrecht: Reidel Publishing Company.

Priest, E.R. 1985a, *Rep. Prog. Phys.* **48**, 955.

Priest, E.R. 1985b, IAU Symp. **107**, 233.

Priest, E.R., Parnell, C.E., and Martin, S.F. 1994, *Astrophys. J.* **427**, 459.

Priest, E.R. and Forbes, T. 2000, "Magnetic Reconnection. MHD Theory and Applications", Cambridge: Cambridge University Press.

Pritchett, P.L. and Wu, C.C. 1979, *Phys.Fluids* **22**, 2140.

Pritchett, P.L. 1984, *J.Geophys.Res.* **89**(A10), 8957.

Pritchett, P.L. 1986, *Phys.Fluids* **29**(9), 2919.

Ramaty, R. 1979, in *Particle Acceleration Mechanisms in Astrophysics*, (eds. J. Arons, C. Max, and C. McKee), New York: AIP Press, p.135.

Ramaty, R. 1986, in The Physics of The Sun (ed. P.A. Sturrock), Dordrecht:Reidel, p.293.

Ramaty, R. and Mandzhavidze, N. 1994, in High Energy Solar Phenomena - A new era of spacecraft measurements (eds. J.M. Ryan and W.T. Vestrand), AIP **294**, 26.

Ramaty, R., Mandzhavidze, N., Kozlovsky, B., and Murphy, R. 1995, ApJ **455**, L193.

Ramaty, R. and Mandzhavidze, N. 2000, in Highly Energetic Physical Processes and Mechanisms for Emission from Astrophysical Plasmas, IAU Symp. **195**, (eds. P.C.H. Martens and S. Tsuruta), p.123.

Ramaty, R. and Mandzhavidze, N. 2001, in Encyclopedia of Astronomy and Astrophysics, Nature Publishing Group and Institute of Physics Publishing, UK.

Reames, D.V., Richardson, I.G., and Wenzel, K.P. 1992, *Astrophys. J.* **387**, 715.

Reames, D.V., Meyer, J.P. and VonRosenvinge, T.T. 1994, *Astrophys. J. Suppl. Ser.* **90**, 649.

Reames, D.V. 1996, in High Energy Solar Physics (eds. R. Ramaty, N. Mandzhavidze, X.-M. Hua), AIP **374**, 35.

Reames, D.V. 1999, *Space Science Revs.* **90**, 3/4, p.413.

Reames, D.V. 2000, in "High Energy Solar Physics: Anticipating HESSI", Astron.Soc.Pac. Conf.Ser. Vol. **206**, (R. Ramaty, and N. Mandzhavidze, eds.) p.102.

Rieger, E. and Marschhäuser, H. 1990, in Max 91' Workshop #3, Estes park, CO., (ed. R. Winglee, A. Kiplinger), p.68.

Robinson, P.A. 1986, *J. Plasma Physics* **36**, 63.

Robinson, P.A. 1988, *Phys.Fluids* **31**(3), 525.
Robinson, P.A. 1989, *Astrophys. J.* **341**, L99.
Robinson, P.A. 1991a, *Solar Phys.* **134**, 299.
Robinson, P.A. 1991b, *Solar Phys.* **136**, 343.
Robinson, P.A. and Benz, A.O. 2000, *Solar Phys.* **194**, 345.
Rosner, R. and Vaiana, G.S. 1978, *Astrophys. J.* **222**, 1104.
Rudawy, P., Falewicz, R., Mandrini, C.H. and Siarkowski, M. 2001, *Astron. Astronomy* **372**, 1030.
Rutherford, E. 1911, *Phil.Mag.* **21**, 699.
Rust, D.M. and Kumar, A. 1996, *Astrophys. J.* **464**, L199.
Ryan, J.M. and Lee, M.A. 1991, *Astrophys. J.* **368**, 316.
Sakai, J.I. and Ohsawa, Y. 1987, *Space Science Rev.* **46**, 113.
Sakai, J.I. and De Jager, C. 1991, *Solar Phys.* **134**, 329.
Sakai, J.I. and De Jager, C. 1996, *Space Science Rev.* **77**, 1.
Sakao, T., Kosugi, T., Masuda, S., Inda, M., Makishima, K., Canfield, R.C., Hudson, H.S. Metcalf, T.R., Wülser, J.P., Acton, L.W., and Ogawara, Y. 1992, *Proc. Astr. Soc. Japan* **44**, L83.
Sakao, T. 1994, PhD Thesis, "Characteristics of Solar Flare Hard X-ray Sources as Revealed with the Hard X-ray Telescope aboard the Yohkoh Satellite". Natl. Astron. Obs., Mitaka, Tokyo, Japan.
Sakao, T. 1999, in "Solar Physics with Radio Observations", Proc. Nobeyama Symp., Kiyosato, Japan, (eds. T. Bastian, N. Gopalswamy, and K. Shibasaki), NRO Report **479**, p.321.
Schmidt, G. 1979, Physics of High Temperature Plasmas, (2nd ed.; New York: Academic Press.
Schmieder, B., Aulanier, G., Demoulin, P., Van Driel-Gesztelyi, L., Roudier, T., Nitta, N. and Cauzzi, G. 1997, *Astron. Astronomy* **325**, 1213.
Schumacher, J. and Kliem, B. 1997, *Phys.Plasmas* **4**(10), 3533.
Schumacher, J., Kliem, B., and Seehafer,N. 2000, *Phys.Plasmas* **7**(1), 108.
Share, G.H. and Murphy, R.J. 1995, *Astrophys. J.* **452**, 933.
Share, G.H., and Murphy, R.J. 1997, *Astrophys. J.* **485**, 409.
Share, G.H., Murphy, R.J., and Ryan, J. 1997, in Proc. 4th Compton Symposium, (eds. C.D. Dermer, M,S. Strickman, and J.D. Kurfess), AIP, p.17.
Share, G.H. and Murphy, R.J. 2000, in High Energy Solar Physics: Anticipating HESSI, ASP Conf.Ser. **206** (eds. R. Ramaty and N. Mandzhavidze), p.377.
Sharma, R.R., Vlahos, L., and Papadopoulos, K. 1982, *Astron. Astronomy* **112**, 377.
Sharma, R.R. and Vlahos, L. 1984, *Astrophys. J.* **280**, 405.
Shibata, K., Ishido, Y., Acton, L.W., Strong, K.T., Hirayama, T., Uchida, Y., McAllister, A.H., Matsumoto, R., Tsuneta, S., Shimizu, T., Hara, H., Sakurai, T., Ichimoto, K., Nishino, Y., and Ogawara, Y. 1992, PASJ **44**, L173.
Shibata, K., Shimojo, M., Yokoyama, T., and Ohyama, M. 1996, in Proc. Yohkoh Conference on Observations of Magnetic Reconnection in the Solar Atmosphere, Bath, England, March 20-22, 1996, (eds. R. Bentley and J. Mariska), Astron.Soc.Pac. Conf. Ser. **111**, p.29.
Shibata, K. and Tanuma, S. 2001, *Earth, Planets, and Space* **53**, 473.
Simnett, G. 1986, *Solar Phys.* **106**, 165.
Simnett, G. and Benz, A.O. 1986, *Astron. Astronomy* **165**, 227.
Simnett, G. and Haines, M.G. 1990, *Solar Phys.* **130**, 253.
Smith, D.F. and Benz, A.O. 1991, *Solar Phys.* **131**, 351.
Smith, D.F. and Brecht, S.H. 1993, *Astrophys. J.* **406**, 298.
Somov, B.V. and Verneta, A.I. 1989, *Solar Phys.* **120**, 93.
Somov, B.V. and Kosugi, T. 1997, *Astrophys. J.* **485**, 859.
Spicer, D.S. 1977a, *Solar Phys.* **53**, 249.
Spicer, D.S. 1977b, *Solar Phys.* **53**, 305.
Spicer, D.S. 1981, *Solar Phys.* **71**, 115.
Spitzer, L. 1967, The Physics of Fully Ionized Gases (2nd ed.; New York: Interscience).
Steinacker, J. and Miller, J.A. 1992, *Astrophys. J.* **393**, 764.
Stewart, R.T. 1965, *Australian J. Phys.* **18**, 67.
Stewart, R.T. 1978, *Solar Phys.* **58**, 121.
Stix, T.H. 1992, Waves in Plasmas, New York: AIP Press. 2001, *Astron. Astronomy* **372**, 1030.
Stone, R.G. and Fainberg, J. 1971, *Solar Phys.* **20**, 106.
Sturrock, P.A. 1973, in Coronal disturbances, IAU-Symposium No. **57**, 437.
Sturrock, P.A. 1966, *Nature* **211**, No.5050, 695.
Sturrock, P.A. 1994, Plasma Physics, Cambridge: Cambridge University Press.
Sweet, P.A. 1958a, in Electromagnetic Phenomena in Cosmical Physics, IAU Symp. 6, (ed. B. Lehnert), Cambridge University Press, London, p.123.

Sweet, P.A. 1958b, *Nuovo Cimento Suppl.* **8**, Ser. X, 188.
Tajima, T., and Sakai, J. 1986, *IEEE Trans. Plasma Sci.* PS-**14**, 929.
Tajima, T., Sakai, J., Nakajima, H., Kosugi, T., Brunel, F., Kundu, M.R. 1987, *Astrophys. J.* **321**, 1031.
Titov, V.S. and Demoulin, P. 1999, *Astron. Astronomy* **351**, 707
Tomczak, M. 2001, *Astron. Astronomy* **366**, 294.
Trottet, G., Pick, M., and Heyvaerts, J. 1979, *Astron. Astronomy* **79**, 164.
Trottet, G., Barat, C., Ramaty, R., Vilmer, N., Dezalay, J.P., Kuznetsov, A., Mandzhavidze, N., Sunyaev, R., Talon, R., Terekhov, O. 1996, in High Energy Solar Physics, (eds. R. Ramaty, N. Mandzhavidze, X.-M. Hua), AIP **374**, 153.
Trottet, G. and Vilmer, N. 1997, *Lecture Notes in Physics* **489**, 219.
Trubnikov, B.A. 1965, *Rev. Plasma Phys.* **1**, 105.
Tsuneta, S. 1985, *Astrophys. J. Lett.* **290**, 353.
Tsuneta, S. 1995, *Proc. Astr. Soc. Japan* **47**, 691.
Tsuneta, S. 1996, *Astrophys. J.* **456**, 840.
Tsuneta, S., Hara, H., Shimizu, T., Acton, L.W., Strong, K.T., Hudson, H.S., and Ogawara, Y. 1992, *Proc. Astr. Soc. Japan* **44**, L63.
Tsuneta, S. and Naito, T. 1998, *Astrophys. J.* **495**, L67.
Uchida, Y., McAllister, A., Khan, J., Sakurai, T., and Jockers, K. 1994, in "X-ray solar physics from YOHKOH", (eds. Uchida, Y., Watanabe, T., Shibata, K., and Hudson, H.S.), Tokyo:Universal Academy Press Inc., p.161.
Uchida, Y., Morita, S., Torii, M. 1999a, in "Solar Physics with Radio Observations", Nobeyama Radio Observatory NRO Report No. **479**, Proc. of Nobeyama Symp. 1998, (eds. T. Bastian, N. Gopalswamy, and K. Shibasaki), p.397.
Uchida, Y., Hirose, S., Cable, S., Morita, S., Torii, M., Uemura, S., and Yamaguchi, T., 1999b, *Proc. Astr. Soc. Japan* **51**, 553.
Van Ballegooijen, A,A. and Martens, P.C.H. 1989, *Astrophys. J.* **343**, 971.
Vestrand, W.T., Forrest, D.J., Chupp, E.L., Rieger, E., and Share,G.H. 1987, *Astrophys. J.* **322**, 1010.
Vestrand, W.T. and Forrest,D.J. 1993, *Astrophys. J.* **409**, L69.
Vestrand, W.T. and Miller, J.A. 1999, in "The Many Faces of the Sun, A Summary of the Results from NASA's Solar Maximum Mission", (eds. K.T. Strong, J.L.R. Saba, B.M. Haisch, J.T. Schmelz), Berlin: Springer, chapter 7, p.231.
Vilmer, N., Trottet, G., and MacKinnon, A.L. 1986, *Astron. Astronomy* **156**, 64.
Vilmer, N. 1987, *Solar Phys.* **111**, 207.
Vilmer, N., Trottet, G., Barat, C., Schwartz, R.A., Enome, S., Kuznetsov, A., Sunyaev, R., and Terekhov, O. 1999, *Astron. Astronomy* **342**, 575.
Vlahos, L. and Sprangle, Ph. 1987, *Astrophys. J.* **322**, 463.
Vlahos, L. 1987, *Solar Phys.* **111**, 155.
Volwerk, M. and Kuijpers, J. 1994, *Astrophys. J. Suppl. Ser.* **90**, 589.
Volwerk, M. 1993, PhD Thesis, *Strong Double Layers in Astrophysical Plasmas*, Univ.Utrecht, The Netherlands.
Vršnak, B., Rudjak, V., and Rompolt, B. 1991, *Solar Phys.* **136**, 151.
Wang, H., Gary, D.E., Zirin, H., Schwartz, R.A., Sakao, T., Kosugi, T., and Shibata, K. 1995, *Astrophys. J.* **453**, 505.
Wang, H., Qiu, J., Denker, C., Spirock, T., Chen, H., and Goode, P.R. 2000, *Astrophys. J.* **542**, 1080.
Wang, J.X. and Shi, Z.X. 1993, *Solar Phys.* **143**, 119.
Warren, H.P., Bookbinder, J.A., Forbes, T.G., Golub,L., Hudson,H.S., Reeves, K., and Warshall, A. 1999, *Astrophys. J.* **527**, L121.
Warren, H.P. and Reeves, K. 2001, *Astrophys. J.* **554**, L103.
Warren, H.P. and Warshall, A.D. 2001, *Astrophys. J.* **560**, L87-L90.
Wheatland, M.S. and Melrose, D.B. 1995, *Solar Phys.* **158**, 283.
White, R.B. 1983, Handbook of Plasma Physics, (eds. M.N.Rosenbluth and R.Z.Sagdeev, Vol.1: Basic Plasma Physics I, p.611 (eds. A.A. Galeev and R.N. Sudan).
White, S.M., Melrose, D.B., and Dulk, G.A. 1983, *Proc.ASA* **5**(2), 188.
Wild, J.P., Sheridan, K.V., and Neylan, A.A. 1959, *Australian J. Phys.* **12**, 369.
Wild, J.P. and Smerd, S.F.1972, *Annu. Rev. Astron. Astrophys.* **10**, 159
Willes, A.J., Robinson, P.A. 1996, *Astrophys. J.* **467**, 465.
Willson, R.F., Lang, K.R., and Gary, D. 1993, *Astrophys. J.* **418**, 490.
Winglee, R.M. 1985a, *Astrophys. J.* **291**, 160.
Winglee, R.M. 1985b, *J.Geophys.Res.* **90**/A10, 9663.
Winglee, R.M. and Dulk, G.A. 1986a, *Astrophys. J.* **307**, 808.
Winglee, R.M. and Dulk, G.A. 1986b, *Astrophys. J.* **310**, 432.

Winglee, R.M. and Dulk, G.A. 1986c, *Solar Phys.* **104**, 93.
Wu, C.S. 1984, *J.Geophys.Res.* **89**, 8857.
Wu, C.S. 1985, *Space Science Reviews* **41**, 215.
Wülser, J.P., Zarro, D.M. and Canfield, R.C. 1992, *Astrophys. J.* **384**, 341.
Xie, R.X., Fu, Q.J., Wang, M., and Liu, Y.Y. 2000, *Solar Phys.* **197**, 375.
Yokoyama, T. and Shibata, K. 1996, *Proc. Astr. Soc. Japan* **48**, 353.
Yokoyama, T., Akita, K., Morimoto, T., Inoue, K., and Newmark, J. 2001, *Astrophys. J.* **546**, L69.
Yurovsky, Y. and Magun, A. 1998, *Solar Phys.* **180**, 409.
Zhang, J., Wang, J.X., Deng, Y.Y., and Wu, D. 2001, *Astrophys. J.* **548**, L99.
Zhou, Y. and Matthaeus, W.H. 1990, *J.Geophys.Res.* **95**, 14881.

GPSR Compliance

The European Union's (EU) General Product Safety Regulation (GPSR) is a set of rules that requires consumer products to be safe and our obligations to ensure this.

If you have any concerns about our products, you can contact us on ProductSafety@springernature.com

In case Publisher is established outside the EU, the EU authorized representative is:

Springer Nature Customer Service Center GmbH
Europaplatz 3
69115 Heidelberg, Germany

Batch number: 10614940

Printed by Printforce, the Netherlands